Meriem Gharbi

Proximity Moving Horizon Estimation

Logos Verlag Berlin

Bibliographic information published by the Deutsche Nationalbibliothek

The Deutsche Nationalbibliothek lists this publication in the Deutsche
Nationalbibliografie; detailed bibliographic data are available
on the Internet at http://dnb.d-nb.de

D93

ISBN 978-3-8325-5456-9

Logos Verlag Berlin GmbH
Georg-Knorr-Str. 4, Geb. 10,
D-12681 Berlin
Germany

Tel.:   +49 (0)30 / 42 85 10 90
Fax: +49 (0)30 / 42 85 10 92
http://www.logos-verlag.de

Proximity Moving Horizon Estimation

Von der Fakultät Konstruktions-, Produktions- und Fahrzeugtechnik
der Universität Stuttgart zur Erlangung der Würde eines
Doktor-Ingenieurs (Dr.-Ing.) genehmigte Abhandlung

Vorgelegt von

Meriem Gharbi

aus Bizerte, Tunesien

Hauptberichter: Prof. Dr.-Ing. Christian Ebenbauer
Mitberichter*innen: Prof. Dr.-Ing. Rolf Findeisen
                                   Prof. Dr. rer. nat. Nicole Radde

Tag der mündlichen Prüfung: 17.12.2021

Institut für Systemtheorie und Regelungstechnik

Universität Stuttgart

2021

Acknowledgements

This thesis is the outcome of my scientific work at the Institute for Systems Theory and Automatic Control (IST) at the University of Stuttgart in the period between October 2017 and September 2021. I enjoyed and learned a lot from my time at the IST and would like to take this opportunity to thank all the people who have accompanied, supported and encouraged me during this interesting but also challenging time.

First and foremost, I would like to express my special thanks and gratitude to Prof. Dr.-Ing. Christian Ebenbauer for the excellent supervision of my work and for his guidance and support during my Masters and PhD. I learned a lot from our regular scientific discussions and thank him not only for the helpful and critical suggestions, but also for granting freedom in doing research, supporting my trips to many international conferences, and the opportunity to carry out stimulating teaching activities. Furthermore, I would like to sincerely thank the members of my doctoral examination committee Prof. Dr.-Ing. Rolf Findeisen, Prof. Dr. rer. nat. Nicole Radde, and Prof. Dr. C. David Remy for their valuable time and for taking interest in my work.

Special thanks and appreciation go to Prof. Dr.-Ing Frank Allgöwer for ensuring a productive, cooperative, and stimulating social and working environment at the IST. Furthermore, I want to express my gratitude to Prof. Bahman Gharesifard for the very pleasant and fruitful collaboration as well as his helpful feedback.

A big thank you is also due to all current and former IST members for the many interesting discussions, conference trips and fun moments we had together. I would like to especially thank my office colleagues Raffaele Soloperto and Zoltan Tuza, who made my PhD journey a much more enjoyable experience. Moreover, I am indebted to the diligent proofreaders Johannes Köhler, Raffaele Soloperto, Julian Berberich, and Anne Koch for their time, dedication, and valuable comments.

Finally, I am grateful to my family who made it possible to embark on this journey and whose support and guidance played a major role in concluding it. A big thank you goes also to my friends who have always motivated and supported me. Last but not least, I am infinitely grateful to my husband Michael for the unconditional support, patience, and understanding in all phases of my PhD and for finding the right words to keep me on track.

Stuttgart, January 2022
Meriem Gharbi

Table of Contents

Abstract

In this thesis, we consider the state estimation problem of constrained discrete-time dynamical systems. More specifically, we develop and analyze moving horizon estimation (MHE) schemes and algorithms within the novel framework of *proximity moving horizon estimation*. In particular, we address the theoretical analysis of the corresponding MHE formulations and develop numerically efficient algorithms for the online implementation.

The first main goal of this thesis is to provide constructive MHE design procedures with desirable guarantees such as nominal stability and robustness of the underlying estimation error. These design procedures are tailored to the considered class of dynamical systems, covering time-invariant and time-varying linear and nonlinear systems, and the guarantees do not require any assumption on the estimation horizon length. In addition to the theoretical results, we study the design and tuning of the performance criteria in proximity MHE from a probabilistic perspective and demonstrate how to ensure a satisfactory performance given prior knowledge on the system disturbances.

The second main goal of this thesis is to develop computationally efficient and reliable MHE algorithms that do not require the exact solution of the constrained MHE problem at each time instant, but rather yield suboptimal state estimates after a limited number of iterations. By taking the dynamics of the employed optimization algorithm into account in the theoretical analysis, we derive sufficient conditions under which desirable stability and performance properties of the estimator hold for any number of optimization algorithm iterations, including the case of a single iteration per time instant. This leads to a novel class of *anytime MHE algorithms* where rigorous guarantees are ensured independently of the number of internal iterations, allowing for a trade-off between computational effort and accuracy of the state estimate.

The obtained theoretical results and practical benefits of the proposed proximity MHE approaches are demonstrated with different numerical examples from the literature.

Deutsche Kurzfassung

Diese Arbeit befasst sich mit der Zustandsschätzung von zeitdiskreten dynamischen Systemen mit Beschränkungen. Insbesondere werden Verfahren und Algorithmen der Zustandsschätzung mit bewegtem Horizont (engl. Moving Horizon Estimation, MHE) innerhalb des neuartigen Grundgerüsts der *Proximity Moving Horizon Estimation* entwickelt und analysiert. Dazu werden theoretische Analysen der entsprechenden MHE-Formulierungen eingesetzt und numerisch effiziente Algorithmen für die Online-Implementierung entwickelt.

Das erste Hauptziel dieser Arbeit besteht darin, konstruktive MHE-Entwurfsverfahren mit relevanten Eigenschaften wie nomineller Stabilität und Robustheit des Schätzfehlers bereitzustellen. Diese Entwurfsverfahren sind speziell auf die betrachtete Klasse dynamischer Systeme, welche zeitinvariante und zeitvariante lineare und nichtlineare Systeme abdeckt, zugeschnitten. Die sich ergebenden Stabilitäts- und Robustheitsgarantien erfordern keine Annahmen über die Länge des zurückliegenden Messhorizonts. Zusätzlich zu den theoretischen Ergebnissen wird der Entwurf und die Feinabstimmung des Gütekriteriums der Proximity-MHE aus einer probabilistischen Perspektive untersucht. Damit wird gezeigt, wie zufriedenstellende Schätzergebnisse bei Vorwissen über die Systemstörungen sichergestellt werden können.

Das zweite Hauptziel dieser Arbeit ist die Entwicklung rechentechnisch effizienter und zuverlässiger MHE-Algorithmen, die keine exakte Lösung des beschränkten MHE-Problems in jedem Zeitschritt erfordern, sondern eine suboptimale Schätzung der Systemzustände nach einer begrenzten Anzahl von Iterationen liefern. Indem die Dynamik des verwendeten Optimierungsalgorithmus in der theoretischen Analyse berücksichtigt wird, lassen sich hinreichende Bedingungen ableiten, unter denen wünschenswerte Stabilitäts- und Performance-Eigenschaften des Schätzers für eine beliebige Anzahl an Iterationen des Optimierungsalgorithmus gelten. Dies schließt den Fall einer einzigen Iteration pro Zeitschritt ein. Damit ergibt sich eine neuartige Klasse von *Anytime-MHE-Algorithmen*, bei denen rigorose Garantien unabhängig von der Anzahl an internen Iterationen gewährleistet sind, was eine Abwägung zwischen Rechenaufwand und Genauigkeit der Zustandsschätzung ermöglicht.

Die erzielten theoretischen Ergebnisse und praktischen Vorteile der vorliegenden Proximity-MHE-Ansätze werden anhand verschiedener numerischer Beispiele aus der Literatur demonstriert.

Chapter 1

Introduction

1.1 Motivation

State estimation is the process of reconstructing an unmeasurable state of a dynamical system from a given model in combination with the available input and measurement data. State estimation is of paramount importance in many engineering domains such as signal processing, geodesy, astronomy, communication, control, and learning, where the success of many of the underlying applications directly relies on the quality of the estimated states. The problem of interpreting observations and computing estimates and predictions is not of recent origin, and the early history dates back to Galileo's statistical analysis of astronomical observations in 1632 (Kailath (1974)). Gauss apparently first discovered the least-squares method in 1795, though it was officially independently introduced and published by Legendre in 1805 (Sorenson (1970)). In his book on planetary orbits published in 1809, Gauss discussed the estimation of the six parameters that determine an elliptical orbit on the basis of more than six observations (Sprott (1978)). The invention of least-squares estimation has provided the foundation for many estimation theories and techniques in the 19th and 20th centuries (Sorenson (1970)). One of the most widely-used state estimators which had a huge impact on the field of state estimation and forms the backbone of numerous applications is the Kalman filter. In his pioneering work, Kalman (1960b) provided a computationally attractive recursive solution to the minimum variance estimation problem for linear systems subject to Gaussian disturbances. Since then, many extensions and generalizations of the Kalman filter were developed, which is revealed by the existing vast and rich literature (see, e.g., Särkkä (2013)). In addition to the fact that it allows for a fast implementation in real-time, the linear Kalman filter exhibits important theoretical guarantees such as stability of the estimation error under standard assumptions (Jazwinski (1970)). However, these guarantees as well as a good practical performance of the Kalman filter are not ensured when nonlinear systems are considered or Gaussian assumptions are violated. Moreover, prior knowledge on the system in form of physical state constraints cannot be directly incorporated into the estimation process.

One successful approach that overcomes the aforementioned issues is moving horizon estimation (MHE). MHE is an optimization-based approach that estimates the state of a dynamical system by solving a suitable optimization problem at each time instant. Hereby, MHE takes a fixed and limited number of the most recent input and measurement data into account and the considered horizon of data is moved forward in time (in a receding horizon manner) when a new measurement becomes available. The main benefits of MHE are (i) its capability to exploit physical constraints on states and disturbances by including them directly in the underlying optimization problem, which may lead to an improved performance compared to the Kalman filter (Muske et al. (1993)), (ii) the handling of general nonlinear systems, (iii) the ability to consider suitable performance criteria designed specifically for the problem at hand, and (iv) tractable online computational complexity compared to the full information estimation problem (Rawlings et al. (2017)). Since the original idea of MHE was presented by Jazwinski (1968) and stability properties were first explored by Ling and Lim (1999); Michalska and Mayne (1993, 1995), MHE has gathered more attention during the last decades from both theoretical and practical viewpoints. In fact, there is a growing number of application examples of MHE, such as automotive applications (Andersson and Thiringer (2018)), mobile and humanoid robots (Bae and Oh (2017); Liu et al. (2016)), state of charge estimation (Hu et al. (2017); Shen et al. (2016)), spacecraft attitude estimation (Huang et al. (2017); Qin and Chen (2013); Vandersteen et al. (2013)), water level regulation in inland navigation (Segovia et al. (2019)), and optimal dosing of cancer therapy (Chen et al. (2012)). Moreover, design methods of MHE have been proposed in the literature which take into account large scale systems (Farina et al. (2010); Haber and Verhaegen (2013); Schneider et al. (2015); Zhang and Liu (2013)), hybrid systems (Ferrari-Trecate et al. (2002)), systems with multi-rate measurements (Gu et al. (2019)), and binary sensors (Battistelli et al. (2017)).

Although the extent of the literature validates the benefits of MHE and reports a significant progress on its theoretical analysis, design, and application, we enumerate the following fundamental challenges in the MHE research that might reduce its practical usefulness.

- Optimization algorithm of MHE:
  In many demanding applications such as in the field of automation, a crucial requirement on the estimator is to process and manipulate large amounts of measurement data in a very fast and reliable fashion. However, MHE can be computationally intensive since it involves the online solution of an optimization problem each time a new measurement becomes available. Thus, an efficient and real-time capable implementation of MHE with theoretical guarantees is required for practical and safety critical applications which fulfills the constraints imposed by the limited computational resources. By contrast, most of the MHE approaches with stability guarantees do not consider the numerical implementation of the underlying optimization problem in the theoretical analysis. More specifically, stability or convergence results are usually presented with respect to the exact solution of the constrained optimization problem and do not take into account the internal iterations of the employed optimization algorithm. Moreover, there is a lack of theoretical results that demonstrate how the accuracy of the suboptimal estimate, which can be characterized by the number of executed optimization algorithm iterations, affects the performance of MHE algorithms. Although these issues have drawn more attention in the recent years (Alessandri and Gaggero (2020); Findeisen et al. (2018); Kong and Sukkarieh (2018); Schiller et al. (2020); Zou et al. (2020)), theoretical studies that address both stability as well as performance of MHE algorithms under rather mild assumptions are rare in the literature and still require investigation.

- Stability theory of MHE:
  As already mentioned, a reliable state estimation strategy with important theoretic properties such as stability of the estimation error is viable for the success of the considered application. Especially for nonlinear systems, the problem of the synthesis

of a state estimator for stability purposes is significant and difficult. This challenge has been addressed in the literature and numerous contributions have been made for the development of MHE design methods in order to guarantee stability of the underlying estimation error. In a notable and rather standard MHE approach (Rawlings et al. (2017)), this problem is tackled by providing a sufficient condition on the approximation of the so-called arrival cost, which summarizes the past history of data not included in the MHE problem. However, the derived condition is rather restrictive and hard to meet for a general nonlinear system. Except for the special case of linear systems, quadratic cost, and convex constraints, developing constructive methods for its satisfaction remains a key issue as pointed out by Rawlings et al. (2017). Nevertheless, there exist alternative MHE approaches which ensure stability and robustness even in the nonlinear case thanks to offline design procedures, see among others the interesting works of Alessandri et al. (2010); Müller (2017); Sui and Johansen (2014) and the detailed discussions in Chapter 2. However, many of the established results either rely on a sufficiently long estimation horizon length, or hold due to sufficient conditions which are limited to a quadratic cost and explicitly depend on global properties of the considered dynamical system, which are often hard to verify in the general nonlinear case. Hence, there is a need for conceptually simple and constructive design procedures of nonlinear and linear MHE schemes which ensure desirable theoretical guarantees.

- Robustness of MHE with respect to disturbances:
  The ability to cope with various disturbances such as measurement outliers, which occur for instance due to sensor malfunction or failures of data transmitters, can be a crucial requirement in many applications (Alessandri and Awawdeh (2016); Ben-Gal (2005); Hawkins (1980); Rousseeuw and Leroy (2005)). Hence, it is desirable to include in the MHE problem meaningful stage costs that describe statistical characteristics of the system disturbances in order to accurately estimate the state of the underlying system. For example, if it is known that outliers occur in the measurements, instead of penalizing the output residuals using the standard ℓ_2-norm, the so-called Huber penalty function can be used as stage cost. This penalty is quadratic for small residuals but linear for large residuals, which makes it less sensitive to outliers. Furthermore, the sparsity promoting ℓ_1-norm can be employed, which has been highly successful in many signal processing applications (Candes et al. (2008); Donoho (2006)). However, there has been little discussion about MHE approaches with stability guarantees in which the commonly-used least-squares formulation is not adapted. This issue is also highlighted in several recent developments (Chu et al. (2012); Geebelen et al. (2013); Haverbeke (2011); Kouzoupis et al. (2016)), which propose to use the ℓ_1-norm or the Huber penalty function instead of the traditional ℓ_2-norm in MHE in order to account for outliers. Despite the fact that these estimators exhibit improved performance, the stability properties of the underlying estimation error are not investigated. Hence, it is desirable to develop MHE approaches that allow for flexible performance criteria and for which theoretical guarantees still hold.

These challenges create an incentive for developing MHE procedures which are easy to implement, easy to design, and with system theoretic guarantees that hold for a broad variety of system classes and performance criteria. More specifically, the goal of this thesis is to provide a unified framework for the design, analysis and implementation of moving horizon estimators, which are theoretically sound with desirable guarantees, computationally efficient, and allow for a flexible formulation of the underlying optimization problem.

1.2 Contributions and outline of the thesis

In this thesis, we introduce the novel framework of *proximity moving horizon estimation* of constrained discrete-time linear and nonlinear systems. It is based on the general conceptual idea of employing a stabilizing analytical a priori solution and combining it with an online optimization in order to obtain an improved performance without jeopardizing the stability properties. Based thereon, we devote our attention to the theoretical analysis of MHE as well as to the practical challenge of computational efficiency. More specifically, we first address MHE theory with a focus on MHE *schemes*, where stability is investigated under the assumption that a solution of the optimization problem is available at each time instant. The underlying cost function consists of two parts: the first part includes a general, possibly nonsmooth convex stage cost which can handle outliers and other data specific characteristics; the second part is a suitable proximity measure to a stabilizing a priori estimate, from which stability can be explicitly inherited. Second, we address MHE implementation with a focus on MHE *algorithms* or *iteration schemes*, where the dynamics of the optimization algorithm are taken into account in the stability analysis. The underlying optimization algorithm computes the next iterate as the solution of a simple convex optimization problem, where a first-order approximation of the MHE stage cost and a suitable proximity measure to the previous iterate are minimized. The optimization algorithm terminates after a limited number of iterations and is warm-started by a stabilizing a priori estimate, from which stability can be implicitly inherited.

In the proposed MHE approaches, the proximity term is characterized by the so-called Bregman distance, which constitutes a measure of distance between two points and is constructed in terms of a strictly convex function (Censor and Zenios (1992)). Moreover, the a priori estimate is generated from a model-based and convergent recursive estimator for which a suitable Lyapunov function of the error dynamics is known. While the stability components consisting of the Bregman distance and the a priori estimate can be designed *offline*, the performance of the estimator is specified through the stage cost and the most recent batch of input and measurement data in the *online* optimization. A graphical illustration of this concept is presented in Figure 1.1.

The methodological core for the design and analysis within the proximity MHE framework is from the field of state estimation and the powerful class of proximal methods for solving nonsmooth convex optimization problems (Beck and Teboulle (2003); Güler (1991); Parikh and Boyd (2014); Rockafellar (1976); Teboulle (1992)). The theory therein is very rich and serves as an efficient mathematical tool for investigating important theoretic properties of the developed proximity MHE approaches. Given the central role played by proximal methods in deriving many of the results presented in this thesis, and since the proposed approaches use the Bregman distance as a proximity measure through which stability can be explicitly or implicitly inherited from the stabilizing a priori estimate, we refer to our framework as "proximity" MHE. Based on this discussion, we summarize in the following the main contributions and present in more detail the outline of this thesis.

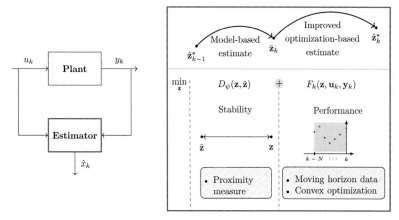

Figure 1.1: Proximity moving horizon estimation: At each time instant k, a stabilizing a priori estimate $\bar{\mathbf{z}}_k$ is computed by updating (predicting) the previous MHE solution $\hat{\mathbf{z}}_{k-1}^*$ through a model-based analytical estimator. Given the calculated a priori estimate, an improved optimization-based estimate $\hat{\mathbf{z}}_k^*$ based on which $\hat{x}_k$ is generated is obtained by solving the proximity MHE problem. In the associated optimization problem, a suitable Bregman distance D_ψ is used as proximity measure to $\tilde{\mathbf{z}}$ and combined in an online optimization with the performance criterion $F_k(\mathbf{z}, \mathbf{u}_k, \mathbf{y}_k)$, which is based on the horizon of most recent input and measurement data $\mathbf{u}_k, \mathbf{y}_k$. In chapter 3, where a solution of the proximity-based MHE problem is computed at each time instant, $\tilde{\mathbf{z}}$ is exactly the a priori estimate $\bar{\mathbf{z}}_k$. In chapter 4, where a fixed number of iterations is executed at each time instant, an iterative optimization update is carried out, in which the Bregman distance keeps $\mathbf{z}$ close to the previous iterate $\tilde{\mathbf{z}}$ and $\bar{\mathbf{z}}_k$ is only used as a warm start strategy. In both cases, the combination ensures both stability and performance of MHE.

Main contributions

This thesis consists of two central contributions, each of which is related to the two main chapters of the thesis.

- The first main contribution of the thesis, Chapter 3, is concerned with the design and analysis of proximity MHE *schemes* for discrete-time linear and nonlinear systems subject to convex constraints. In this chapter, we present the novel proximity-based formulation of the MHE optimization problem, where the cost function includes a convex stage cost as well as a Bregman distance centered around a stabilizing a priori estimate. The main objective is to derive sufficient conditions on the Bregman distance and the a priori estimate such that stability properties of the underlying estimation error can be established. Based on these conditions, we propose explicit design procedures which are tailored to the considered system class. A technical contribution is a novel and unified Lyapunov-based approach for the stability analysis, which allows MHE to inherit the stability properties of the employed a priori estimate by choosing the Bregman distance as a Lyapunov function. Furthermore, we embed the proposed MHE formulation into a stochastic framework in order to obtain a Bayesian interpretation, where the stabilizing model-based estimator is considered as a priori knowledge obtained offline, before observed data from the most recent measurements are included. Based thereon, we describe how the design and tuning of the stage cost can be guided by the statistics of the disturbances.

- The second main contribution of the thesis, Chapter 4, is concerned with the design and analysis of proximity MHE *algorithms* for discrete-time linear and nonlinear systems subject to convex constraints. In this chapter, we present a novel proximity-based MHE iteration scheme, which reduces computational burden by performing only a limited number of optimization algorithm iterations each time a new measurement is received. More specifically, a suitable gradient-based proximal algorithm for minimizing the sum of convex stage cost is initialized based on a stabilizing a priori estimate and used to deliver a suboptimal estimate after single or multiple iterations per time instant. By means of a rigorous Lyapunov analysis, we derive conditions under which stability of the underlying estimation errors is ensured for any arbitrary number of optimization algorithm iterations. Thereby, we obtain a so-called *anytime MHE algorithm*, where stability is guaranteed independently of the executed number of internal iterations, and which allows the user to achieve a trade-off between computational complexity and accuracy of the state estimate. Although this anytime property is induced from the stabilizing a priori estimate, the (suboptimal) bias of the employed recursive estimator used to construct the a priori estimate is fading away with each iteration since it is only used to warm start the optimization algorithm. This is an implicit stabilizing regularization approach of the a priori estimate, which is conceptually different from the explicit stabilizing regularization used in Chapter 3 to ensure stability. Furthermore, we study the performance of the proximity MHE algorithm in terms of a regret analysis, which is widely used in the field of online convex optimization to characterize performance. By adapting this notion of regret to our setting, we provide performance guarantees in terms of rigorously derived regret bounds. The established bounds allow to measure the real-time regret of the proposed algorithm that carries out only finitely many optimization iterations (due to limited computing power and/or minimum required sampling rate) relative to a comparator algorithm that gets instantaneously an optimal solution from some oracle.

Thesis outline

Chapter 2: Background. In this chapter, we present the basic principle of moving horizon estimation and a short review on some of the relevant and related literature, which serves as a basis for comparison with the results derived in this thesis. Moreover, we provide a brief overview on proximal methods for convex optimization.

Chapter 3: Proximity-based MHE schemes In this chapter, we first start by presenting the problem setup and formulating the proximity-based MHE optimization problem. Second, we focus on the class of linear time-varying systems, for which we establish global uniform exponential stability of the estimation error in the absence of disturbances, as well as input-to-state stability when additive process and measurement disturbances affect the system. A Bayesian interpretation of the proximity MHE scheme is also established and the relationship to Kalman filtering is investigated. For the special class of linear time-invariant systems, we show that the theoretical results hold under very mild assumptions, where only detectability of the system matrices is needed. Third, we extend our results to nonlinear systems and establish local uniform exponential stability of the underlying estimation error under suitable assumptions. Moreover, we consider a class of nonlinear systems that can be transformed into systems that are affine in the unmeasured state. For these systems, we formulate the proximity MHE problem as a convex optimization problem and show that the estimation error is globally uniformly exponentially stable. Due to the simple proximity-based design, the theoretical guarantees derived in this chapter hold for any horizon length and irrespectively of the convex stage cost being used. Finally, we illustrate the benefits of the proposed MHE approaches using numerical examples from the literature.

The results of this chapter are based on (Gharbi and Ebenbauer (2018, 2019a,b, 2020); Gharbi et al. (2020a)).

Chapter 4: Anytime proximity-based MHE algorithms In this chapter, we first present the estimation problem of interest. Second, we describe the proximity-based MHE iteration scheme in detail. At each time instant, the underlying optimization algorithm is warm-started by a stabilizing a priori estimate and delivers a state estimate in real-time. Third, we consider linear systems and prove that global uniform exponential stability of the resulting estimation error can be ensured for any number of optimization algorithm iterations. In addition, we establish performance guarantees of the proposed MHE iteration scheme in terms of regret upper bounds. Our results show that both exponential stability and a sublinear regret can be guaranteed, where the latter can be rendered smaller by increasing the number of optimization iterations. Fourth, we show that the stability results derived in the linear case can be extended to nonlinear systems by proving local uniform exponential stability of the underlying estimation error. Thanks to the simple proximity-based design of the MHE algorithm, the stability and performance guarantees derived in this chapter hold for any number of optimization algorithm iterations, any horizon length, and for a rather general convex stage cost that is not necessarily quadratic. Finally, we use numerical examples to illustrate the obtained stability and regret results as well as the computational efficiency of the proposed iteration scheme.

The results of this chapter are based on (Gharbi and Ebenbauer (2021); Gharbi et al. (2020b, 2021)).

Chapter 6: Conclusions. In this chapter, we summarize the main results of this thesis and provide perspectives for future research.

Appendices. In Appendix A, we provide stability definitions of discrete-time systems which are employed throughout this thesis. Moreover, we recall how stability properties can be established using Lyapunov functions.

Chapter 2

Background

In this chapter, we give a brief review on optimization-based state estimation by introducing full information estimation (FIE) and moving horizon estimation (MHE) and presenting some approaches from the literature that guarantee stability of the estimators. Furthermore, we provide a short background on proximal methods for solving convex optimization problems which serves as a starting point towards the proximity-based formulation and analysis of MHE in the subsequent chapters of the thesis.

2.1 Moving horizon estimation

In this section, we briefly present standard stability results for FIE and MHE in a deterministic setting, where the considered system class as well as the main assumptions are based on (Rawlings et al., 2017, Chapter 4).

Consider a discrete-time nonlinear system of the form

$$x_{k+1} = f(x_k, w_k), \tag{2.1a}$$
$$y_k = h(x_k) + v_k, \tag{2.1b}$$

where $k \in \mathbb{N}$ denotes the discrete time instant, $x_k \in \mathbb{R}^n$ the state vector, $y_k \in \mathbb{R}^p$ the measurement vector, and $w_k \in \mathbb{R}^{m_w}$ and $v_k \in \mathbb{R}^p$ account for unknown process and measurement disturbances, respectively. Moreover, the initial condition $x_0 \in \mathbb{R}^n$ of system (2.1) is unknown. The functions $f : \mathbb{R}^n \times \mathbb{R}^{m_w} \to \mathbb{R}^n$ and $h : \mathbb{R}^n \to \mathbb{R}^p$ are assumed to be continuous. The state and disturbances are known to verify the following constraints

$$x_k \in \mathcal{X}_k \subseteq \mathbb{R}^n, \quad w_k \in \mathcal{W}_k \subseteq \mathbb{R}^{m_w}, \quad v_k \in \mathcal{V}_k \subseteq \mathbb{R}^p, \quad k \in \mathbb{N}, \tag{2.2}$$

where the sets $\mathcal{X}_k$, $\mathcal{W}_k$, and $\mathcal{V}_k$ are nonempty and closed, with $0 \in \mathcal{W}_k$, and $0 \in \mathcal{V}_k$. We let $x(k; x_0, \mathbf{w}_k)$ refer to the solution of system (2.1) at time k with initial state x_0 and disturbances $\mathbf{w}_k = \{w_0, \cdots, w_{k-1}\}$. Note that for simplicity of presentation, and analogous to the problem setup in (Rawlings et al., 2017, Chapter 4), inputs u_k are not considered in system (2.1). Nevertheless, the subsequent assumptions and results can be extended to systems of the form $x_{k+1} = f(x_k, u_k, w_k)$ without any particular conceptual difficulties.

The goal is to find at each time instant k an estimate $\hat{x}_k$ of the state x_k given the model (2.1), the constraints (2.2), and the available measurements $\{y_0, y_1, \ldots, y_{k-1}\}$. In FIE, the state estimation problem is tackled by solving at each time instant k a suitable constrained optimization problem that incorporates all the available past measurements.

More specifically, the FIE problem is

$$\min_{\hat{x}_0, \hat{w}_0, \dots, \hat{w}_{k-1}} \quad \sum_{i=0}^{k-1} l_i(\hat{w}_i, \hat{v}_i) + \Gamma_0\left(\hat{x}_0\right) \tag{2.3a}$$

$$\text{s.\,t.} \quad \hat{x}_{i+1} = f(\hat{x}_i, \hat{w}_i), \qquad i = 0, \dots, k-1 \tag{2.3b}$$

$$y_i = h(\hat{x}_i) + \hat{v}_i, \qquad i = 0, \dots, k-1 \tag{2.3c}$$

$$\hat{x}_i \in \mathcal{X}_i, \qquad i = 0, \dots, k \tag{2.3d}$$

$$\hat{w}_i \in \mathcal{W}_i, \quad \hat{v}_i \in \mathcal{V}_i, \qquad i = 0, \dots, k-1, \tag{2.3e}$$

where the variables $\hat{x}_i \in \mathbb{R}^n$ with $i = 0, \dots, k$ and $\hat{w}_i \in \mathbb{R}^{m_w}$, $\hat{v}_i \in \mathbb{R}^p$ with $i = 0, \dots, k-1$ represent estimates of the states, the model disturbances and the output residuals measuring the residual between the true measurement y_i and the estimated output $h(\hat{x}_i)$, respectively. We set the initial state $\hat{x}_0$ and the model disturbance sequence $\{\hat{w}_0, \dots, \hat{w}_{k-1}\}$ as decision variables of problem (2.3) since the remaining variables $\hat{x}_i$ and $\hat{v}_i$ depend implicitly on $\hat{x}_0$ and $\hat{w}_i$ through the model (2.3b), (2.3c). In (2.3a), the function $l_i : \mathbb{R}^{m_w} \times \mathbb{R}^p \to \mathbb{R}_+$ refers to the stage cost for penalizing the model disturbance and the output residual, and the function $\Gamma_0 : \mathbb{R}^n \to \mathbb{R}_+$ refers to a suitable prior weighting. Assuming that a solution $(\hat{x}_0^*, \hat{w}_0^*, \dots, \hat{w}_{k-1}^*)$ to problem (2.3) exists, the state estimate $\hat{x}_k$ can be computed using the system dynamics (2.3b). Note that by assuming linear system dynamics and a quadratic cost function, and by discarding the inequality constraints in problem (2.3), the Kalman filter can be derived by solving the FIE problem using forward dynamic programming (Rawlings et al. (2017)).

In several works on nonlinear FIE and MHE (e.g. Müller (2017); Rao et al. (2003); Rawlings et al. (2017)), the desired stability property of the estimator is given by global asymptotic stability (GAS) in the disturbance-free case and robust global asymptotic stability (RGAS) in the presence of disturbances. In these works, saying that an estimator is GAS implies that the dynamics of the underlying estimation error $x_k - \hat{x}_k$ is GAS. Moreover, an estimator is RGAS implies that the dynamics of the estimation error is input-to-state stability (ISS) with respect to the process and measurement disturbances. The class of systems for which RGAS is investigated is specified by the notion of incremental input/output-to-state stability (i-IOSS), which was introduced by Sontag and Wang (1997) in order to characterize detectability of nonlinear systems.

Definition 2.1 (i-IOSS). *The system $x^+ = f(x, w)$, $y = h(x)$ is i-IOSS if there exist functions $\beta \in \mathcal{KL}$ and $\gamma_1, \gamma_2 \in \mathcal{K}$ such that for every two initial conditions z_1 and z_2, and any two disturbance sequences $\mathbf{w}_1$ and $\mathbf{w}_2$, the following holds for all $k \in \mathbb{N}$:*

$$\|x(k; z_1, \mathbf{w}_1) - x(k; z_2, \mathbf{w}_2)\| \leq \beta(\|z_1 - z_2\|, k) \tag{2.4}$$
$$+ \gamma_1(\|\mathbf{w}_1 - \mathbf{w}_2\|_{[0:k-1]}) + \gamma_2(\|\mathbf{h}(\mathbf{x}_1) - \mathbf{h}(\mathbf{x}_2)\|_{[0:k-1]}),$$

where $\|\mathbf{w}_{[i:k]}\| := \sup_{i \leq j \leq k} \{\|w_j\|\}$ and $\mathbf{h}(\mathbf{x}_1) := \{h(x(0; z_1, \mathbf{w}_1)), h(x(1; z_1, \mathbf{w}_1)), \cdots\}$.

The i-IOSS condition (2.4) requires that the difference between two state trajectories is bounded in terms of the differences in their initial conditions, their process disturbances, and their nominal outputs. More detail on i-IOSS can be found in (Allan et al. (2020)).

If system (2.1) is i-IOSS and problem (2.3) satisfies the following mild assumption, GAS of the estimation error generated by the FIE scheme can be ensured.

Assumption 2.1 (Positive definite stage cost). *The prior weighting* Γ_0 *and the stage cost* l_i *are continuous functions and there exist* K_∞-functions $\underline{\gamma}_x, \underline{\gamma}_w, \underline{\gamma}_v, \overline{\gamma}_x, \overline{\gamma}_w, \overline{\gamma}_v$ *such that the following holds*

$$\underline{\gamma}_x(\|x - \bar{x}_0\|) \leq \Gamma_0(x) \leq \overline{\gamma}_x(\|x - \bar{x}_0\|) \tag{2.5}$$

$$\underline{\gamma}_w(\|w\|) + \underline{\gamma}_v(\|v\|) \leq l_i(w, v) \leq \overline{\gamma}_w(\|w\|) + \overline{\gamma}_v(\|v\|) \qquad i \in \mathbb{N} \tag{2.6}$$

for all $x \in \mathbb{R}^n$, $w \in \mathbb{R}^{m_w}$ *and* $v \in \mathbb{R}^p$. *Furthermore,* $\bar{x}_0 \in \mathbb{R}^n$ *which denotes the initial guess is assumed to satisfy* $\bar{x}_0 \in \mathcal{X}_0$.

Assumption 2.1 ensures that the prior weighting Γ_0 is positive definite with respect to the initial guess $\bar{x}_0$ and that the stage cost is positive definite with respect to zero. Moreover, it ensures that a solution to problem (2.3) always exists by the Weierstrass theorem, which states that the set of minimizers of a continuous function on a compact set is nonempty and bounded (Rawlings et al., 2017, Proposition A.7). More specifically, since Γ_0 and l_i are continuous for all $i \in \mathbb{N}$, and f and h are continuous, the FIE cost function is continuous in the decision variables. Moreover, a minimizer to problem (2.3) has to lie in a bounded and closed set. This is due to the fact that the lower bounds in Assumption 2.1 imply that the cost function is radially unbounded in the decision variables, and hence all its sublevel sets are bounded, and that the feasible set is closed and nonempty since the true system satisfies the inequality constraints.

Theorem 2.1 (Rawlings et al. (2017)). *Consider an i-IOSS system* (2.1) *with* $w_k = 0$, $v_k = 0$, $k \in \mathbb{N}$, *and the FIE problem* (2.3). *Let Assumption 2.1 hold. Then, the estimation error of the full information estimator is GAS.*

This result is a direct consequence of (Rawlings et al., 2017, Theorem 4.27), which establishes that the estimator is RGAS with respect to convergent disturbances under an additional technical assumption, which requires that there exists a K-function σ such that, for all convergent disturbances $(\mathbf{w}, \mathbf{v})$,

$$\sum_{i=0}^{\infty} l_i(w_i, v_i) \leq \sigma(\|(\mathbf{w}, \mathbf{v})\|). \tag{2.7}$$

Obviously, in Theorem 2.1, this assumption automatically holds since $w_k = 0, v_k = 0, k \in \mathbb{N}$. In (Ji et al. (2015)), RGAS is shown for bounded disturbances when an extra weighted max-term of l_i is added to the sum of stage cost.

Even though FIE enjoys these important theoretical properties under mild assumptions, it is computationally intractable since the complexity of the underlying optimization problem increases whenever a new measurement becomes available. In order to bound the size of the problem, constrained MHE constitutes a suitable alternative. In MHE, a state estimate $\hat{x}_k$ at time instant k is computed by solving an optimization problem with a fixed horizon of length $N \in \mathbb{N}_+$ and based on the most recent N measurements $\{y_{k-N}, \dots, y_{k-1}\}$. The horizon is shifted forward in time as a new measurement becomes available, which allows for a fixed number of decision variables at each time instant. A graphical illustration of MHE is depicted in Figure 2.1. A typical MHE problem can be formulated as follows:

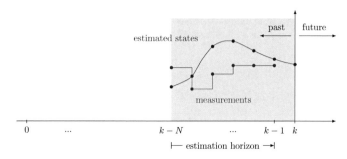

Figure 2.1: A description of the basic idea of MHE

$$\min_{\hat{x}_{k-N}, \hat{\mathbf{w}}_k} \quad \sum_{i=k-N}^{k-1} l_i(\hat{w}_i, \hat{v}_i) + \Gamma_{k-N}\left(\hat{x}_{k-N}\right) \tag{2.8a}$$

$$\text{s.\,t.} \quad \hat{x}_{i+1} = f(\hat{x}_i, \hat{w}_i), \qquad i = k-N, \ldots, k-1 \tag{2.8b}$$

$$y_i = h(\hat{x}_i) + \hat{v}_i, \qquad i = k-N, \ldots, k-1 \tag{2.8c}$$

$$\hat{x}_i \in \mathcal{X}_i, \qquad i = k-N, \ldots, k \tag{2.8d}$$

$$\hat{w}_i \in \mathcal{W}_i, \ \hat{v}_i \in \mathcal{V}_i, \qquad i = k-N, \ldots, k-1, \tag{2.8e}$$

where $\hat{\mathbf{w}}_k := \{\hat{w}_{k-N}, \cdots, \hat{w}_{k-1}\}$ denotes the estimated process disturbances over the estimation horizon. In (2.8a), the prior weighting $\Gamma_{k-N} : \mathbb{R}^n \to \mathbb{R}_+$ needs to be appropriately designed in order to ensure stability of MHE, as will be discussed below. Assuming that a solution $(\hat{x}_{k-N}^*, \hat{\mathbf{w}}_k^*)$ to problem (2.8) exists, where $\hat{\mathbf{w}}_k^* := \{\hat{w}_{k-N}^*, \cdots, \hat{w}_{k-1}^*\}$, we obtain the state estimate $\hat{x}_k = x\left(k; \hat{x}_{k-N}^*, \hat{\mathbf{w}}_k^*\right)$. The resulting estimation error is given by $x_k - \hat{x}_k = x\left(k; x_{k-N}, \mathbf{w}_k\right) - x\left(k; \hat{x}_{k-N}^*, \hat{\mathbf{w}}_k^*\right)$. For the time instants $k \leq N$, we set $N = k$ in problem (2.8), which reduces to the FIE problem (2.3). The MHE scheme can be summarized by Algorithm 2.1.

Algorithm 2.1 Basic MHE scheme

Offline: Specify the stage cost l_i and the estimation horizon N, and choose a suitable prior weighting Γ_{k-N}.

Online:

1: **for** $k = 0, 1, 2, \cdots$ **do**
2:      get the measurements $\{y_{k-N}, \ldots, y_{k-1}\}$
3:      compute a solution $(\hat{x}_{k-N}^*, \hat{\mathbf{w}}_k^*)$ to problem (2.8)
4:      obtain the state estimate $\hat{x}_k = x\left(k; \hat{x}_{k-N}^*, \hat{\mathbf{w}}_k^*\right)$
5: **end for**

In MHE, one of the main challenges is how to provide sufficient conditions on the prior weighting Γ_{k-N} in order to ensure stability of the estimation error. Setting $\Gamma_{k-N}(\cdot) = 0$ means that the state estimate is reconstructed as the best fit to the measurement sequence with length N. In particular, we assume that system (2.1) is observable in the sense that there exists $N_0 \in \mathbb{N}_+$ and $\gamma_w, \gamma_v \in \mathcal{K}$ such that for every two initial conditions z_1 and z_2, and any two disturbance sequences $\mathbf{w}_1$ and $\mathbf{w}_2$, the following holds for all $k \geq N_0$:

$$\left\| z_1 - z_2 \right\| \leq \gamma_w \left(\left\| \mathbf{w}_1 - \mathbf{w}_2 \right\|_{[0:k-1]} \right) + \gamma_v \left(\left\| \mathbf{y}_{z_1,\mathbf{w}_1} - \mathbf{y}_{z_2,\mathbf{w}_2} \right\|_{[0:k-1]} \right), \qquad (2.9)$$

where $\mathbf{y}_{z_1,\mathbf{w}_1} := \{ h(x(0; z_1, \mathbf{w}_1)), h(x(1; z_1, \mathbf{w}_1)), \cdots \}$ (see, e.g., (Rawlings et al., 2017, Definition 4.13)). If the horizon length is chosen such that $N \geq N_0$ and Assumption 2.1 is satisfied, the estimator is GAS (Rawlings et al., 2017, Theorem 4.18). However, since N has to be chosen large enough (for linear systems, this implies that N has to be greater than the state dimension n), the size of the nonlinear optimization problem to be solved at each $k \in \mathbb{N}$ can be rather large, which yields to a high computational complexity.

In order to be able to consider detectable systems, the prior weighting Γ_{k-N} can be chosen as the so-called full information arrival cost that accounts for measurements which are not included in the optimization problem.

Definition 2.2 (Arrival cost). *The full information arrival cost is defined as*

$$Z_k(p) = \min_{\hat{x}_0, \hat{w}_0, \ldots, \hat{w}_{k-1}} \sum_{i=0}^{k-1} l_i(\hat{w}_i, \hat{v}_i) + \Gamma_0(\hat{x}_0) \qquad (2.10)$$

$$\begin{aligned}
\text{s.t.} \quad & \hat{x}_{i+1} = f(\hat{x}_i, \hat{w}_i), && i = 0, \ldots, k-1 \\
& y_i = h(\hat{x}_i) + \hat{v}_i, && i = 0, \ldots, k-1 \\
& \hat{x}_i \in \mathcal{X}_i, && i = 0, \ldots, k \\
& \hat{w}_i \in \mathcal{W}_i, \ \hat{v}_i \in \mathcal{V}_i, && i = 0, \ldots, k-1, \\
& x(k; \hat{x}_0, \hat{\mathbf{w}}_k) = p.
\end{aligned}$$

By forward dynamic programming, it can be shown that the MHE problem (2.8) with $\Gamma_k(\cdot) = Z_k(\cdot)$ for all $k \in \mathbb{N}$ yields the same state estimate as the one obtained by solving the FIE problem (2.3). An analytical expression for the exact arrival cost (2.10) can be computed if the inequality constraints are discarded and in the special case of linear systems

$$f(x, w) := A\,x + G\,w, \qquad h(x) := C\,x, \qquad (2.11)$$

quadratic stage cost

$$l_i(w, v) := \|w\|_{Q_i^{-1}}^2 + \|v\|_{R_i^{-1}}^2, \qquad i \in \mathbb{N}, \qquad (2.12)$$

where $Q_i \in \mathbb{S}_{++}^{m_w}$ and $R_i \in \mathbb{S}_{++}^p$ for all $i \in \mathbb{N}$, and quadratic prior weighting

$$\Gamma_0(x) := \|x - \bar{x}_0\|_{P_0^{-1}}^2, \qquad (2.13)$$

where $P_0 \in \mathbb{S}_{++}^n$ and $\bar{x}_0 \in \mathbb{R}^n$ is a given initial guess. More specifically, the arrival cost in this special case is

$$Z_k(p) = \|p - \hat{x}_k\|_{P_k^{-1}}^2 + V_k^0, \qquad (2.14a)$$

where V_k^0 is the optimal FIE cost at time instant k and P_k denotes the Kalman filter covariance matrix that satisfies the discrete-time Riccati equation

$$P_{k+1} = AP_kA^\top - AP_kC^\top \left(C\,P_k\,C^\top + R_k \right)^{-1} CP_kA^\top + GQ_kG^\top. \qquad (2.14b)$$

Hence, unconstrained linear MHE with stage cost (2.12) and prior weighting $\Gamma_k\left(\cdot\right) = Z_k(\cdot)$ given by the exact arrival cost (2.14) is GAS due to Theorem 2.1. Except for this special case, an analytical expression for the exact arrival cost is hard to obtain.

For constrained linear and nonlinear MHE, there are conceptually two approaches in the literature for the design of the prior weighting with which stability of the estimator can be established (Alessandri et al. (2010)). Either suitable conditions for approximating the arrival cost are devised or a specific structure is assigned to the prior weighting and suitable conditions on its parameters are determined.

Arrival cost approximations

For detectable linear systems subject to convex constraints, the prior weighting

$$\Gamma_k\left(p\right) = \|p - \hat{x}_k\|_{P_k^{-1}}^2 \tag{2.15}$$

becomes an approximation of the arrival cost. Here, $\hat{x}_k$ refers to the MHE estimate at time instant k and P_k is computed via (2.14b). In the nominal case, this approximation is sufficient for ensuring GAS of MHE with stage cost (2.12). Moreover, in the presence of convergent disturbances for which condition (2.7) holds, linear constrained MHE with prior weighting (2.15) is RGAS (Rawlings et al., 2017, Corollary 4.31). For general nonlinear systems, the following condition on Γ_k is sufficient for GAS of MHE.

Assumption 2.2 (Prior weighting). *The prior weighting Γ_k is continuous and satisfies the following inequalities for all $k \geq N$*

$$\hat{V}_k^0 + \underline{\gamma}_p(\|p - \hat{x}_k\|) \leq \Gamma_k(p) \leq \min_{\hat{x}_{k-N}, \hat{\mathbf{w}}_k} \sum_{i=k-N}^{k-1} l_i(\hat{w}_i, \hat{v}_i) + \Gamma_{k-N}\left(\hat{x}_{k-N}\right) \tag{2.16}$$

$$\text{s.\,t.} \quad \begin{aligned} \hat{x}_{i+1} &= f(\hat{x}_i, \hat{w}_i), & i &= k-N, \ldots, k-1 \\ y_i &= h(\hat{x}_i) + \hat{v}_i, & i &= k-N, \ldots, k-1 \\ \hat{x}_i &\in \mathcal{X}_i, & i &= k-N, \ldots, k \\ \hat{w}_i &\in \mathcal{W}_i, \ \hat{v}_i \in \mathcal{V}_i, & i &= k-N, \ldots, k-1, \\ x\left(k; \hat{x}_{k-N}, \hat{\mathbf{w}}_k\right) &= p, \end{aligned}$$

where $\underline{\gamma}_p \in \mathcal{K}_\infty$ and $\hat{V}_k^0$ denotes the MHE optimal cost at time instant k.

Theorem 2.2 (Rawlings et al. (2017)). *Consider an i-IOSS system (2.1) with $w_k = 0$, $v_k = 0$, $k \in \mathbb{N}$, and the MHE problem (2.8) with stage cost satisfying Assumption 2.1 and prior weighting satisfying Assumption 2.2. Then, the estimation error of the moving horizon estimator is GAS.*

This result is a direct consequence of (Rawlings et al., 2017, Theorem 4.28), which establishes that the estimator is RGAS under convergent disturbances satisfying condition (2.7). Assumption 2.2 and in particular continuity of Γ_k and the lower bound in (2.16) imply that similar arguments to the FIE case can be used to establish the existence of a solution to the resulting MHE problem. The upper bound in (2.16) is interpreted by Rao et al. (2003) as making sure that the approximate arrival cost Γ_k does not add any new information to the problem. Although the carried out stability analysis is rather elegant, this condition is hard to verify for general nonlinear systems and does not provide a constructive way for determining analytical approximations of the arrival cost. In fact, Rawlings et al. (2017) noted that for the general nonlinear case, ensuring satisfaction of the upper bound in (2.16) "remains a key technical challenge for MHE research".

In order to facilitate the design of nonlinear MHE, Rao (2000) discusses using the quadratic approximation of the arrival cost (2.15) as a prior weighting, where P_k refers to the covariance matrix estimates from the extended Kalman filter (EKF). More specifically, with

$$A_k = \frac{\partial f}{\partial x}\bigg|_{(\hat{x}_k,0)}, \quad G_k = \frac{\partial f}{\partial w}\bigg|_{(\hat{x}_k,0)}, \quad C_k = \frac{\partial h}{\partial x}\bigg|_{\hat{x}_k}, \tag{2.17}$$

the weight P_k is computed by recursively solving

$$P_{k+1} = A_k P_k A_k^\top - A_k P_k C_k^\top \left(C_k P_k C_k^\top + R_k\right)^{-1} C_k P_k A_k^\top + G_k Q_k G_k^\top, \tag{2.18}$$

where $Q_k \in \mathbb{S}_{++}^{m_w}$ and $R_k \in \mathbb{S}_{++}^{p}$ for all $k \in \mathbb{N}$. This approximation of the arrival cost is referred to as filtering update, since the prior weighting Γ_{k-N} employed at time k uses the filtered MHE estimate $\hat{x}_{k-N}$ obtained by solving the MHE problem at time $k - N$ given the measurement sequence $\{y_{k-2N}, \ldots, y_{k-N-1}\}$. Note that nonlinear MHE with filtering update and horizon length $N = 1$ is shown to be equivalent to the EKF when the measurement equation is linear (Robertson et al. (1996)) in the sense that they yield the same state estimates. An alternative approximation of the arrival cost is the smoothing update (Rao (2000)), where the prior weighting Γ_{k-N} employed at time k uses the smoothed MHE estimate $x(k - N; \hat{x}_{k-N-1}^\star, \hat{\mathbf{w}}_{k-N}^\star)$ obtained by solving the MHE problem at time $k-1$ given the measurement sequence $\{y_{k-N-1}, \ldots, y_{k-2}\}$. However, both approximations are guaranteed to satisfy Assumption 2.2 when linear systems, quadratic stage cost and convex constraints are considered (Rao et al. (2001)).

Alternative techniques in the literature for approximating the arrival cost of MHE include sampling-based methods (Ungarala (2009)), unscented Kalman filter (Qu and Hahn (2009)), constrained particle filters (López-Negrete et al. (2011)), and a local approximation based on a gradient condition for the arrival cost (Baumgärtner et al. (2020)). However, these methods do not investigate the stability properties of the associated estimators.

Design methods for the prior weighting

As mentioned above, based on the sufficient condition (2.16) for stability of MHE, it might not be possible to compute the prior weighting Γ_k offline. In the following, we briefly discuss other results form the literature where Γ_k can be designed a priori.

By assuming that Γ_k is positive definite with respect to the MHE estimate $\hat{x}_k$, Müller (2017) shows that there exists $N_0 \in \mathbb{N}$, such that for all $N \geq N_0$, RGAS of MHE holds under bounded disturbances for certain i-IOSS nonlinear systems. More specially, the $\mathcal{KL}$ function in the i-IOSS condition (2.4) is assumed to satisfy $\beta(r, s) \leq c_\beta r^p \Psi(s)$, for all $r \in \mathbb{R}_{++}$ and $s \in \mathbb{N}_+$, where Ψ is a nonincreasing function satisfying $\lim_{s\to\infty} \Psi(s) = 0$. Furthermore, the author shows that the estimation error converges to zero for decaying disturbances. These results are established for the first time for MHE of detectable nonlinear systems under bounded disturbances with and without an additional max-term of the stage cost l_i. However, the value of N_0 depends on a priori bounds on the initial error and disturbances.

In (Knüfer and Müller (2018)), a stronger i-IOSS condition for exponentially detectable nonlinear systems is considered and a penalty on the difference to a previously estimated state sequence is added to the quadratic cost function in order to ensure robust exponential stability of the estimation error. Using Lyapunov-based approaches as suggested by the authors in order to establish whether the considered systems verify these technical i-IOSS conditions is however difficult and not fully clear.

Specific attention was also devoted to constructive methods for the design of the prior weighting in order to preserve nominal and robust stability of MHE with quadratic stage cost (2.12). For nonlinear observable systems satisfying (2.9), Alessandri and Awawdeh (2016); Alessandri et al. (2008, 2010, 2012) consider a least squares formulation of the MHE problem and choose a quadratic prior weighting

$$\Gamma_k(p) = \|p - \bar{x}_k\|_P^2 \tag{2.19}$$

that is centred around a suitable a priori estimate $\bar{x}_k \in \mathbb{R}^n$, which they choose as the smoothed MHE estimate computed at the previous time instant $k-1$. They provide sufficient conditions on the weight $P = pI_n$, where $p \in \mathbb{R}_{++}$, such that global exponential stability of the estimation error can be ensured in the disturbance-free case. More specifically, the parameter p has to fulfill a suitable inequality that depends on global Lipschitz and observability constants of the considered system. In the linear case for instance, this condition depends on $\|A\|$ and the minimum eigenvalue of the observability matrix $\begin{bmatrix} C^\top & (CA)^\top & \cdots & (CA^N)^\top \end{bmatrix}^\top$. In the presence of additive bounded process and measurement disturbances, a bounded estimation error is guaranteed.

For observable linear systems (2.11), Sui and Johansen (2014); Sui et al. (2010) consider also a least-squares estimation problem with stage cost (2.12) and quadratic prior weighting (2.19). What is particularly interesting in their approach is that a stabilizing pre-estimating Luenberger observer is employed in the a priori estimate. In particular, the a priori estimate $\bar{x}_{k-N}$ is computed based on the last MHE solution as

$$\bar{x}_{k-N} = A\,\hat{x}^*_{k-N-1} + L(y_{k-N-1} - C\hat{x}^*_{k-N-1}). \tag{2.20}$$

Here, the observer gain $L \in \mathbb{R}^{n \times p}$ is designed such that all eigenvalues of the matrix $A - LC$ are strictly within the unit circle. Due to this formulation, ISS of the estimation error with respect to additive process and measurement disturbances can be guaranteed if the weight matrices $P \in \mathbb{S}^n_{++}$ and $R_i \in \mathbb{S}^p_{++}$ verify the following linear matrix inequalities

$$(A - LC)^\top P(A - LC) - P \prec -Q_1 \tag{2.21}$$

$$P - F_N^\top \tilde{R} F_N \prec -Q_2 \tag{2.22}$$

for some $Q_1, Q_2 \in \mathbb{S}^n_{++}$, where $F_N = \begin{bmatrix} C^\top & (C\tilde{A})^\top & \cdots & (C\tilde{A}^N)^\top \end{bmatrix}^\top$ with $\tilde{A} = A - LC$. Moreover, in the stage cost (2.12), the weight matrices R_i employed in the MHE problem are such that $\operatorname{diag}(R_{k-N}, \cdots, R_{k-1}) =: \tilde{R}_k = \tilde{R}_{k+1} =: \tilde{R}$ for all $k \in \mathbb{N}$. Note that the second linear matrix inequality (2.22) results from the fact that the pre-estimating observer is used in the forward predictions of the MHE problem as well.

MHE algorithms

All the previously discussed MHE approaches rely on the assumption that a solution to the underlying optimization problem is obtained at each time instant k, thereby reducing the applicability of MHE in engineering applications where a reliable state estimate has to be computed in real-time. This real-time challenge related to the online solution of optimization-based estimation strategies has in fact drawn special attention in recent MHE research. For instance, fast optimization strategies based on interior-point methods (Haverbeke et al. (2009); Jørgensen et al. (2004)) and Nesterov's fast gradient method (Morabito et al. (2015)) are proposed. However, no stability analysis is given there. Alessandri et al. (2008) consider approximation schemes, in which suboptimal solutions for minimizing quadratic cost functions with a given accuracy are allowed and upper bounds on the estimation errors are derived under observability assumptions. However, no optimization algorithm is specified. A similar convergence analysis is carried by Zavala (2010) for the MHE algorithm presented in (Zavala et al. (2008)), where a nominal background problem is solved based on predicted future measurements. When the true measurement arrives, the actual state is computed using a fast online correction step. To show that the generated estimation errors remain bounded, the associated approximate cost and resulting suboptimality are taken into account in the analysis. Similar to our work, Schiller et al. (2020) propose a suboptimal MHE scheme whose stability properties are inherited from an auxiliary observer and hold independently of the horizon length and even if no optimization is performed. Also in this approach, the employed optimization algorithm is not specified. Particularly interesting are works which explicitly consider the dynamics of the optimization algorithm in the convergence analysis. In (Alessandri and Gaggero (2017)), a fast MHE implementation is achieved by performing single or multiple iterations of gradient or Newton methods to minimize least-squares cost functions (2.12). For linear systems, global exponential stability of the estimation errors is shown based on an explicit representation of the error dynamics. However, the required observability assumption restricts the choice of the horizon length N and implies that it has to be greater than the state dimension n. Moreover, variants of the so-called real-time iteration scheme (Kühl et al. (2011)) are proposed, which performs a single Gauss-Newton iteration per time instant. Each iteration is split up into a preparation and an estimation phase, where the latter phase is shown to be in the range of milliseconds. The local convergence results derived by Wynn et al. (2014) are established for unconstrained nonlinear systems, i.e., no inequality constraints are considered, and hold under the assumptions of system observability in N steps (2.9). Real-time implementations of MHE are also successfully carried out in real-world applications, such as structural vibration applications (Abdollahpouri et al. (2017)), battery management systems (Morabito et al. (2017)), induction machines (Favato et al. (2019)), and industrial separation processes (Küpper et al. (2009)). However, theoretical studies that consider both stability as well as performance of MHE schemes under rather mild assumptions are to the best of the authors' knowledge rarely addressed in the literature.

Summarizing, the stability properties of the estimator can be guaranteed by design based on how to incorporate information on old measurements that are not accounted for explicitly in the receding-horizon optimization problem. Nevertheless, the design of stable and robust nonlinear MHE approaches is a topic which is still under active development (Raković and Levine (2018)), given the rather restrictive and in part non-verifiable assumptions that the considered nonlinear system and / or the MHE scheme have to fulfill. For linear systems, the aforementioned issues are circumvented but the established stability results are limited to a least-squares formulation of the optimization problem. Moreover, the theoretical results in most of the MHE approaches, expect for few works (see the discussion above), require optimal solutions and are not provided for the case where only a limited number of optimization algorithm iterations can be performed at each time instant. In this thesis, we tackle these issues within the framework of proximity-based MHE, which is inspired by connecting the problem of moving horizon state estimation to proximal methods for solving convex optimization problems.

2.2 Proximal methods

In the field of convex optimization, proximal methods are powerful algorithms for solving nonsmooth, constrained, convex optimization problems as an alternative to classical smooth optimization algorithms like the gradient descent or the Newton method (Parikh and Boyd (2014)). This class of algorithms appears to originate in the 60s, where Moreau (1965) introduced the so-called proximal operator to regularize the original function to be minimized by a quadratic proximity term. Based on a successive evaluation of the proximal operator, Rockafellar (1970) developed the proximal point algorithm for solving the original problem and explored its convergence properties. Bregman (1967) considered replacing the quadratic proximity term by a non-Euclidean distance of measure which is given by the so-called Bregman distance. This led to a generalized form of the proximal point algorithm which Censor and Zenios (1992) called proximal minimization with D-functions. Moreover, Nemirovskiĭ and Yudin (1983) introduced a generalized form of the projected subgradient algorithm called the mirror descent algorithm, where the Bregman distance is used to 'adjust gradient updates to fit problem geometry'. Since then, proximal methods have found many extensions and applications and are still an active area of research (Ryu and Boyd (2016)). In this brief review of proximal methods, we focus on three special instances relevant to our thesis, which are the proximal minimization with D-functions, the proximal point algorithm, and the mirror descent algorithm. Since we cannot cover all details of each method in this section, we refer the interested reader to the original publications and to the monograph of Parikh and Boyd (2014) and the textbook of Beck (2017) for more detail and thorough discussion of the presented approaches.

Let us begin with a few basic facts about subgradients, which are employed in the design and analysis of the subsequent algorithms. For a nondifferentiable convex function $f : \mathbb{R}^n \to \mathbb{R} \cup \{+\infty\}$, the notion of derivative is generalized by the subdifferential as follows.

Definition 2.3 (Subgradient and subdifferential). *A vector $y \in \mathbb{R}^n$ is a subgradient of a convex function $f : \mathbb{R}^n \to \mathbb{R} \cup \{+\infty\}$ at $x \in \mathbb{R}^n$ if, for all $z \in \mathbb{R}^n$,*

$$f(z) \geq f(x) + y^\top (z - x). \tag{2.23}$$

The set $\partial f(x)$ denotes the subdifferential of f at x, which is the set of all its subgradients at that point. If f is differentiable, then $\partial f(x)$ consists of one point which is the gradient of f at x.

For example, if $f : \mathbb{R} \to \mathbb{R}$ is the ℓ_1-norm, i.e., $f(x) = \|x\|_1$, its subdifferential $\partial f(x)$ is precisely the gradient for all $x \neq 0$. For $x = 0$, f is not differentiable, and we have $\partial f(0) = [-1, 1]$. If $f(x) = I_C(x)$, where I_C is the so-called indicator function of a nonempty closed and convex set $C \subseteq \mathbb{R}^n$, i.e.,

$$I_C := \begin{cases} 0 & x \in C, \\ \infty & x \notin C, \end{cases} \tag{2.24}$$

then the subdifferential of the indicator function is the normal cone operator N_C (Ryu and Boyd (2016)), which is defined as

$$N_C(x) := \begin{cases} \{y | y^\top (z - x) \leq 0 \quad \forall z \in C\} & x \in C, \\ \emptyset & x \notin C. \end{cases} \tag{2.25}$$

In other words, $\partial I_C(x) = N_C(x)$. Note that if $x \in \text{int}(C)$, then $N_C(x) = \{0\}$.

Since f is a convex function, the first-order necessary and sufficient condition for optimality can be characterized by the subdifferential as (Parikh and Boyd (2014))

$$x^* \text{ minimizes } f \quad \Leftrightarrow \quad 0 \in \partial f(x^*), \tag{2.26}$$

i.e., x^* is a global minimizer to the problem $\min_x f(x)$ if and only if (2.26) is satisfied. Consider now the following convex optimization problem

$$\min_{x \in \mathbb{R}^n} \quad F(x) = f(x) + g(x), \tag{2.27}$$

where we assume the following.

Assumption 2.3 (Composite objective function).
A) $f, g : \mathbb{R}^n \to \mathbb{R} \cup \{+\infty\}$ *are proper, closed, and convex functions.*
B) *There exists a constant $L_f \in \mathbb{R}_{++}$, such that, at any point $x \in \mathbb{R}^n$ and any subgradient $f'(x) \in \partial f(x)$, it holds that $\|f'(x)\| \leq L_f$.*
C) *The set of minimizers of problem (2.27) is nonempty and referred to by X^*.*

Assumption A) implies that the epigraphs of f and g are nonempty closed convex sets, and that f and g may take on the extended value $+\infty$ and hence may contain implicit constraints. For example, if we consider the indicator function I_C of a nonempty closed and convex set $C \subseteq \mathbb{R}^n$ as defined in (2.24), then (2.27) amounts to the constrained problem

$$\min_{x \in C} \quad f(x). \tag{2.28}$$

In Assumption B), we assume that all the subgradients of f are bounded. Concerning Assumption C), we denote the optimal value of problem (2.27) by F^{opt}.

The goal is to use suitable iterative first-order procedures for finding a solution to problem (2.27). More specifically, we consider iteration schemes of the form

$$x_{k+1} = \arg\min_{x \in \mathbb{R}^n} \left\{ f'(x_k)^\top x + g(x) + \frac{1}{\eta_k} D_\psi(x, x_k) \right\}, \tag{2.29}$$

where $k \in \mathbb{N}$ denotes the iteration index, x_k the k-th iterate of the algorithm initialized by $x_0 \in \mathbb{R}^n$, $\eta_k \in \mathbb{R}_{++}$ an appropriately chosen step size, and $f'(x_k) \in \partial f(x_k)$. Note that the choice of the subgradient $f'(x_k)$ from the subdifferential $\partial f(x_k)$ is arbitrary. Algorithms that perform the above update step are called mirror-C methods, where "C" refers to the fact that we are considering a composite objective function (Beck (2017)). In (2.29), the iterate x_{k+1} is obtained by minimizing the sum of a linear approximation of the function f, the function g and a proximity term given by the Bregman distance $D_\psi : \mathbb{R}^n \times \mathbb{R}^n \to \mathbb{R}_+$, which is a non-Euclidean distance defined as follows.

Definition 2.4 (Bregman distance). *Let $\psi : \mathbb{R}^n \to \mathbb{R}$ be a continuously differentiable and strictly convex function. Then the Bregman distance constructed from ψ is defined as*

$$D_\psi(x,y) = \psi(x) - \psi(y) - (x-y)^\top \nabla \psi(y), \qquad \forall x,y \in \mathbb{R}^n. \tag{2.30}$$

As can be seen from the definition, D_ψ computes the difference between the value of the function ψ at a point x and the first-order Taylor approximation of ψ around a point y evaluated at x. In the following, we give some background information on central properties of Bregman distances (see among others Censor and Zenios (1992); Teboulle (1992, 2018) for more detail).

Although Bregman distances are similar to distance functions and measure the proximity between two points x and y, they are in general not symmetric, i.e., $D_\psi(x,y) \neq D_\psi(y,x)$ and do not satisfy the triangle inequality. However, by the strict convexity of ψ, it follows that $D_\psi(x,y) \geq 0$ for all $x,y \in \mathbb{R}^n$ and $D_\psi(x,y) = 0$ if and only if $x = y$. Moreover, $D_\psi(x,y)$ is by definition convex in the first argument x.

A very common choice for the Bregman distance is the weighted squared Euclidean distance $D_\psi(x,y) = \frac{1}{2}\|x-y\|_P^2$, where P is a positive definite matrix. This distance is induced from the function $\psi(z) = \frac{1}{2}\|z\|_P^2$, which can be directly shown by substituting ψ in the definition (2.30). If $g = I_C$ in the composite problem (2.27), the choice of an appropriate Bregman distance can be guided by the geometry of the set C. For example, if we consider the unit simplex $C = \{x \in \mathbb{R}_{++}^n : \sum_{i=1}^n x = 1\}$, the so-called Kullback-Leibler divergence distance

$$D_\psi(x,y) = \sum_{i=1}^n x_i \log(x_i/y_i) \qquad \forall x,y \in C \tag{2.31}$$

is recommended as a suitable Bregman distance (Beck (2017)). In this case, D_ψ is generated from $\psi(x) = \sum_{i=1}^n x_i \log x_i$, $x \in \mathbb{R}_{++}^n$, which refers to the so-called entropy function over the positive orthant (Beck and Teboulle (2003)). A list of common Bregman distances can be found in (Dhillon and Tropp (2008)). The following properties satisfied by the Bregman distance will also be essential for deriving many of the results in the subsequent chapters.

Definition 2.5 (Bregman projection). *Let the set $C \subseteq \mathbb{R}^n$ be nonempty, closed and convex. The Bregman projection $\Pi_C^\psi(\bar{x})$ onto C is the closest point in C to $\bar{x}$ with respect to the Bregman distance D_ψ:*

$$\Pi_C^\psi(\bar{x}) = \arg\min_{x \in C} \quad D_\psi(x, \bar{x}). \tag{2.32}$$

If $\psi(z) = \frac{1}{2}\|z\|^2$, then D_ψ-projections are classical orthogonal projections. The next key identity can be proven by directly using the definition of D_ψ.

Lemma 2.1 (Beck and Teboulle (2003)). *Let the function D_ψ denote a Bregman distance induced from ψ. Then for any $a,b,c \in \mathbb{R}^n$, the following three-points identity holds*

$$D_\psi(c,a) + D_\psi(a,b) - D_\psi(c,b) = (\nabla \psi(b) - \nabla \psi(a))^\top (c-a). \tag{2.33}$$

Moreover, it has been shown that Bregman distances possess the following useful property (Censor and Zenios, 1992, Proposition 3.5).

Lemma 2.2 (Censor and Zenios (1992)). *Let the set $C \subseteq \mathbb{R}^n$ be nonempty, closed and convex. Suppose $x \in C$. Then,*

$$D_\psi\big(\Pi_C^\psi(\bar{x}), \bar{x}\big) \leq D_\psi\big(x, \bar{x}\big) - D_\psi\big(x, \Pi_C^\psi(\bar{x})\big). \qquad (2.34)$$

In the following, we present three special cases of the mirror-C algorithm (2.29) which are relevant to our work and that are based on special classes of the optimization problem (2.27) as well as on special choices of the Bregman distance D_ψ.

2.2.1 Proximal minimization algorithm with D-functions

Setting $f \equiv 0$ in the composite objective function (2.27) gives the problem

$$\min_{x \in \mathbb{R}^n} \quad F(x) = g(x). \qquad (2.35)$$

The iteration step (2.29) in this case takes the form

$$x_{k+1} = \arg\min_{x \in \mathbb{R}^n} \left\{ g(x) + \frac{1}{\eta_k} D_\psi(x, x_k) \right\}. \qquad (2.36)$$

Following the terminology in (Censor and Zenios (1992); Chen and Teboulle (1993)), the associated algorithm is refered to as proximal minimization with D-functions (PMD), whose basic convergence properties can be summarized in the following theorem.

Theorem 2.3 (Chen and Teboulle (1993)). *Consider problem (2.35) and let Assumption 2.3 hold. Let $\{x_k\}_{k \in \mathbb{N}_+}$ be the sequence generated by the PMD based on the iteration step (2.36). Then, for any $t \in \mathbb{N}_+$,*

$$F(x_t) - F^{\text{opt}} \leq \frac{D_\psi(x^*, x_0)}{\sum_{k=0}^{t-1} \eta_k} \qquad \forall x^* \in X^*. \qquad (2.37)$$

More specifically, we can see that the PMD converges, i.e., $F(x_t) - F^{\text{opt}} \to 0$ as $t \to \infty$ if $\lim_{t \to \infty} \sum_{k=0}^t \eta_k = \infty$. Hence, the algorithm is guaranteed to converge for any constant step size $\eta_k = \eta \in \mathbb{R}_{++}$, $k \in \mathbb{N}$. Theorem 2.3 can be proven by either using the three-points identity in Lemma 2.1 or Lemma 2.2 to first show that, for all $x^* \in X^*$, the following holds

$$D_\psi(x^*, x_{k+1}) \leq D_\psi(x^*, x_k), \qquad \forall k \in \mathbb{N}. \qquad (2.38)$$

This inequality amounts to saying that the Bregman distance to a minimizer x^* decreases at each step of the algorithm. In the next chapter, we will use the same tools, i.e., Lemma 2.1 and Lemma 2.2, to establish similar intermediate results to (2.38) in the context of proximity-based MHE.

The proximal point algorithm

In the following, we consider the PMD algorithm with a specific choice of the Bregman distance. More specifically, we construct D_ψ in (2.36) based on the function $\psi(x) = \frac{1}{2}\|x\|^2$. We obtain

$$x_{k+1} = \arg\min_{x \in \mathbb{R}^n} \left\{ g(x) + \frac{1}{2\eta_k}\|x - x_k\|^2 \right\}. \qquad (2.39)$$

An algorithm that is based on this iteration step is refered to as the proximal point algorithm (PPA) or the proximal minimization algorithm (Parikh and Boyd (2014)). Since the PMD generalizes the PPA, we can directly deduce the convergence properties of the PPA from Theorem 2.3. For instance, if we use a constant step size $\eta \in \mathbb{R}_{++}$, it holds that the sequence $\{x_k\}_{k \in \mathbb{N}_+}$ generated by the PPA converges to some point $x^* \in X^*$ and that, for any $t \in \mathbb{N}_+$

$$F(x_t) - F^{\text{opt}} \leq \frac{\|x^* - x_0\|^2}{2\,t\,\eta} \qquad \forall x^* \in X^*. \tag{2.40}$$

Due to the fact that the Bregman distance is quadratic, convergence can be also established by using proximal operator theory. The proximal operator $\text{prox}_{\eta g}(v) : \mathbb{R}^n \to \mathbb{R}^n$ of a function g with parameter $\eta \in \mathbb{R}_{++}$, is defined as (Parikh and Boyd (2014))

$$\text{prox}_{\eta g}(v) = \arg\min_{x \in \mathbb{R}^n} \left\{ g(x) + \frac{1}{2\eta}\|x - v\|^2 \right\}. \tag{2.41}$$

We can see that the point $\text{prox}_{\eta g}(v)$ is a compromise between minimizing the function g and being in proximity to v, where η is a trade-off parameter between the two terms. Note that the function we want to minimize on the right-hand side of (2.41) is strongly convex, so it possesses a unique minimizer for every $v \in \mathbb{R}^n$. Hence, the proximal operator is well-defined. For example, if $g = I_C$ is the indicator function defined in (2.24), where C is a closed nonempty convex set, the associated proximal operator is

$$\text{prox}_{\eta g}(v) = \arg\min_{x \in C} \|x - v\|, \tag{2.42}$$

which is the Euclidean projection onto C. A fundamental property of the proximal operator is that minimizers of g are fixed points of $\text{prox}_{\eta g}$, i.e., we have

$$x^* \text{ minimizes } g \quad \Leftrightarrow \quad x^* = \text{prox}_{\eta g}(x^*). \tag{2.43}$$

A proof of this property can be found in (Parikh and Boyd, 2014, page 131). Since we can minimize g by finding a fixed point of its proximal operator, a fixed point iteration of the form

$$x_{k+1} = \text{prox}_{\eta g}(x_k) \tag{2.44}$$

can be used to obtain x^*. This is in fact an iteration step of the PPA, where (2.44) reduces to (2.39) by using the definition of the proximal operator (2.41) and by setting $\eta_k = \eta$ for all $k \in \mathbb{N}$.

In the following, we discuss how this link between proximal operators and fixed point theory can be useful for establishing convergence of the PPA from this viewpoint. We do this by first showing that the proximal operator associated to g is the so-called resolvent of the subdifferential operator ∂g with parameter η, which is defined as the point-to-point mapping $R_{\partial g} := (I + \eta \partial g)^{-1}$. Let $x^+ = \text{prox}_{\eta g}(v)$. We have in view of (2.26) that

$$x^+ = \arg\min_{x \in \mathbb{R}^n} \left\{ g(x) + \frac{1}{2\eta}\|x - v\|^2 \right\} \iff 0 \in \eta \partial g(x^+) + x^+ - v. \tag{2.45}$$

Hence, with the above relation, it can be demonstrated that the proximal operator $\text{prox}_{\eta g}$ is the resolvent of ∂g since

$$x^+ = \text{prox}_{\eta g}(v) \iff v \in \eta \partial g(x^+) + x^+$$
$$\iff x^+ \in (I + \eta \partial g)^{-1}(v). \qquad (2.46)$$

Moreover, it is established that the resolvent of the subdifferential operator is an averaged operator (Parikh and Boyd (2014)), i.e., it can be written as a weighted average of I, the identity mapping, and a nonexpansive operator G as $R_{\partial g} = (1-\theta)I + \theta G$ for some $\theta \in (0,1)$. Here, the nonexpansiveness property refers to Lipschitz continuity with constant 1. Note that $R_{\partial g}$ and hence also the proximal operator are nonexpansive as well, i.e.,

$$\|\text{prox}_{\eta g}(u) - \text{prox}_{\eta g}(v)\| \leq \|u - v\|, \quad \forall u, v \in \mathbb{R}^n. \qquad (2.47)$$

Since the proximal operator $\text{prox}_{\eta g}$ is the resolvent $R_{\partial g}$, the latter is an averaged operator, and fixed point iterations associated to averaged operators converge to a fixed point if one exists (Ryu and Boyd, 2016, Section 5.2), the PPA converges under the assumption that a minimizer exists.

2.2.2 Mirror descent algorithm

Setting $g = I_C$ in the composite objective function (2.27) gives the problem

$$\min_{x \in C} \quad F(x) = f(x), \qquad (2.48)$$

where $C \subseteq \mathbb{R}^n$ is a nonempty closed and convex set. The iteration step (2.29) becomes

$$x_{k+1} = \arg\min_{x \in C} \left\{ f'(x_k)^\top x + \frac{1}{\eta_k} D_\psi(x, x_k) \right\}, \qquad (2.49)$$

and an algorithm that is based on (2.49) is refered to as the mirror descent algorithm (MDA). If $C = \mathbb{R}^n$, the function f is differentiable (i.e., $f'(x) = \nabla f(x)$ for all $x \in \mathbb{R}^n$), and the Bregman distance D_ψ is chosen as $D_\psi(x, x_k) = \frac{1}{2}\|x - x_k\|_P^2$, where P is a positive definite matrix, taking the first-order optimality condition of (2.49) at the solution x_{k+1} yields

$$0 = \nabla f(x_k) + \frac{1}{\eta_k} P(x_{k+1} - x_k). \qquad (2.50)$$

Hence, we can obtain in this case an analytical expression for the MDA iteration step:

$$x_{k+1} = x_k - \eta_k P^{-1} \nabla f(x_k). \qquad (2.51)$$

Furthermore, if $P = I_n$, we recover the gradient descent step for minimizing the function f,

$$x_{k+1} = x_k - \eta_k \nabla f(x_k). \qquad (2.52)$$

Hence, similar to how the PMD algorithm generalizes the PPA, the MDA generalizes the gradient method to the non-Euclidean setting.

As mentioned above, the choice of an appropriate Bregman distance can be guided by the geometry of the constraint set C. For example, if C is the unit simplex introduced above and the Kullback-Leiber divergence (2.31) is used as a Bregman distance, (2.49) can be also expressed analytically to form the iteration step of the so-called entropic descent algorithm (Beck and Teboulle (2003))

$$x_{k+1}^j = \frac{x_k^j \exp^{-\eta_k f'(x_k)^j}}{\sum_{j=1}^n x_k^j \exp^{-\eta_k f'(x_k)^j}}, \qquad j = 1, \cdots, n \tag{2.53}$$

where $f'(x_k)^j$ is the j-th component of the subgradient $f'(x_k) \in \partial f(x_k)$.
Recall that the choice of the subgradient $f'(x_k)$ in (2.49) is arbitrary, therefore, the direction of $-f'(x_k)$ is not necessarily a descent direction (see for instance (Beck, 2017, Example 8.3)). Hence, the sequence of function values $\{F(x_k)\}_{k \in \mathbb{N}_+}$ generated by the MDA is not necessarily monotone. Convergence of the method is therefore established based on the best achieved function value defined as

$$F^{\text{best},t} := \min_{k=0,1,\ldots,t} F(x_k). \tag{2.54}$$

Based on the three-points identity presented in Lemma 2.1, Beck and Teboulle (2003) establish in the following theorem an upper bound on the distance of $F^{\text{best},t}$ to F^{opt}.

Theorem 2.4 (Beck and Teboulle (2003)). *Consider problem (2.48) and let Assumption 2.3 hold. Suppose that ψ is strongly convex with parameter $\sigma \in \mathbb{R}_{++}$. Let $\{x_k\}_{k \in \mathbb{N}_+}$ be the sequence generated by the MDA based on the iteration step (2.49). Then, for any $t \in \mathbb{N}_+$,*

$$F^{\text{best},t} - F^{\text{opt}} \leq \frac{D_\psi(x^*, x_0) + \frac{L_f^2}{2\sigma} \sum_{k=0}^{t-1} \eta_k^2}{\sum_{k=0}^{t-1} \eta_k} \qquad \forall x^* \in X^*, \tag{2.55}$$

where $F^{\text{best},t}$ is defined in (2.54).

More specifically, we can see that the MDA converges, i.e., $F^{\text{best},t} - F^{\text{opt}} \to 0$ as $t \to \infty$ if the step sizes verify

$$\frac{\sum_{k=0}^t \eta_k^2}{\sum_{k=0}^t \eta_k} \to 0, \qquad t \to \infty. \tag{2.56}$$

If we fix the total number of iterations $t \in \mathbb{N}_+$, one simple choice for the step size that verifies the above condition is (Beck, 2017, Theorem 9.16)

$$\eta_k = \frac{\sqrt{2\sigma D_\psi(x^*, x_0)}}{L_f \sqrt{t}}, \qquad k = 0, 1, \cdots, t-1. \tag{2.57}$$

With this constant step size, (2.55) becomes

$$F^{\text{best},t} - F^{\text{opt}} \leq \frac{\sqrt{2 D_\psi(x^*, x_0)} L_f}{\sqrt{\sigma} \sqrt{t}}. \tag{2.58}$$

In order to be able to compute η_k based on (2.57), Beck and Teboulle (2003) discuss ways to get around the unknown quantity $D_\psi(x^*, x_0)$ since it depends on the unknown minimizer x^* of problem (2.48). For example, if $D_\psi(x, x_0)$ is bounded over the set C, a quantity $R(x_0)$ satisfying $R(x_0) \geq \max_{x \in C} D_\psi(x, x_0)$ can replace $D_\psi(x^*, x_0)$ in the upper bound in (2.55) and hence also in the step size (2.57). Note that convergence of the MDA is also established under a dynamic (non-constant) step size rule, where the total number of iterations t need not be fixed a priori (Beck, 2017, Theorem 9.18).

2.3 Summary

In this chapter, we briefly reviewed some of the standard MHE strategies from the literature, where stability of the estimator is enforced through a suitable design of the prior weighting. Moreover, we gave a brief overview on three different proximal methods for solving nonsmooth convex minimization problems, for which we presented basic convergence results. More specifically, in all the methods in question, the idea is to find a minimizer of a nonsmooth convex function by iteratively solving a proximal subproblem, where the smooth Bregman distance is employed as a proximity regularizing term to the previous iterate. Coming back to the main topic of this thesis, we will embed this idea in the context of MHE in the next chapter. In particular, the underlying optimization problem will consist of two parts: a nonsmooth convex stage cost and a Bregman distance employed as a proximity term to a stabilizing a priori estimate. Equipped with the presented tools for the analysis of proximal methods, this novel concept of proximity MHE will provide a unified framework for the design, analysis and implementation of moving horizon estimators with theoretical guarantees.

Chapter 3

Proximity-based MHE schemes

In this chapter, we address the moving horizon state estimation problem of constrained discrete-time linear and nonlinear systems and introduce a novel proximity-based formulation of the underlying optimization problem. More specifically, the cost function in proximity MHE (pMHE) consists of a performance measure designed based on a convex, possibly nonsmooth stage cost, as well as a suitable Bregman distance to an a priori estimate that is constructed based on a classical model-based stabilizing estimator, from which stability can be inherited. A major theme in this chapter is to tailor the design of the pMHE problem to the considered class of dynamical systems and devise suitable sufficient conditions for the exponential stability of the resulting estimation error by means of a Lyapunov analysis. The obtained stability results can be ensured by a rather simple design of the Bregman distance and the a priori estimate, without jeopardizing the user's freedom in selecting suitable stage costs and keeping the horizon length small.

This chapter is structured as follows. In Section 3.1, we introduce the problem setup and the proximity-based formulation of the MHE optimization problem in its general form. In Section 3.2, we specifically consider linear systems, for which we investigate nominal and robust stability properties of the resulting estimation error, and present a Bayesian interpretation of pMHE. In Section 3.3, we establish the stability properties of nonlinear pMHE for a broad class of discrete-time nonlinear systems. Moreover, we focus on a special class of nonlinear systems that can be transformed into systems which are affine in the unmeasured state and for which the resulting pMHE problem is convex. In Section 3.4, we use various simulation examples to illustrate our theoretical results and show an improved performance compared to standard MHE schemes (cf. Chapter 2) in case of non-Gaussian disturbances. Finally, we conclude this chapter in Section 3.5 with a summary.

The results presented in this chapter are based on (Gharbi and Ebenbauer (2018, 2019a,b, 2020); Gharbi et al. (2020a)).

3.1 Problem setup and proximity MHE

In this section, we consider the state estimation problem of general discrete-time nonlinear systems of the form

$$x_{k+1} = f_k(x_k, u_k, w_k), \tag{3.1a}$$

$$y_k = h_k(x_k) + v_k, \tag{3.1b}$$

where $k \geq k_0 \in \mathbb{N}$ denotes the discrete time instant, $x_k \in \mathbb{R}^n$ the state vector, $u_k \in \mathbb{R}^m$ the input vector, and $y_k \in \mathbb{R}^p$ the measurement vector. We assume for simplicity of presentation that the time instant at which we start the observations is zero, i.e., $k_0 = 0$. The vectors $w_k \in \mathbb{R}^{m_w}$ and $v_k \in \mathbb{R}^p$ represent unknown process and measurement disturbances. Furthermore, the initial condition $x_0 \in \mathbb{R}^n$ of system (3.1) is unknown. The state and disturbances are known to satisfy the following constraints

$$x_k \in \mathcal{X}_k \subseteq \mathbb{R}^n, \quad w_k \in \mathcal{W}_k \subseteq \mathbb{R}^{m_w}, \quad v_k \in \mathcal{V}_k \subseteq \mathbb{R}^p, \qquad k \in \mathbb{N}. \tag{3.2}$$

We impose the following assumptions.

Assumption 3.1 (System functions). *For any $k \in \mathbb{N}$, the functions $f_k : \mathbb{R}^n \times \mathbb{R}^m \times \mathbb{R}^{m_w} \to \mathbb{R}^n$ and $h_k : \mathbb{R}^n \to \mathbb{R}^p$ are twice continuously differentiable and Lipschitz continuous with Lipschitz constants $c_f \in \mathbb{R}_{++}$ and $c_h \in \mathbb{R}_{++}$, respectively, in all of their arguments and uniformly over $k \in \mathbb{N}$.*

Assumption 3.2 (Constraint sets). *For any $k \in \mathbb{N}$, the sets $\mathcal{X}_k$, $\mathcal{W}_k$, and $\mathcal{V}_k$ are closed and convex, $0 \in \mathcal{W}_k$, and $0 \in \mathcal{V}_k$.*

For the sake of convenience, we briefly adapt the notation we used in the previous chapter to system (3.1). Let $x\,(k; x_i, i, \mathbf{u}_k, \mathbf{w}_k)$ denote the solution of system (3.1) at time instant k with initial state x_i at time i and input and disturbance sequences $\mathbf{u}_k = \{u_i, \cdots, u_{k-1}\}$ and $\mathbf{w}_k = \{w_i, \cdots, w_{k-1}\}$, respectively. This notation is simplified to $x\,(k; x_i, i, \mathbf{u}_k)$ in the disturbance-free case where $w_k = 0$ for all $k \in \mathbb{N}$.

Our aim is to compute an estimate $\hat{x}_k$ of the state x_k in a moving horizon fashion, given the model (3.1), the constraints (3.2) and a constant number $N \in \mathbb{N}_+$ of past inputs $\{u_{k-N}, \ldots, u_{k-1}\}$ and measurements $\{y_{k-N}, \ldots, y_{k-1}\}$. This is the same problem statement as in the previous chapter, for which we reported various MHE formulations in the literature, many of which are based on solving an optimization problem of the form (2.8), where an appropriate prior weighting Γ_k, often quadratic, is used to ensure stability. In this thesis, we propose a proximity-based MHE formulation that is motivated by linking MHE to proximal methods presented in Section 2.2. To see this link, and for simplicity of presentation, let us consider as starting point the MHE problem (2.8) with quadratic prior weighting (2.19) and zero process disturbances $\hat{w}_i = 0$. Let us also discard the inequality constraints. For a fixed time instant $k \in \mathbb{N}$, the resulting MHE problem can be written more compactly as

$$\hat{x}_{k-N}^* = \arg\min_{x \in \mathbb{R}^n} \left\{ \hat{f}(x) + \frac{1}{2}\|x - \bar{x}_{k-N}\|_P^2 \right\}, \tag{3.3}$$

where $\hat{f} : \mathbb{R}^n \to \mathbb{R}$ accounts for the sum of stage cost and $\bar{x}_{k-N} \in \mathbb{R}^n$ denotes a suitable a priori estimate. We compare this formulation to a standard proximal point algorithm (PPA) introduced in Section 2.2.1 and given by

$$x_{k+1}^* = \operatorname{prox}_{\eta f}(x_k^*). \tag{3.4a}$$

The proximal operator is defined as

$$\operatorname{prox}_{\eta f}(v) = \arg\min_{x \in \mathbb{R}^n} \left\{ f(x) + \frac{1}{2}\|x - v\|_{\eta I}^2 \right\}, \tag{3.4b}$$

where $f : \mathbb{R}^n \to \mathbb{R}$ is convex function and $\eta \in \mathbb{R}_{++}$ denotes a suitable step size. Recall that the PPA iteratively finds a minimizer of a convex and potentially nondifferentiable function f by appending a quadratic proximity measure to the current iterate and solving the resulting regularized proximal subproblem instead. We can observe that the MHE formulation in which the quadratic prior weighting is added to the sum of stage cost fits very well in this proximity setup. In particular, notice that (3.3) is rather similar to (3.4), i.e., we can write the estimator (3.3) as

$$\hat{x}^*_{k-N} = \hat{\text{prox}}_{\hat{f}}(\bar{x}_{k-N}), \tag{3.5a}$$

where the modified proximal operator is defined as

$$\hat{\text{prox}}_{\hat{f}}(v) = \arg\min_{x \in \mathbb{R}^n} \left\{ \hat{f}(x) + \frac{1}{2}\|x - v\|^2_P \right\}. \tag{3.5b}$$

Moreover, notice that using a Luenberger observer as a stabilizing pre-estimating observer to generate the a priori estimate in (3.5) as is the case for the MHE approach proposed by Sui and Johansen (2014) yields

$$\hat{x}^*_{k-N} = \hat{\text{prox}}_{\hat{f}}\left((A - LC)\hat{x}^*_{k-N-1} + Bu_{k-N-1} + Ly_{k-N-1}\right). \tag{3.6}$$

Hence, by contrast to the standard proximal formulation given in (3.4) in which the previous solution is directly substituted into the proximal operator, in the modified formulation, the previous MHE solution $\hat{x}^*_{k-N-1}$ enters via the a priori estimation step as can be seen in (3.6). This can be interpreted as introducing an additional prediction step in the proximal algorithm.

The established similarities between MHE and proximal methods motivate the main conceptual idea behind the proximity-based MHE formulation, which is to formulate the underlying optimization problem with a rather general and flexible convex stage cost in order to ensure a satisfactory performance, *and* a proximity prior term to a stabilizing a priori estimate $\bar{x}_{k-N}$ from which stability can be inherited. Hence, in pMHE, the convex stage cost need not be quadratic and can be nonsmooth allowing for instance to employ the ℓ_1-norm which is known to be robust against outliers. Furthermore, motivated by the fact that the proximal minimization algorithm with D-functions (PMD) introduced in Section 2.2.1 generalizes the PPA by replacing the quadratic proximity term with the more general Bregman distance, we propose to employ in the pMHE cost function a suitable Bregman distance as a generalized proximity measure to a stabilizing a priori estimate.

In light of this discussion, and according to the presented problem setup, we introduce in the following the pMHE scheme for the state estimation problem of system (3.1) with constraints (3.2). While the formulation of the underlying optimization problem is introduced on a conceptual level, specific design methods will follow in the subsequent sections. We consider again the general MHE problem introduced in (2.8). In pMHE, we choose the stage cost l_i as the sum of convex and nonnegative functions $r_i : \mathbb{R}^p \to \mathbb{R}_+$ and $q_i : \mathbb{R}^{m_w} \to \mathbb{R}_+$, which penalize the output residual $\hat{v}_i$ and process disturbance $\hat{w}_i$, respectively. More specifically, we set

$$l_i(\hat{w}_i, \hat{v}_i) := r_i(\hat{v}_i) + q_i(\hat{w}_i), \qquad i = k - N, \ldots, k - 1, \tag{3.7}$$

and formulate the pMHE problem as follows

$$\min_{\hat{\mathbf{z}}_k} \quad \sum_{i=k-N}^{k-1} r_i(\hat{v}_i) + q_i(\hat{w}_i) + D_{\psi_k}\left(\hat{\mathbf{z}}_k, \bar{\mathbf{z}}_k\right) \tag{3.8a}$$

$$\text{s.t.} \quad \hat{x}_{i+1} = f_i(\hat{x}_i, u_k, \hat{w}_i), \qquad i = k-N, \ldots, k-1 \tag{3.8b}$$

$$y_i = h_i(\hat{x}_i) + \hat{v}_i, \qquad i = k-N, \ldots, k-1 \tag{3.8c}$$

$$\hat{x}_i \in \mathcal{X}_i, \qquad i = k-N, \ldots, k \tag{3.8d}$$

$$\hat{w}_i \in \mathcal{W}_i, \ \hat{v}_i \in \mathcal{V}_i, \qquad i = k-N, \ldots, k-1, \tag{3.8e}$$

where we collect the decision variables $(\hat{x}_{k-N}, \hat{\mathbf{w}}_k)$ in the vector $\hat{\mathbf{z}}_k$, i.e.,

$$\hat{\mathbf{z}}_k := \begin{bmatrix} \hat{x}_{k-N}^\top & \hat{\mathbf{w}}_k^\top \end{bmatrix}^\top = \begin{bmatrix} \hat{x}_{k-N}^\top & \hat{w}_{k-N}^\top & \cdots & \hat{w}_{k-1}^\top \end{bmatrix}^\top \in \mathbb{R}^{Nm_w+n}. \tag{3.9}$$

Note that with a slight abuse of notation, we use $\hat{\mathbf{w}}_k$ in order to refer to the disturbance sequence $\{\hat{w}_{k-N}, \cdots, \hat{w}_{k-1}\}$ or to the column vector $\begin{bmatrix} \hat{w}_{k-N}^\top & \cdots & \hat{w}_{k-1}^\top \end{bmatrix}^\top \in \mathbb{R}^{Nm_w}$. In the pMHE cost function (3.8a), $D_{\psi_k} : \mathbb{R}^{Nm_w+n} \times \mathbb{R}^{Nm_w+n} \to \mathbb{R}_+$ denotes the Bregman distance induced at time instant $k \in \mathbb{N}$ from the function $\psi_k : \mathbb{R}^{Nm_w+n} \to \mathbb{R}$ as

$$D_{\psi_k}(\mathbf{z}_1, \mathbf{z}_2) = \psi_k(\mathbf{z}_1) - \psi_k(\mathbf{z}_2) - (\mathbf{z}_1 - \mathbf{z}_2)^\top \nabla \psi_k(\mathbf{z}_2). \tag{3.10}$$

A brief overview on the useful properties of Bregman distances is provided in Section 2.2. Note that we choose to endow the function ψ_k in (3.8a) with the time index k and not $k - N$ in order to refer to the Bregman distance D_{ψ_k} in the pMHE cost function to be minimized at time instant k. By eliminating the system dynamics (3.8b) and (3.8c), the pMHE problem (3.8) can be written more compactly as

$$\min_{\hat{\mathbf{z}}_k \in \mathcal{S}_k} \quad J_k(\hat{\mathbf{z}}_k) = F_k(\hat{\mathbf{z}}_k) + D_{\psi_k}\left(\hat{\mathbf{z}}_k, \bar{\mathbf{z}}_k\right), \tag{3.11a}$$

where $F_k : \mathbb{R}^{Nm_w+n} \to \mathbb{R}_+$ denotes the sum of stage cost

$$F_k(\hat{\mathbf{z}}_k) = \sum_{i=k-N}^{k-1} r_i\left(y_i - h_i\left(x\left(i; \hat{x}_{k-N}, k-N, \mathbf{u}, \hat{\mathbf{w}}\right)\right)\right) + q_i(\hat{w}_i) \tag{3.11b}$$

with $\mathbf{u} = \{u_{k-N}, \cdots, u_{k-1}\}$ and $\hat{\mathbf{w}} = \{\hat{w}_{k-N}, \cdots, \hat{w}_{i-1}\}$ for each $i = k-N, \cdots, k-1$ and the set $\mathcal{S}_k \subseteq \mathbb{R}^{Nm_w+n}$ represents the pMHE feasible set in the sense that

$$\mathcal{S}_k := \Big\{ \hat{\mathbf{z}}_k = \begin{bmatrix} \hat{x}_{k-N}^\top & \hat{\mathbf{w}}_k^\top \end{bmatrix}^\top : x\left(i; \hat{x}_{k-N}, k-N, \mathbf{u}, \hat{\mathbf{w}}\right) \in \mathcal{X}_i, \qquad i = k-N, \ldots, k \tag{3.11c}$$

$$\hat{w}_i \in \mathcal{W}_i, \qquad i = k-N, \ldots, k-1$$

$$y_i - h(x\left(i; \hat{x}_{k-N}, k-N, \mathbf{u}, \hat{\mathbf{w}}\right)) \in \mathcal{V}_i, \quad i = k-N, \ldots, k-1 \Big\}.$$

Since the true system satisfies the inequality constraints and therefore $(x_{k-N}, \mathbf{w}_k)$ lies in the set $\mathcal{S}_k$, the pMHE problem is feasible at all times.

We impose the following assumptions on the stage cost and the Bregman distance, with which we can establish the existence of a solution to the pMHE problem (3.8).

Assumption 3.3 (pMHE stage cost). *For any $i \in \mathbb{N}$, the functions r_i and q_i are continuous uniformly over i, convex and nonnegative, and attain their minimum zero at zero.*

Assumption 3.4 (Bregman distance). *For any $k \in \mathbb{N}$, the function ψ_k is continuously differentiable, strongly convex with constant $\sigma_k \in \mathbb{R}_{++}$ and strongly smooth with constant $\gamma_k \in \mathbb{R}_{++}$, which implies the following for the associated Bregman distance*

$$\frac{\sigma_k}{2}\|\mathbf{z}_1 - \mathbf{z}_2\|^2 \leq D_{\psi_k}(\mathbf{z}_1, \mathbf{z}_2) \leq \frac{\gamma_k}{2}\|\mathbf{z}_1 - \mathbf{z}_2\|^2 \tag{3.12}$$

for all $\mathbf{z}_1, \mathbf{z}_2 \in \mathbb{R}^{Nm_w+n}$. Moreover, these properties hold uniformly over k. In particular, there exist $\sigma \in \mathbb{R}_{++}$ and $\gamma \in \mathbb{R}_{++}$ such that $\sigma_k \geq \sigma$ and $\gamma_k \leq \gamma$ for all $k \in \mathbb{N}$.

Some comments on Assumptions 3.3 and 3.4 are in order. First, Assumption 3.3 highlights the importance of a convex stage cost in pMHE, which will prove useful in the proximity-based stability analyses even in the nonlinear case. Note that most MHE formulations in the literature, including the works that establish stability under the rather general Assumption 2.1 on the stage cost, use least squares formulations when specific applications are considered. Concerning Assumption 3.4, even though the class of Bregman distances for which the quadratic lower and upper bounds (3.12) hold might seem rather restrictive, it includes the important special case of quadratic distances $D_{\psi_k}(\mathbf{z}_1, \mathbf{z}_2) = \frac{1}{2}\|\mathbf{z}_1 - \mathbf{z}_2\|^2_{P_k}$ induced from the function $\psi_k(\mathbf{z}) = \frac{1}{2}\|\mathbf{z}\|^2_{P_k}$, with $P_k \in \mathbb{S}^{Nm_w+n}_{++}$, as discussed in Section 2.2. In fact, quadratic distances are widely used as prior weighting in MHE as we have seen in the previous chapter and above discussions in order to ensure stability of the estimation error (cf. (2.15)). In our setting, we can consider in addition any function of the form $\psi_k(\mathbf{z}) = \frac{1}{2}\|\mathbf{z}\|^2_{P_k} + B_k(\mathbf{z})$, where $B_k : \mathbb{R}^{Nm_w+n} \to \mathbb{R}$ is a convex and strongly smooth function, i.e., a convex function whose gradient is uniformly Lipschitz continuous.

Theorem 3.1. *Consider system (3.1) and let Assumptions 3.1-3.4 hold. Then, the pMHE optimization problem (3.8) has a bounded and nonempty set of minimizers, one of which we denote by $\hat{\mathbf{z}}^*_k$.*

Proof. First, due to the continuity of the system functions f_k, h_k, the functions r_i and q_i in the stage cost as well as of the Bregman distance, the pMHE cost function J_k in (3.11) is continuous in the decision variable $\hat{\mathbf{z}}_k$. Moreover, given the nonnegativity of r_i and q_i in Assumption 3.3, J_k can be lower bounded by the Bregman distance D_{ψ_k}. Furthermore, due to the lower bound on D_{ψ_k} in Assumption 3.4, it holds that $J_k(\hat{\mathbf{z}}_k) \geq \frac{\sigma}{2}\|\hat{\mathbf{z}}_k - \bar{\mathbf{z}}_k\|^2$ for all $\hat{\mathbf{z}}_k \in \mathbb{R}^{Nm_w+n}$ and fixed $\bar{\mathbf{z}}_k$. Hence, the cost function J_k is radially unbounded in $\hat{\mathbf{z}}_k$, and all its sublevel sets are therefore bounded. In view of Assumption 3.2, the feasible set $\mathcal{S}_k$ is closed and nonempty, since the true system satisfies the inequality constraints. Hence, any existing minimizers lie in the intersection of $\mathcal{S}_k$ and a closed and bounded sublevel set containing the true state. This nonempty intersection is therefore closed and bounded, and hence compact. Applying the Weierstrass Theorem stating that the set of minimizers of a continuous function on a compact set is nonempty and bounded establishes the desired result. A more rigorous and detailed discussion can be found in the proof of (Bayer, 2019, Theorem 4.1). □

In the pMHE cost function (3.8a), the Bregman distance D_{ψ_k} is used as a proximity measure to a stabilizing a priori estimate, which we refer to as $\bar{\mathbf{z}}_k \in \mathbb{R}^{Nm_w+n}$. This a priori estimate evolves based on a simple, model-based, and recursive state estimator whose dynamics can be described by an operator $\Phi_k : \mathbb{R}^{Nm_w+n} \to \mathbb{R}^{Nm_w+n}$. More specifically, given the pMHE solution $\hat{\mathbf{z}}_k^*$ obtained at time instant k, the a priori estimate for the next time instant $k + 1$ is calculated as

$$\bar{\mathbf{z}}_{k+1} = \Phi_k\left(\hat{\mathbf{z}}_k^*\right). \tag{3.13}$$

Hence, for the propagation of the pMHE scheme at time k to the next time instant $k + 1$, only the solution $\hat{\mathbf{z}}_k^*$ has to be retained. Note that Φ_k is time varying since it generates the a priori estimate based on inputs and measurements of the system. For the sake of clarity, let us summarize the following notations which will be used throughout the chapter:

$$\hat{\mathbf{z}}_k^* := \begin{bmatrix} \hat{x}_{k-N}^{*\top} & \hat{w}_{k-N}^{*\top} & \dots & \hat{w}_{k-1}^{*\top} \end{bmatrix}^\top \in \mathbb{R}^{Nm_w+n}, \tag{3.14a}$$

$$\bar{\mathbf{z}}_k := \begin{bmatrix} \bar{x}_{k-N}^{\top} & \bar{w}_{k-N}^{\top} & \dots & \bar{w}_{k-1}^{\top} \end{bmatrix}^\top \in \mathbb{R}^{Nm_w+n}, \tag{3.14b}$$

$$\mathbf{z}_k := \begin{bmatrix} x_{k-N}^{\top} & w_{k-N}^{\top} & \dots & w_{k-1}^{\top} \end{bmatrix}^\top \in \mathbb{R}^{Nm_w+n}. \tag{3.14c}$$

At time instant k, we denote with $\hat{\mathbf{z}}_k^*$ a solution of the pMHE optimization problem (3.8), with $\bar{\mathbf{z}}_k$ the a priori estimate and with $\mathbf{z}_k$ the true state x_{k-N} with the true process disturbance sequence $\mathbf{w}_k = \{w_{k-N}, \cdots, w_{k-1}\}$. Given the definition of $\mathbf{z}_k$, we now require the following assumption on the a priori estimate operator Φ_k.

Assumption 3.5 (A priori estimate operator). *In the disturbance-free case where* $\mathbf{z}_k = \begin{bmatrix} x_{k-N}^{\top} & 0 & \dots & 0 \end{bmatrix}^\top$ *and* $\mathbf{z}_{k+1} = \begin{bmatrix} x_{k-N+1}^{\top} & 0 & \dots & 0 \end{bmatrix}^\top$, *it holds that* $\Phi_k(\mathbf{z}_k) = \mathbf{z}_{k+1}$.

Assumption 3.5 states that, in the absence of disturbances, if we substitute the true state in the a priori estimate operator Φ_k, we should obtain the next true state. This reflects the intuitive fact that, in this case, the stabilizing estimator based on which the a priori estimate is constructed should follow the true system dynamics.

Based on the pMHE solution $\hat{\mathbf{z}}_k^*$ and a forward prediction of the dynamics (3.8b), we obtain the state estimate at time instant k as

$$\hat{x}_k = x\left(k; \hat{x}_{k-N}^*, k-N, \mathbf{u}_k, \hat{\mathbf{w}}_k^*\right). \tag{3.15}$$

We define the pMHE error for system (3.1) as

$$\mathbf{z}_k - \hat{\mathbf{z}}_k^* = \begin{bmatrix} x_{k-N} \\ \mathbf{w}_k \end{bmatrix} - \begin{bmatrix} \hat{x}_{k-N}^* \\ \hat{\mathbf{w}}_k^* \end{bmatrix} = \begin{bmatrix} e_{k-N} \\ \mathbf{w}_k - \hat{\mathbf{w}}_k^* \end{bmatrix}, \tag{3.16}$$

where $e_{k-N} := x_{k-N} - \hat{x}_{k-N}^*$, and the estimation error as

$$x_k - \hat{x}_k = x\left(k; x_{k-N}, k-N, \mathbf{u}_k, \mathbf{w}_k\right) - x\left(k; \hat{x}_{k-N}^*, k-N, \mathbf{u}_k, \hat{\mathbf{w}}_k^*\right). \tag{3.17}$$

Overall, the pMHE scheme can be summarized by Algorithm 3.1.

Remark 3.1. For $k \leq N$, we consider the pMHE problem (3.8) with $N = k$. More specifically, we have $\hat{\mathbf{z}}_k^* = \begin{bmatrix} \hat{x}_0^{*\top} & \hat{w}_0^{*\top} & \dots & \hat{w}_{k-1}^{*\top} \end{bmatrix}^\top$ and set $\bar{\mathbf{z}}_k = \begin{bmatrix} \bar{x}_0^{\top} & \bar{w}_0^{\top} & \dots & \bar{w}_{k-1}^{\top} \end{bmatrix}^\top$, where $\hat{\mathbf{z}}_k^*, \bar{\mathbf{z}}_k \in \mathbb{R}^{km_w+n}$ and $\bar{x}_0$ in the a priori estimate denotes the initial guess. In other words, this amounts to solving for the first N time instants the FIE problem introduced in Section 2.1, where all the available measurements are taken into account in the optimization process at each time $k \leq N$.

Algorithm 3.1 General pMHE scheme

Offline: Specify the functions r_i, q_i, the estimation horizon N, and design a suitable Bregman distance D_{ψ_k} and a priori estimate operator Φ_k. Choose an initial guess $\bar{x}_0 \in \mathbb{R}^n$ and set $\bar{\mathbf{z}}_0 = \bar{x}_0$.

Online:

1: **for** $k = 0, 1, 2, \cdots$ **do**
2:      get the inputs $\{u_{k-N}, \ldots, u_{k-1}\}$ and measurements $\{y_{k-N}, \ldots, y_{k-1}\}$
3:      compute a solution $\hat{\mathbf{z}}_k^*$ to problem (3.8) or equivalently solve

$$\hat{\mathbf{z}}_k^* = \underset{\hat{\mathbf{z}}_k \in \mathcal{S}_k}{\arg\min} \ \{F_k(\hat{\mathbf{z}}_k) + D_{\psi_k}(\hat{\mathbf{z}}_k, \bar{\mathbf{z}}_k)\}$$

4:      obtain the state estimate $\hat{x}_k$ based on $\hat{\mathbf{z}}_k^*$
5:      compute the a priori estimate $\bar{\mathbf{z}}_{k+1} = \Phi_k(\hat{\mathbf{z}}_k^*)$ for the next time instant
6: **end for**

In pMHE, the Bregman distance D_{ψ_k} is added to the time-varying, convex and potentially non-differentiable sum of stage cost F_k. The latter consists of the functions r_i and q_i, which are designed to attain their respective minimum values when the output residuals and process disturbances are zero, i.e., when the true nominal value of the state is achieved. The goal is to make sure that the solution of the resulting optimization problem and consequently the state estimate converge to the true state. In the following sections, we will see how this can be enforced in the framework of pMHE through simple design approaches of the Bregman distance D_{ψ_k} and the a priori estimate operator Φ_k, whose design can be fixed a priori and is based on the system class under consideration. More specifically, choosing appropriate stability components D_{ψ_k} and Φ_k will ensure desirable theoretical properties like nominal stability of the estimation error or its robustness with respect to additive process and measurement disturbances, which hold for a rather general convex stage cost and for any horizon length. Moreover, we will show by establishing a stochastic point of view on linear pMHE as well as via simulation examples how a rather general and not necessarily quadratic performance criterion enables to handle non-Gaussian disturbances such as outliers and to consider other specific data characteristics.

3.2 Proximity MHE for linear systems

In this section, we assume that system (3.1) is a discrete-time linear time-varying (LTV) system of the form

$$x_{k+1} = A_k\, x_k + B_k\, u_k + w_k, \tag{3.18a}$$

$$y_k = C_k\, x_k + v_k, \tag{3.18b}$$

with the specific dimension $m_{\mathrm{w}} = n$ and where all matrices A_k, B_k, C_k are of compatible dimensions. The resulting constrained pMHE optimization problem to be solved at each time instant k is

$$\min_{\hat{\mathbf{z}}_k} \quad \sum_{i=k-N}^{k-1} r_i(\hat{v}_i) + q_i(\hat{w}_i) + D_{\psi_k}(\hat{\mathbf{z}}_k, \bar{\mathbf{z}}_k) \tag{3.19a}$$

$$\text{s.t.} \quad \hat{x}_{i+1} = A_i\,\hat{x}_i + B_i\,u_i + \hat{w}_i, \qquad i = k-N, \dots, k-1 \tag{3.19b}$$

$$y_i = C_i\,\hat{x}_i + \hat{v}_i, \qquad i = k-N, \dots, k-1 \tag{3.19c}$$

$$\hat{x}_i \in \mathcal{X}_i, \qquad i = k-N, \dots, k \tag{3.19d}$$

$$\hat{w}_i \in \mathcal{W}_i, \ \hat{v}_i \in \mathcal{V}_i, \qquad i = k-N, \dots, k-1. \tag{3.19e}$$

Given the linear dynamics, the convex constraints and convex objective, the overall pMHE optimization problem is convex. Moreover, since the Bregman distance D_{ψ_k} is constructed from a strongly convex function ψ_k as specified by Assumption 3.4, there exists a unique solution $\hat{\mathbf{z}}_k^*$ to (3.19), based on which the state estimate defined in (3.15) can be explicitly computed as

$$\hat{x}_k = x\left(k; \hat{x}_{k-N}^*, k-N, \mathbf{u}_k, \hat{\mathbf{w}}_k^*\right) \tag{3.20}$$

$$= \Phi(k, k-N)\,\hat{x}_{k-N}^* + \sum_{i=k-N}^{k-1} \Phi(k, i+1)\,(B_i\,u_i + \hat{w}_i^*).$$

Here, $\Phi(k, k_0) := A_{k-1} \dots A_{k_0}$ denotes the multiplication of the transition matrices from some state x_{k_0} to x_k for the homogeneous part of system (3.18). Note that $\Phi(k+1, k) = A_k$ and we let $\Phi(k, k) = I_n$.

In Section 3.2.1, the stability properties of the linear pMHE scheme are analyzed when no disturbances act on the system (i.e., $w_k = 0$, $v_k = 0$, $k \in \mathbb{N}$), followed by a corresponding robustness analysis in Section 3.2.2. For both these settings, we adapt a deterministic view on the estimation problem (3.19) before we switch to a probabilistic point of view in Section 3.2.3. In the latter section, we show how pMHE can be interpreted as a Bayesian estimator with stability guarantees. In Section 3.2.4, we focus on pMHE design approaches specific to the important special case of linear time-invariant (LTI) systems. The aforementioned sections are based on and taken in parts literally from (Gharbi and Ebenbauer (2018))[1], (Gharbi and Ebenbauer (2019a))[2], and (Gharbi and Ebenbauer (2019b))[3].

3.2.1 Nominal stability of the estimation error

In the following, we derive for the disturbance-free case ($w_k = 0, v_k = 0, k \in \mathbb{N}$) sufficient conditions on the Bregman distance and a priori estimate operator for the global uniform exponential stability (GUES) of the estimation error generated by the pMHE scheme based on problem (3.19). The stability analysis essentially relies on useful properties of Bregman

---

[1]M. Gharbi and C. Ebenbauer. A proximity approach to linear moving horizon estimation. In *Proc. 6th IFAC Conference on Nonlinear Model Predictive Control*, volume 51, pages 549–555, 2018 © 2018 Elsevier Ltd.

[2]M. Gharbi and C. Ebenbauer. A proximity moving horizon estimator based on Bregman distances and relaxed barrier functions. In *Proc. 18th European Control Conference (ECC)*, pages 1790–1795. IEEE, 2019a © 2019 IEEE.

[3]M. Gharbi and C. Ebenbauer. Proximity moving horizon estimation for linear time-varying systems and a Bayesian filtering view. In *Proc. 58th Conference on Decision and Control (CDC)*, pages 3208–3213. IEEE, 2019b © 2019 IEEE.

distances such as Lemma 2.2 and tools from the theory of PMD algorithms introduced in Section 2.2.1. Let us first state the following key lemma which holds for the pMHE problem (3.19) independently of the specific choices of the Bregman distance and the a priori estimate operator.

Lemma 3.1. *Consider system* (3.18) *with* $w_k = 0, v_k = 0, k \in \mathbb{N}$, *and the pMHE problem* (3.19) *at a given time instant* k. *Let Assumptions 3.1-3.4 hold. Then,*

$$D_{\psi_k}\Big(\mathbf{z}_k, \hat{\mathbf{z}}_k^*\Big) \leq D_{\psi_k}\Big(\mathbf{z}_k, \Phi_{k-1}\big(\hat{\mathbf{z}}_{k-1}^*\big)\Big), \tag{3.21}$$

where $\mathbf{z}_k$ *denotes the true system state,* $\hat{\mathbf{z}}_k^*$ *the pMHE solution, and* Φ_k *the a priori estimate operator as introduced in* (3.13).

Proof. The proof of this result follows similar steps for establishing (2.38) in the convergence analysis of the PMD algorithm we presented in Section 2.2.1. Consider the compact formulation of the pMHE problem in (3.11). Due to the linear setup, the cost function J_k is strongly convex and the feasible set $\mathcal{S}_k$ is convex, closed and nonempty (see the proof of Theorem 3.1). By optimality of the unique solution $\hat{\mathbf{z}}_k^*$, it holds for all $\mathbf{z} \in \mathcal{S}_k$ that $J_k(\hat{\mathbf{z}}_k^*) \leq J_k(\mathbf{z})$, i.e.,

$$F_k(\hat{\mathbf{z}}_k^*) + D_{\psi_k}(\hat{\mathbf{z}}_k^*, \bar{\mathbf{z}}_k) \leq F_k(\mathbf{z}) + D_{\psi_k}(\mathbf{z}, \bar{\mathbf{z}}_k), \qquad k \in \mathbb{N}. \tag{3.22}$$

Hence, for all $\mathbf{z} \in \mathcal{S}_k$ with $F_k(\mathbf{z}) \leq F_k(\hat{\mathbf{z}}_k^*)$, we have

$$D_{\psi_k}(\hat{\mathbf{z}}_k^*, \bar{\mathbf{z}}_k) \leq D_{\psi_k}(\mathbf{z}, \bar{\mathbf{z}}_k). \tag{3.23}$$

In view of Definition 2.5, we can therefore say that $\hat{\mathbf{z}}_k^*$ is the unique Bregman projection of $\bar{\mathbf{z}}_k$ onto the convex set Ω_k defined as

$$\Omega_k := \{\mathbf{z} \in \mathcal{S}_k : F_k(\mathbf{z}) \leq F_k(\hat{\mathbf{z}}_k^*)\}, \tag{3.24}$$

i.e., it holds that

$$\hat{\mathbf{z}}_k^* = \Pi_{\Omega_k}^{\psi_k}(\bar{\mathbf{z}}_k) = \arg\min_{\mathbf{z} \in \Omega_k} D_{\psi_k}(\mathbf{z}, \bar{\mathbf{z}}_k). \tag{3.25}$$

Due to Assumption 3.3 and the fact that $w_k = 0, v_k = 0, k \in \mathbb{N}$ in system (3.18), substituting the true system state $\mathbf{z}_k$ in (3.11b) yields

$$F_k(\mathbf{z}_k) = \sum_{i=k-N}^{k-1} r_i\left(y_i - C_i\, x\left(i; x_{k-N}, k-N, \mathbf{u}\right)\right) + q_i\left(0\right) = \sum_{i=k-N}^{k-1} r_i\left(0\right) + q_i\left(0\right) = 0. \tag{3.26}$$

Moreover, Assumption 3.3 implies that the function F_k is nonnegative. Thus, it holds that $0 = F_k(\mathbf{z}_k) \leq F_k(\hat{\mathbf{z}}_k^*)$ for any $k \in \mathbb{N}$. Furthermore, due to Assumption 3.2, $\mathbf{z}_k \in \mathcal{S}_k$. Hence, we deduce that $\mathbf{z}_k \in \Omega_k$. Based on Lemma 2.2, (3.25), and the nonnegativity of Bregman distances, we therefore get

$$0 \leq D_{\psi_k}(\hat{\mathbf{z}}_k^*, \bar{\mathbf{z}}_k) \leq D_{\psi_k}(\mathbf{z}_k, \bar{\mathbf{z}}_k) - D_{\psi_k}(\mathbf{z}_k, \hat{\mathbf{z}}_k^*) \tag{3.27}$$

and hence $D_{\psi_k}(\mathbf{z}_k, \hat{\mathbf{z}}_k^*) \leq D_{\psi_k}(\mathbf{z}_k, \bar{\mathbf{z}}_k)$. By setting $\bar{\mathbf{z}}_k = \Phi_{k-1}\big(\hat{\mathbf{z}}_{k-1}^*\big)$ as specified in (3.13), we obtain the desired inequality. $\qquad \square$

What we have shown in Lemma 3.1 is that, if we solve the optimization problem (3.19) at a given time instant k, the Bregman distance between the disturbance-free true system state $\mathbf{z}_k$ and the pMHE optimal solution $\hat{\mathbf{z}}_k^*$ is guaranteed to be smaller than the distance between $\mathbf{z}_k$ and the stabilizing a priori estimate $\bar{\mathbf{z}}_k$.

Remark 3.2. We can observe the following conceptual difference between the PMD algorithm presented in Section 2.2.1 and the linear pMHE scheme. In the proximity formulation given in (2.36), the iterate x_k is directly substituted into the Bregman distance in order to compute the next iterate x_{k+1}, and the result (2.38) with respect to a minimizer x^* holds. In the pMHE problem (3.11), the solution $\hat{\mathbf{z}}_k^*$ enters via the a priori estimate operator Φ_k as specified in (3.13) in order to obtain the next solution $\hat{\mathbf{z}}_{k+1}^*$, and the result (3.21) with respect to the true system state $\mathbf{z}_k$ holds. As previously discussed, this can be interpreted as introducing an additional prediction step in the PMD algorithm, which is given by the operator Φ_k.

We state in the following theorem the stability properties of the estimation error generated from the pMHE scheme based on problem (3.19).

Theorem 3.2. *Consider system* (3.18) *with* $w_k = 0, v_k = 0, k \in \mathbb{N}$, *and the pMHE problem* (3.19). *Let Assumptions 3.1-3.5 hold. Suppose there exist* $M \in \mathbb{N}_+$, $c \in \mathbb{R}_{++}$, *such that the Bregman distance* D_{ψ_k} *and the a priori estimate operator* Φ_k *satisfy*

$$D_{\psi_k}\left(\Phi_{k-1}(\mathbf{z}), \Phi_{k-1}(\hat{\mathbf{z}})\right) - D_{\psi_{k-1}}(\mathbf{z}, \hat{\mathbf{z}}) \leq -c\left\|\mathbf{z} - \hat{\mathbf{z}}\right\|^2 \tag{3.28}$$

for all $k \geq M$ *and* $\mathbf{z}, \hat{\mathbf{z}} \in \mathbb{R}^{(N+1)n}$. *Then, the estimation error* (3.17) *is GUES.*

Proof. Let us first prove that the pMHE error (3.16) is GUES by showing that there exists a continuous time-varying Lyapunov function V_k which globally satisfies conditions (A.6a) and (A.6b) in Theorem A.2 in Appendix A. The idea of this proof is to choose the Bregman distance D_{ψ_k} as a candidate Lyapunov function, i.e., we set

$$V_k(\mathbf{z}_k, \hat{\mathbf{z}}_k^*) = D_{\psi_k}(\mathbf{z}_k, \hat{\mathbf{z}}_k^*). \tag{3.29}$$

Due to the uniform quadratic lower and upper bounds implied by (3.12) in Assumption 3.4, it holds that

$$\frac{\sigma}{2}\|\mathbf{z}_1 - \mathbf{z}_2\|^2 \leq V_k(\mathbf{z}_1, \mathbf{z}_2) \leq \frac{\gamma}{2}\|\mathbf{z}_1 - \mathbf{z}_2\|^2 \tag{3.30}$$

for any $k \in \mathbb{N}$ and $\mathbf{z}_1, \mathbf{z}_2 \in \mathbb{R}^{(N+1)n}$. Hence, condition (A.6a) is satisfied with $c_1 = \frac{\sigma}{2}$ and $c_2 = \frac{\gamma}{2}$. Furthermore, exploiting (3.21) in Lemma 3.1 leads to

$$\Delta V_k := V_k(\mathbf{z}_k, \hat{\mathbf{z}}_k^*) - V_{k-1}(\mathbf{z}_{k-1}, \hat{\mathbf{z}}_{k-1}^*) \tag{3.31}$$
$$\leq D_{\psi_k}\left(\mathbf{z}_k, \Phi_{k-1}\left(\hat{\mathbf{z}}_{k-1}^*\right)\right) - D_{\psi_{k-1}}\left(\mathbf{z}_{k-1}, \hat{\mathbf{z}}_{k-1}^*\right).$$

Since $\mathbf{z}_k = \Phi_{k-1}(\mathbf{z}_{k-1})$ by Assumption 3.5, it holds that

$$D_{\psi_k}\left(\mathbf{z}_k, \Phi_{k-1}\left(\hat{\mathbf{z}}_{k-1}^*\right)\right) = D_{\psi_k}\left(\Phi_{k-1}\left(\mathbf{z}_{k-1}\right), \Phi_{k-1}\left(\hat{\mathbf{z}}_{k-1}^*\right)\right). \tag{3.32}$$

Thus, by substituting (3.32) into the Lyapunov difference (3.31) and using the sufficient condition (3.28), we arrive at

$$\Delta V_k \leq D_{\psi_k}\left(\Phi_{k-1}\left(\mathbf{z}_{k-1}\right), \Phi_{k-1}\left(\hat{\mathbf{z}}_{k-1}^*\right)\right) - D_{\psi_{k-1}}(\mathbf{z}_{k-1}, \hat{\mathbf{z}}_{k-1}^*) \tag{3.33}$$
$$= -c\|\mathbf{z}_{k-1} - \hat{\mathbf{z}}_{k-1}^*\|^2$$

for all $k \geq M$, which establishes condition (A.6b) with $c_3 = c$ and thereby GUES of the pMHE error (3.16). From this, stability of the estimation error (3.17) follows directly by simple Lipschitz arguments. By the triangle inequality, we have

$$\left\|x_k - \hat{x}_k\right\| \leq \left\|x\left(k; x_{k-N}, k-N, \mathbf{u}_k\right) - x\left(k; \hat{x}_{k-N}^*, k-N, \mathbf{u}_k\right)\right\| \tag{3.34}$$
$$+ \left\|x\left(k; \hat{x}_{k-N}^*, k-N, \mathbf{u}_k\right) - x\left(k; \hat{x}_{k-N}^*, k-N, \mathbf{u}_k, \hat{\mathbf{w}}_k^*\right)\right\|.$$

By Assumption 3.1, the function f_k is Lipschitz continuous in all of its arguments for any $k \in \mathbb{N}$ with the uniform Lipschitz constant $c_f \in \mathbb{R}_{++}$. Hence, using the notation $e_{k-N} = x_{k-N} - \hat{x}_{k-N}^*$ introduced in (3.16), we get

$$\left\|x\left(k; x_{k-N}, k-N, \mathbf{u}_k\right) - x\left(k; \hat{x}_{k-N}^*, k-N, \mathbf{u}_k\right)\right\| \tag{3.35}$$
$$= \left\|A_{k-1}\, x\left(k-1; x_{k-N}, k-N, \mathbf{u}_{k-1}\right) - A_{k-1}\, x\left(k-1; \hat{x}_{k-N}^*, k-N, \mathbf{u}_{k-1}\right)\right\|$$
$$\leq c_f \left\|x\left(k-1; x_{k-N}, k-N, \mathbf{u}_{k-1}\right) - x\left(k-1; \hat{x}_{k-N}^*, k-N, \mathbf{u}_{k-1}\right)\right\|$$
$$\vdots$$
$$\leq c_f^N \left\|e_{k-N}\right\|,$$

since the input terms cancel out. Similarly, we can compute

$$\left\|x\left(k; \hat{x}_{k-N}^*, k-N, \mathbf{u}_k\right) - x\left(k; \hat{x}_{k-N}^*, k-N, \mathbf{u}_k, \hat{\mathbf{w}}_k^*\right)\right\| \leq \sum_{i=k-N}^{k-1} c_f^{k-i}\left\|\hat{w}_i^*\right\|. \tag{3.36}$$

Substituting the above upper bounds in (3.34) yields

$$\|x_k - \hat{x}_k\| \leq c_f^N \|e_{k-N}\| + \sum_{i=k-N}^{k-1} c_f^{k-i}\|\hat{w}_i^*\| \tag{3.37}$$
$$\leq c_f^N \|\mathbf{z}_k - \hat{\mathbf{z}}_k^*\| + \sum_{i=k-N}^{k-1} c_f^{k-i}\|\mathbf{z}_k - \hat{\mathbf{z}}_k^*\|.$$

Since the pMHE error is GUES, according to Definition A.5 in Appendix A, there exist constants $\alpha \in \mathbb{R}_{++}$ and $\beta \in (0,1)$ such that

$$\|\mathbf{z}_k - \hat{\mathbf{z}}_k^*\| \leq \alpha\beta^k\|\mathbf{z}_0 - \hat{\mathbf{z}}_0^*\| \tag{3.38}$$

for all $k \in \mathbb{N}_+$ and $\mathbf{z}_0, \hat{\mathbf{z}}_0^* \in \mathbb{R}^{(N+1)n}$. Hence, substituting (3.38) in (3.37) yields

$$\|x_k - \hat{x}_k\| \leq \bar{c}\,\alpha\beta^k\|\mathbf{z}_0 - \hat{\mathbf{z}}_0^*\| \tag{3.39}$$

for all $k \in \mathbb{N}_+$ and $\mathbf{z}_0, \hat{\mathbf{z}}_0^* \in \mathbb{R}^{(N+1)n}$, where $\bar{c} := c_f^N + \sum_{i=k-N}^{k-1} c_f^{k-i}$. In view of Remark 3.1, the pMHE problem (3.19) at time instant $k = 0$ reduces to $\hat{\mathbf{z}}_0^* = \arg\min_{\hat{\mathbf{z}}_0 \in \mathcal{S}_0} D_{\psi_0}(\hat{\mathbf{z}}_0, \bar{\mathbf{z}}_0) =$

$\Pi_{\mathcal{S}_0}^{\psi_0}(\bar{\mathbf{z}}_0)$. Invoking Lemma 2.2 yields that $D_{\psi_0}(\mathbf{z}_0, \hat{\mathbf{z}}_0^*) \leq D_{\psi_0}(\mathbf{z}_0, \bar{\mathbf{z}}_0)$, since the true system state satisfies the inequality constraints, i.e., $\mathbf{z}_0 \in \mathcal{S}_0$. By the uniform bounds on the Bregman distance in Assumption 3.4, we get $\|\mathbf{z}_0 - \hat{\mathbf{z}}_0^*\| \leq \sqrt{\gamma/\sigma} \|\mathbf{z}_0 - \bar{\mathbf{z}}_0\|$. Recall that in Algorithm 3.1, we set $\bar{\mathbf{z}}_0 = \bar{x}_0$, where $\bar{x}_0$ denotes the initial guess. Moreover, $\mathbf{z}_0 = x_0$ is the true initial state. We finally obtain in (3.39) that

$$\|x_k - \hat{x}_k\| \leq \tilde{\alpha}\beta^k \|x_0 - \bar{x}_0\| \tag{3.40}$$

for all $k \in \mathbb{N}_+$ and $x_0, \bar{x}_0 \in \mathbb{R}^n$, where $\tilde{\alpha} := \bar{c}\alpha\sqrt{\gamma/\sigma}$, which proves that the estimation error (3.17) is GUES. $\qquad\square$

Some remarks on Theorem 3.2 are in order. First, a crucial issue for the design of the pMHE scheme is the validity of condition (3.28). It states that, given two vectors $\mathbf{z}, \hat{\mathbf{z}} \in \mathbb{R}^{(N+1)n}$ and the Bregman distance $D_{\psi_{k-1}}(\mathbf{z}, \hat{\mathbf{z}})$, a prediction from the a priori estimate operator Φ_{k-1} will yield a contracting Bregman distance D_{ψ_k}. Hence, by requiring the operator to be a contraction mapping that reduces a Bregman distance between a pair of points and using this Bregman distance in the pMHE cost function, we can satisfy (3.28). This gives rise to a key concept in pMHE, which is to design the a priori estimate operator from an estimator with (exponentially) stable dynamics and to select the Bregman distance as the Lyapunov function with which the stability of this estimator can be verified. Simple, model-based, and recursive state estimation strategies for which the construction of suitable Lyapunov functions is known can be used. This includes for instance the computationally efficient discrete-time Kalman filter and the Luenberger observer. Second, note that we do not impose any assumption on the horizon length $N \in \mathbb{N}_+$. Even if $N = 0$, it holds that $\hat{x}_k = \hat{\mathbf{z}}_k^* = \arg\min_{\hat{\mathbf{z}}_k \in \mathcal{X}_k} D_{\psi_k}(\hat{\mathbf{z}}_k, \bar{\mathbf{z}}_k)$ and hence the pMHE estimates correspond to the Bregman projection (see Definition 2.5) of the stabilizing a priori estimate onto the set of state constraints. Finally, from a technical point of view, another advantage of the proposed proximity-based analysis is that it allows to consider a rather general stage cost since the stability of the estimation error can be shown without an explicit representation of the error dynamics and hence to avoid technical stability proofs with strong assumptions. It is also worth noting that the underlying Lyapunov function not only depends on the error between the true state and the estimated state, but also on the difference between the true disturbances and the estimated ones.

As discussed in Section 2.1, exponential stability of classical MHE formulations for linear systems (Alessandri et al. (2003); Rao et al. (2001); Sui and Johansen (2014)) is established for the case where the functions r_i and q_i in (3.19) as well as the prior weighting are chosen as weighted quadratic functions. In our setup, however, r_i, q_i need not be differentiable and are chosen according to desired performance criteria as long as Assumption 3.3 holds true. For example, depending on the problem setting, we can choose any norm $r_i(v) = \|v\|_p$ where $p \in \{1, 2, \infty\}$, or any function satisfying Assumption 3.3 such as the so-called Huber penalty function, which will be defined later in Section 3.2.3. This flexibility of design allows to handle the case where measurements might be affected by outliers by employing the ℓ_1-norm or the Huber penalty, without the need to explicitly alter the MHE scheme to account for or detect outliers (cf. Alessandri and Awawdeh (2016)). We will discuss more thoroughly in Section 3.2.3 how one can choose suitable stage costs given prior knowledge on the statistics of the disturbances.

Φ_k and D_{ψ_k} based on the LTV Kalman filter

We now present a design approach for the a priori estimate operator and the associated Bregman distance, which will be demonstrated to satisfy condition (3.28). Given the time-varying setup and due to its well-established stability properties, it is reasonable to use the LTV Kalman filter for constructing the a priori estimate. Moreover, employing the Kalman filter will enable us to embed the pMHE scheme in a Bayesian framework in Section 3.2.3.

Let $\mathbf{z} = \begin{bmatrix} x^\top & \mathbf{w}^\top \end{bmatrix}^\top$, $\hat{\mathbf{z}} = \begin{bmatrix} \hat{x}^\top & \hat{\mathbf{w}}^\top \end{bmatrix}^\top$, $x, \hat{x} \in \mathbb{R}^n$, $\mathbf{w}, \hat{\mathbf{w}} \in \mathbb{R}^{Nn}$. Given a time instant $k \in \mathbb{N}$ and a horizon length $N \in \mathbb{N}_+$, the a priori estimate operator based on the Kalman filter is

$$\Phi_k(\mathbf{z}) = \begin{bmatrix} A_{k-N}\left(x + K_{k-N}\left(y_{k-N} - C_{k-N}\,x\right)\right) + B_{k-N}\,u_{k-N} \\ \mathbf{0} \end{bmatrix}, \tag{3.41}$$

where $\mathbf{0} \in \mathbb{R}^{Nn}$ refers to the zero vector. In particular, consider the notation introduced in (3.14) and the pMHE solution $\hat{\mathbf{z}}_k^*$ at time k. In view of (3.13), the a priori estimate for the next time instant is

$$\bar{\mathbf{z}}_{k+1} = \Phi_k(\hat{\mathbf{z}}_k^*) = \begin{bmatrix} \bar{x}_{k-N+1}^\top & 0 & \dots & 0 \end{bmatrix}^\top, \tag{3.42a}$$

where we have zero a priori process disturbances, i.e., $\bar{w}_i = 0$, $i = k - N + 1, \dots, k$, and

$$\bar{x}_{k-N+1} = A_{k-N}\left(\hat{x}_{k-N}^* + K_{k-N}\left(y_{k-N} - C_{k-N}\,\hat{x}_{k-N}^*\right)\right) + B_{k-N}\,u_{k-N}. \tag{3.42b}$$

The associated Bregman distance D_{ψ_k} used at time instant k is a weighted quadratic function given by

$$D_{\psi_k}(\mathbf{z}, \hat{\mathbf{z}}) = \frac{1}{2}\|x - \hat{x}\|_{\Pi_{k-N}^-}^2 + \frac{1}{2}\|\mathbf{w} - \hat{\mathbf{w}}\|_{\bar{W}}^2, \tag{3.43}$$

where $\Pi_{k-N}^- \in \mathbb{S}_{++}^n$ refers to the inverse of the Kalman filter covariance matrix and $\bar{W} \in \mathbb{S}_{++}^{Nn}$ is a weight matrix of the form $\bar{W} = \mathrm{diag}(W, \dots, W)$ with $W \in \mathbb{S}_{++}^n$ arbitrary. The time-varying Kalman gain and covariance matrix are computed as

$$K_k = P_k^- C_k^\top \left(C_k P_k^- C_k^\top + R\right)^{-1}, \tag{3.44a}$$

$$P_{k+1}^- = A_k\,P_k^+\,A_k^\top + Q, \qquad \Pi_k^- := \left(P_k^-\right)^{-1}, \tag{3.44b}$$

$$P_k^+ = (I - K_k C_k)\,P_k^-, \qquad \Pi_k^+ := \left(P_k^+\right)^{-1}, \tag{3.44c}$$

where $P_0^- \in \mathbb{S}_{++}^n$, $Q \in \mathbb{S}_{++}^n$ and $R \in \mathbb{S}_{++}^p$. In this deterministic setting, the matrices P_0^-, Q and R can be considered as design parameters that can be chosen arbitrarily. Their design in Section 3.2.3 will be motivated from a probabilistic point of view based on the statistics of the disturbances.

Establishing that the proposed design based on the Kalman filter verifies (3.28) in Theorem 3.2 motivates the following notions of uniform observability and controllability for LTV systems, which are due to Kalman (1960a) and further discussed in (Jazwinski, 1970, Chapter 7).

Definition 3.1 (Complete uniform observability). *Consider an LTV system of the form* (3.18) *in which the matrices A_k are nonsingular. The pair (A_k, C_k) is uniformly completely observable if there exist $N_0 \in \mathbb{N}_+$ and $\alpha, \beta \in \mathbb{R}_{++}$ such that*

$$\alpha I \preceq \sum_{i=k-N_0}^{k} \Phi^{-\top}(k,i) \, C_i^{\top} R \, C_i \, \Phi^{-1}(k,i) \preceq \beta I, \qquad \forall k \geq N_0, \tag{3.45}$$

with some $R \in \mathbb{S}_{++}^{p}$ (typically the covariance matrix of the measurement disturbances in a probabilistic setup).

Definition 3.2 (Complete uniform controllability). *Consider an LTV system of the form* (3.18). *The pair $(A_k, \sqrt{Q})$ is uniformly completely controllable if there exist $N_0 \in \mathbb{N}_+$ and $\alpha, \beta \in \mathbb{R}_{++}$ such that*

$$\alpha I \preceq \sum_{i=k-N_0}^{k} \Phi(k,i+1) \, Q \, \Phi^{\top}(k,i+1) \preceq \beta I, \qquad \forall k \geq N_0, \tag{3.46}$$

with some $Q \in \mathbb{S}_{++}^{n}$ (typically the covariance matrix of the process disturbances in a probabilistic setup).

Assumption 3.6 (Linear system properties). *System* (3.18) *is uniformly completely observable and uniformly completely controllable.*

The following lemma summarizes the results of (Jazwinski, 1970, Lemmas 7.1 and 7.2), which establish uniform lower and upper bounds on the a posteriori covariance matrix P_k^+ computed via the recursion (3.44), and where the corresponding proofs can be found.

Lemma 3.2 (Jazwinski (1970)). *Consider system* (3.18) *and let Assumption 3.6 hold, in which R and Q refer to the matrices used in* (3.44). *Then, there exists $M \in \mathbb{N}_+$ such that P_k^+ in* (3.44c) *is uniformly lower and upper bounded for all $k \geq M$, i.e., there exist $\underline{p}, \overline{p} \in \mathbb{R}_{++}$ such that $\underline{p} I \preceq P_k^+ \preceq \overline{p} I$ for all $k \geq M$.*

We now state that the proposed Kalman filter design approach of the a priori estimate operator and the Bregman distance satisfies amongst others the sufficient condition (3.28) in Theorem 3.2.

Proposition 3.1. *Consider system* (3.18) *with $w_k = 0$, $v_k = 0$, $k \in \mathbb{N}$ and let Assumptions 3.1 and 3.6 hold. Then, the Bregman distance* (3.43) *and the a priori estimate operator* (3.41) *verify Assumptions 3.4 and 3.5. Moreover, there exist $M \in \mathbb{N}_+$, $c \in \mathbb{R}_{++}$ such that they satisfy condition* (3.28) *for all $\mathbf{z}, \hat{\mathbf{z}} \in \mathbb{R}^{(N+1)n}$ and $k \geq M$.*

Proof. Since $w_k = 0$, $v_k = 0$, $k \in \mathbb{N}$, evaluating the a priori estimate operator (3.41) at $\mathbf{z}_k = \begin{bmatrix} x_{k-N}^{\top} & 0 & \dots & 0 \end{bmatrix}^{\top}$, i.e., the true state x_{k-N} with zero process disturbances yields

$$\Phi_k(\mathbf{z}_k) = \begin{bmatrix} A_{k-N} \left(x_{k-N} + K_{k-N} \left(y_{k-N} - C_{k-N} \, x_{k-N} \right) \right) + B_{k-N} \, u_{k-N} \\ 0 \end{bmatrix} \tag{3.47}$$

$$= \begin{bmatrix} A_{k-N} x_{k-N} + B_{k-N} \, u_{k-N} \\ 0 \end{bmatrix} = \begin{bmatrix} x_{k-N+1} \\ 0 \end{bmatrix} = \mathbf{z}_{k+1},$$

since $y_{k-N} = C_{k-N} x_{k-N}$. Hence, Assumption 3.5 is satisfied.

In view of Assumption 3.6 and Lemma 3.2, there exist $M \in \mathbb{N}_+$ and $\underline{p}, \overline{p} \in \mathbb{R}_{++}$ such that $\underline{p}I \preceq P_k^+ \preceq \overline{p}I$ for all $k \geq M$. Due to Assumption 3.1, $\|A_k\|$ is uniformly bounded by c_{f}. This is due to the fact that, by the uniform Lipschitz continuity assumption, it holds that $\frac{\|A_k(x_1-x_2)\|}{c_{\mathrm{f}}\|x_1-x_2\|} \leq 1$ for all $x_1, x_2 \in \mathbb{R}^n$ with $x_1 \neq x_2$. In particular, this inequality should also hold for $\max_{z=x_1-x_2 \neq 0} \frac{\|A_k z\|}{c_{\mathrm{f}}\|z\|} = \frac{\|A_k\|}{c_{\mathrm{f}}} \leq 1$. Hence, $\|A_k\| \leq c_{\mathrm{f}}$ for all $k \in \mathbb{N}$. In addition, since $Q \in \mathbb{S}_{++}^n$, a uniform lower and upper bound on P_k^- in (3.44b) can be established as well, i.e., there exist $\underline{p}', \overline{p}' \in \mathbb{R}_{++}$ such that $\underline{p}'I \preceq P_k^- \preceq \overline{p}'I$ for all $k \geq M$.

Let $\mathbf{z} = \begin{bmatrix} x^\top & \mathbf{w}^\top \end{bmatrix}^\top$, $\hat{\mathbf{z}} = \begin{bmatrix} \hat{x}^\top & \hat{\mathbf{w}}^\top \end{bmatrix}^\top$, $x, \hat{x} \in \mathbb{R}^n$, $\mathbf{w}, \hat{\mathbf{w}} \in \mathbb{R}^{Nn}$. We obtain for the Bregman distance (3.43) that

$$\frac{1}{2\overline{p}'}\|x - \hat{x}\|^2 + \frac{1}{2}\lambda_{\min}(\overline{W})\|\mathbf{w} - \hat{\mathbf{w}}\|^2 \leq D_{\psi_k}(\mathbf{z}, \hat{\mathbf{z}}) \leq \frac{1}{2\underline{p}'}\|x - \hat{x}\|^2 + \frac{1}{2}\lambda_{\max}(\overline{W})\|\mathbf{w} - \hat{\mathbf{w}}\|^2. \tag{3.48}$$

Therefore, Assumption 3.4 is fulfilled with $\sigma = \min(\frac{1}{\overline{p}'}, \lambda_{\min}(\overline{W}))$ and $\gamma = \max(\frac{1}{\underline{p}'}, \lambda_{\max}(\overline{W}))$. Given the a priori estimate operator (3.41) and the Bregman distance (3.43), we compute

$$\Delta D_k := D_{\psi_k}(\Phi_{k-1}(\mathbf{z}), \Phi_{k-1}(\hat{\mathbf{z}})) - D_{\psi_{k-1}}(\mathbf{z}, \hat{\mathbf{z}}) \tag{3.49}$$

$$= \frac{1}{2}\|(A_{k-N-1}(I - K_{k-N-1}C_{k-N-1}))(x - \hat{x})\|_{\Pi_{k-N}^-}^2 - \frac{1}{2}\|x - \hat{x}\|_{\Pi_{k-N-1}^-}^2 - \frac{1}{2}\|\mathbf{w} - \hat{\mathbf{w}}\|_{\overline{W}}^2,$$

since we have zero a priori disturbances and the input and measurement terms cancel out in the first term of the right-hand side of (3.49). Let us now derive upper bounds for the first two terms in the right-hand side of (3.49), whose sum amounts to

$$\frac{1}{2}\|(A_{k-N-1}(I - K_{k-N-1}C_{k-N-1}))(x - \hat{x})\|_{\Pi_{k-N}^-}^2 - \frac{1}{2}\|x - \hat{x}\|_{\Pi_{k-N-1}^-}^2 \tag{3.50}$$

$$= \frac{1}{2}(x - \hat{x})^\top \left(\tilde{A}_{k-N-1}^\top \Pi_{k-N}^- \tilde{A}_{k-N-1} - \Pi_{k-N-1}^-\right)(x - \hat{x}),$$

where $\tilde{A}_k := A_k(I - K_k C_k)$ for $k \in \mathbb{N}$. Note that $\tilde{A}_k$ is invertible since $(I - K_k C_k) = P_k^+ \Pi_k^-$ in view of (3.44c) and A_k is nonsingular as implied by Assumption 3.6. The result in (Reif and Unbehauen, 1999, Lemma 6) states that

$$\Pi_{k+1}^- \preceq \tilde{A}_k^{-\top}\left(\Pi_k^- - \Pi_k^-\left(\Pi_k^+ + A_k^\top Q^{-1} A_k\right)^{-1}\Pi_k^-\right)\tilde{A}_k^{-1}. \tag{3.51}$$

Hence, by defining $U_k := \Pi_k^-\left(\Pi_k^+ + A_k^\top Q^{-1} A_k\right)^{-1}\Pi_k^-$, we obtain

$$\tilde{A}_k^\top \Pi_{k+1}^- \tilde{A}_k - \Pi_k^- \preceq -U_k. \tag{3.52}$$

Assumptions 3.1 and 3.6 imply that, for all $k \geq M$,

$$\|U_k\| \geq \frac{1}{\overline{p}'^2(1/\overline{p} + c_{\mathrm{f}}^2/\lambda_{\min}(Q))} =: \tilde{u}. \tag{3.53}$$

Hence, it holds for (3.50) that

$$\frac{1}{2}\|(A_{k-N-1}(I - K_{k-N-1}C_{k-N-1}))(x - \hat{x})\|_{\Pi_{k-N}^-}^2 - \frac{1}{2}\|x - \hat{x}\|_{\Pi_{k-N-1}^-}^2 \leq -\frac{\tilde{u}}{2}\|x - \hat{x}\|^2 \tag{3.54}$$

for all $k \geq M$. Substituting (3.54) in (3.49) yields

$$\Delta D_k \leq -\frac{\tilde{u}}{2}\|x - \hat{x}\|^2 - \frac{1}{2}\|\mathbf{w} - \hat{\mathbf{w}}\|_{\bar{W}}^2. \tag{3.55}$$

Thus, for all $k \geq M$, condition (3.28) holds with $c = \frac{1}{2}\min(\tilde{u}, \lambda_{\min}(\bar{W}))$. □

Overall, we have the following (inherent) stability result for linear pMHE based on the Kalman filter, which is a direct consequence of Theorem 3.2 and Proposition 3.1.

Corollary 3.1. *Consider system* (3.18) *with* $w_k = 0$, $v_k = 0$, $k \in \mathbb{N}$ *and the pMHE problem* (3.19) *with a priori estimate operator* (3.41) *and Bregman distance* (3.43) *based on the Kalman filter. Let Assumptions 3.1-3.3 and 3.6 hold. Then, the estimation error* (3.17) *is GUES.*

Corollary 3.1 ensures GUES of the estimation error generated from the pMHE scheme through the use of the Kalman filter in the design of the stabilizing a priori estimate and the Bregman distance. Although the Kalman filter is the optimal estimator for unconstrained linear systems subject to Gaussian disturbances, the formulation of the pMHE problem enables to additionally incorporate constraints and handle other data specific characteristics such as outliers. We therefore expect pMHE to yield a better state estimate compared to the standard Kalman filter in the non-Gaussian case.

In Section 2.1, we reviewed the linear constrained MHE scheme discussed in (Rawlings et al. (2017)), where the LTI system dynamics are given by (2.11) and the stage cost is quadratic (2.12). If we compare the associated MHE problem with the pMHE problem (3.19) with a priori estimate operator (3.41) and Bregman distance (3.43), we notice that the Kalman filter covariance matrix is employed in both design approaches, i.e., in the quadratic prior weighting (2.15) and in the quadratic Bregman distance (3.43). Nevertheless, due to the proximity-based analysis and design of pMHE, using the stabilizing a priori estimate allows to establish a rather stronger result: GUES of the estimation error of pMHE with convex stage cost, compared to GAS of MHE with quadratic stage cost. However, in our case, Assumption 3.6 is needed in order to be able to inherit the stability properties from the Kalman filter. In other words, we require observability instead of detectability as assumed in (Rawlings et al. (2017)). Nevertheless, we will see in Section 3.2.4 that using the Luenberger observer instead of the Kalman filter in pMHE for LTI systems will allow to consider detectable systems as well.

Remark 3.3. Note that if the state is known to satisfy inequality constraints, there exists a number of methods aimed at modifying the Kalman filter in order to be able to take advantage of this additional knowledge (Simon (2010)). For instance, Simon and Simon (2005) showed theoretically and via simulations that projecting the unconstrained Kalman estimate onto the constraint set yields unbiased state estimates and smaller state error covariance than the unconstrained case. Nevertheless, we argue that in pMHE based on the Kalman filter, inequality constraints can fit easily and more naturally into the structure of the estimator.

Remark 3.4. In the a priori estimate operator (3.41), we choose zero a priori process disturbances since, in the nominal case, the true disturbances to which we want to converge are zero. Alternatively, we could investigate other design approaches for the a priori disturbances such as constructing them based on a suitable disturbance observer. Even though this approach fits very well in our proximity framework, we do not pursue it in this thesis and leave it as future research. Nevertheless, even in the presence of disturbances, this simple design of the operator (3.41) will prove to be sufficient for the inherent robustness of pMHE, as we will see in the next section.

3.2.2 Robust stability of the estimation error

In this section, we consider system (3.18) with unknown additive process and measurement disturbances w_k and v_k and assume convex polytopic state constraints of the form

$$\mathcal{X}_k = \left\{ x \in \mathbb{R}^n : C_{\mathrm{x},k}\, x \le d_{\mathrm{x},k} \right\}, \qquad k \in \mathbb{N}, \tag{3.56}$$

where $C_{\mathrm{x},k} \in \mathbb{R}^{q_{\mathrm{x}} \times n}$ and $d_{\mathrm{x},k} \in \mathbb{R}^{q_{\mathrm{x}}}$ with $q_{\mathrm{x}} \in \mathbb{N}_+$ refering to the number of affine state constraints. Moreover, in (3.2), we set $\mathcal{W}_k = \mathbb{R}^n$ and $\mathcal{V}_k = \mathbb{R}^p$ for any $k \in \mathbb{N}$. We investigate in a deterministic setting the inherent robustness properties of linear pMHE (3.19) with respect to the disturbances in terms of an input-to-state stability (ISS) analysis, for which a brief overview is presented in Appendix A. In particular, we focus on the case where the Bregman distance D_{ψ_k} and the a priori estimate operator Φ_k are constructed based on the Kalman filter, i.e., we choose the a priori estimate (3.42) and the quadratic time-varying Bregman distance (3.43). The resulting pMHE optimization problem is given by

$$\min_{\hat{\mathbf{z}}_k} \quad \sum_{i=k-N}^{k-1} r_i(\hat{v}_i) + q_i(\hat{w}_i) + \frac{1}{2}\|\hat{w}_i\|_W^2 + \frac{1}{2}\|\hat{x}_{k-N} - \bar{x}_{k-N}\|_{\Pi_{k-N}^-}^2 \tag{3.57a}$$

$$\text{s.\,t.} \quad \hat{x}_{i+1} = A_i\,\hat{x}_i + B_i\,u_i + \hat{w}_i, \qquad i = k-N, \dots, k-1 \tag{3.57b}$$

$$y_i = C_i\,\hat{x}_i + \hat{v}_i, \qquad i = k-N, \dots, k-1 \tag{3.57c}$$

$$C_{\mathrm{x},k}\,\hat{x}_i \le d_{\mathrm{x},k}, \qquad i = k-N, \dots, k. \tag{3.57d}$$

The compact formulation of the pMHE problem (3.11) becomes

$$\min_{\hat{\mathbf{z}}_k \in \mathcal{S}_k} \quad J_k\left(\hat{\mathbf{z}}_k\right) = F_k(\hat{\mathbf{z}}_k) + \frac{1}{2}\|\hat{\mathbf{z}}_k - \bar{\mathbf{z}}_k\|_{M_k}^2, \tag{3.58a}$$

where $M_k := \operatorname{diag}(\Pi_{k-N}^-, W, W, \dots, W) \in \mathbb{S}_{++}^{(N+1)n}$ and the convex polytopic set $\mathcal{S}_k$ represents the (stacked) state constraints (3.57d) given by

$$\mathcal{S}_k = \left\{ \hat{\mathbf{z}}_k \in \mathbb{R}^{(N+1)n} : C_{\mathrm{z},k}\,\hat{\mathbf{z}}_k \le d_{\mathrm{z},k} \right\}. \tag{3.58b}$$

The matrix $C_{\mathrm{z},k} \in \mathbb{R}^{(N+1)q_{\mathrm{x}} \times (N+1)n}$ and the vector $d_{\mathrm{z},k} \in \mathbb{R}^{(N+1)q_{\mathrm{x}}}$ can be computed by simple algebraic manipulations. In particular, this can be achieved by expressing each $\hat{x}_i$ in the inequality constraints (3.57d) in terms of the decision variable $\hat{\mathbf{z}}_k$ given the linear system dynamics (3.57b).

In principle, an intuitive approach for establishing that the estimation error is ISS with respect to the disturbances is to extend the result of Lemma 3.1 to the perturbed case and adapt the Lyapunov analysis developed for proving Theorem 3.2 accordingly. However, Lemma 3.1 uses the crucial step that, in the disturbance-free case, the sum of stage cost F_k evaluated at the true system state is smaller than the sum of stage cost at the pMHE solution (cf. (3.26)), which does not necessarily hold true in the presence of disturbances. Hence, we require different tools for establishing the robustness properties of the estimator. Indeed, designing the Bregman distance as a weighted quadratic function enables us to employ alternative techniques based on the theory of the PPA in Section 2.2.1. More specifically, similar to the proximal operator (2.41), let us introduce the (time-varying) pMHE operator $\rho_k : \mathbb{R}^{(N+1)n} \to \mathbb{R}^{(N+1)n}$

$$\rho_k\left(\bar{\mathbf{z}}\right) = \arg\min_{\mathbf{z} \in \mathcal{S}_k} \quad F_k(\mathbf{z}) + \frac{1}{2}\left\|\mathbf{z} - \bar{\mathbf{z}}\right\|_{M_k}^2, \tag{3.59}$$

which is well-defined at each time instant $k \in \mathbb{N}$ due to the fact that the function we want to minimize on the right-hand side of (3.59) is strongly convex, so it possesses a unique minimizer for every $\bar{\mathbf{z}} \in \mathbb{R}^{(N+1)n}$. Note that if we set $\bar{\mathbf{z}} = \bar{\mathbf{z}}_k$, i.e., the pMHE a priori estimate, it holds that $\rho_k\left(\bar{\mathbf{z}}_k\right) = \hat{\mathbf{z}}_k^*$. Moreover, recall from the previous chapter that minimizers of a convex function are fixed points of the associated proximal operator (see (2.43) for more details). In the context of pMHE, for a fixed time instant k and with no disturbances affecting the system (3.18), the true system state $\mathbf{z}_k$ with zero disturbances can be interpreted as a fixed point of the operator ρ_k, i.e., we have $\rho_k\left(\mathbf{z}_k\right) = \mathbf{z}_k$ for any $k \in \mathbb{N}$. This is stated more formally in the following lemma.

Lemma 3.3. *Consider the pMHE operator ρ_k defined in* (3.59). *Let Assumption 3.3 hold. In the disturbance-free case where* $\mathbf{z}_k = \begin{bmatrix} x_{k-N}^\top & 0 & \cdots & 0 \end{bmatrix}^\top$, *if $\mathbf{z}_k \in \text{int}(\mathcal{S}_k)$, then $\rho_k\left(\mathbf{z}_k\right) = \mathbf{z}_k$ for all $k \in \mathbb{N}$.*

Proof. First, we equivalently reformulate (3.59) at the true system state $\mathbf{z}_k$ as

$$\rho_k\left(\mathbf{z}_k\right) = \arg\min_{\mathbf{z} \in \mathbb{R}^{(N+1)n}} \quad F_k(\mathbf{z}) + I_{\mathcal{S}_k}(\mathbf{z}) + \frac{1}{2}\left\|\mathbf{z} - \mathbf{z}_k\right\|_{M_k}^2, \tag{3.60}$$

where $I_{\mathcal{S}_k}$ denotes the indicator function of the convex set $\mathcal{S}_k$ (see (2.24) for the definition of the function). In the following, we show that in the disturbance-free case, $\mathbf{z}_k$ minimizes $F_k(\mathbf{z}) + I_{\mathcal{S}_k}(\mathbf{z})$. In other words, and according to the first-order necessary and sufficient condition for optimality (2.26), we show that $0 \in \partial I_{\mathcal{S}_k}(\mathbf{z}_k) + \partial F_k(\mathbf{z}_k)$ holds. Since $\mathbf{z}_k \in \text{int}(\mathcal{S}_k)$, then $N_{\mathcal{S}_k}(\mathbf{z}_k) = \{0\}$, where $N_{\mathcal{S}_k}$ denotes the normal cone defined in (2.25). Given that the subdifferential of the indicator function is the normal cone operator, it holds that $\partial I_{\mathcal{S}_k}(\mathbf{z}_k) = N_{\mathcal{S}_k}(\mathbf{z}_k) = \{0\}$. Second, we have that $F_k\left(\mathbf{z}_k\right) = \sum_{i=k-N}^{k-1} q_i(0) + r_i(0) = 0$ by Assumption 3.3. Hence, the minimal value of F_k which is zero is attained at the true system state $\mathbf{z}_k$ and $0 \in \partial F_k(\mathbf{z}_k)$. Thus, $0 \in \partial I_{\mathcal{S}_k}(\mathbf{z}_k) + \partial F_k(\mathbf{z}_k)$. We obtain in view of (3.60) and the above arguments that

$$0 \in \partial F_k(\mathbf{z}_k) + \partial I_{\mathcal{S}_k}(\mathbf{z}_k) \Leftrightarrow 0 \in \partial F_k(\mathbf{z}_k) + \partial I_{\mathcal{S}_k}(\mathbf{z}_k) + M_k(\mathbf{z}_k - \mathbf{z}_k) \tag{3.61}$$

$$\Leftrightarrow \mathbf{z}_k = \arg\min_{\mathbf{z} \in \mathbb{R}^{(N+1)n}} \quad F_k(\mathbf{z}) + I_{\mathcal{S}_k}(\mathbf{z}) + \frac{1}{2}\left\|\mathbf{z} - \mathbf{z}_k\right\|_{M_k}^2$$

$$\Leftrightarrow \mathbf{z}_k = \rho_k\left(\mathbf{z}_k\right),$$

which finishes the proof of the lemma. $\qquad\square$

In the following, we establish a modified version of the nonexpansiveness property of the proximal operator (see (2.47) in Section 2.2.1) adapted to the context of pMHE. We have the following result.

Lemma 3.4. *Consider the pMHE operator ρ_k defined in (3.59). Let Assumption 3.3 hold. Then, the following inequality holds*

$$\|\rho_k(\bar{\mathbf{z}}) - \rho_k(\bar{\mathbf{z}}')\|_{M_k}^2 \leq \|\bar{\mathbf{z}} - \bar{\mathbf{z}}'\|_{M_k}^2 \tag{3.62}$$

for all $k \in \mathbb{N}$ and $\bar{\mathbf{z}}, \bar{\mathbf{z}}' \in \mathbb{R}^{(N+1)n}$.

Proof. The steps in this proof are based on common procedures for showing the non-expansiveness of the proximal operator (Ryu and Boyd (2016)). The Lagrangian $L : \mathbb{R}^{(N+1)n} \times \mathbb{R}^{(N+1)q_x} \to \mathbb{R}$ associated with problem (3.59) is

$$L(\mathbf{z}, \lambda) = F_k(\mathbf{z}) + \frac{1}{2}\|\mathbf{z} - \bar{\mathbf{z}}\|_{M_k}^2 + \lambda^\top (C_{\mathbf{z},k}\,\mathbf{z} - d_{\mathbf{z},k}), \tag{3.63}$$

where λ denotes the Lagrange multiplier associated with the inequality constraints (3.58b). Recall that $\partial F_k(\mathbf{z})$ denotes the subdifferential of the (potentially nondifferentiable) function F_k at a point $\mathbf{z}$ (cf. Definition 2.3). The Karush-Kuhn-Tucker (KKT) optimality conditions at the minimizer $\rho_k(\bar{\mathbf{z}})$ for the a priori estimate $\bar{\mathbf{z}}$ are

$$0 \in M_k(\rho_k(\bar{\mathbf{z}}) - \bar{\mathbf{z}}) + \partial F_k(\rho_k(\bar{\mathbf{z}})) + C_{\mathbf{z},k}^\top \lambda \tag{3.64a}$$

$$\lambda^\top (C_{\mathbf{z},k}\,\rho_k(\bar{\mathbf{z}}) - d_{\mathbf{z},k}) = 0 \tag{3.64b}$$

$$C_{\mathbf{z},k}\,\rho_k(\bar{\mathbf{z}}) - d_{\mathbf{z},k} \leq 0 \tag{3.64c}$$

$$\lambda \geq 0. \tag{3.64d}$$

For the a priori estimate $\bar{\mathbf{z}}'$, we can similarly write down the corresponding KKT conditions at the minimizer $\rho_k(\bar{\mathbf{z}}')$ with the Lagrange multiplier $\lambda' \in \mathbb{R}^{(N+1)q_x}$. Subtracting the gradient condition (3.64a) at $\rho_k(\bar{\mathbf{z}}')$ from the gradient condition at $\rho_k(\bar{\mathbf{z}})$ yields

$$M_k(\bar{\mathbf{z}} - \bar{\mathbf{z}}') \in M_k(\rho_k(\bar{\mathbf{z}}) - \rho_k(\bar{\mathbf{z}}')) + \partial F_k(\rho_k(\bar{\mathbf{z}})) - \partial F_k(\rho_k(\bar{\mathbf{z}}')) + C_{\mathbf{z},k}^\top (\lambda - \lambda'). \tag{3.65}$$

We multiply (3.65) from the left by $(\rho_k(\bar{\mathbf{z}}) - \rho_k(\bar{\mathbf{z}}'))^\top$ and obtain

$$(\rho_k(\bar{\mathbf{z}}) - \rho_k(\bar{\mathbf{z}}'))^\top M_k(\bar{\mathbf{z}} - \bar{\mathbf{z}}') \in \|\rho_k(\bar{\mathbf{z}}) - \rho_k(\bar{\mathbf{z}}')\|_{M_k}^2 + (\rho_k(\bar{\mathbf{z}}) - \rho_k(\bar{\mathbf{z}}'))^\top C_{\mathbf{z},k}^\top (\lambda - \lambda')$$
$$+ (\rho_k(\bar{\mathbf{z}}) - \rho_k(\bar{\mathbf{z}}'))^\top (\partial F_k(\rho_k(\bar{\mathbf{z}})) - \partial F_k(\rho_k(\bar{\mathbf{z}}'))). \tag{3.66}$$

Due to the convexity of the function F_k, the following holds (Parikh and Boyd, 2014)

$$(\rho_k(\bar{\mathbf{z}}) - \rho_k(\bar{\mathbf{z}}'))^\top (g - g') \geq 0, \quad \forall g \in \partial F_k(\rho_k(\bar{\mathbf{z}})),\, g' \in \partial F_k(\rho_k(\bar{\mathbf{z}}')). \tag{3.67}$$

Using (3.67) in (3.66) yields

$$(\rho_k(\bar{\mathbf{z}}) - \rho_k(\bar{\mathbf{z}}'))^\top M_k(\bar{\mathbf{z}} - \bar{\mathbf{z}}') \geq \|\rho_k(\bar{\mathbf{z}}) - \rho_k(\bar{\mathbf{z}}')\|_{M_k}^2 + (\rho_k(\bar{\mathbf{z}}) - \rho_k(\bar{\mathbf{z}}'))^\top C_{\mathbf{z},k}^\top (\lambda - \lambda'). \tag{3.68}$$

Given the complementary slackness condition (3.64b), the second term of the right-hand side of (3.68) satisfies

$$\left(\lambda^\top C_{\mathbf{z},k}\,\rho_k(\bar{\mathbf{z}}) - \lambda^\top C_{\mathbf{z},k}\,\rho_k(\bar{\mathbf{z}}')\right)^\top + \left(\lambda'^\top C_{\mathbf{z},k}\,\rho_k(\bar{\mathbf{z}}') - \lambda'^\top C_{\mathbf{z},k}\,\rho_k(\bar{\mathbf{z}})\right)^\top \tag{3.69}$$

$$= \left(\lambda^\top d_{\mathbf{z},k} - \lambda^\top C_{\mathbf{z},k}\,\rho_k(\bar{\mathbf{z}}')\right)^\top + \left(\lambda'^\top d_{\mathbf{z},k} - \lambda'^\top C_{\mathbf{z},k}\,\rho_k(\bar{\mathbf{z}})\right)^\top$$

$$= (d_{\mathbf{z},k} - C_{\mathbf{z},k}\,\rho_k(\bar{\mathbf{z}}'))^\top \lambda + (d_{\mathbf{z},k} - C_{\mathbf{z},k}\,\rho_k(\bar{\mathbf{z}}))^\top \lambda' \geq 0,$$

where the nonngeativity of the last expression holds as a result of (3.64c) and (3.64d). Hence, (3.68) becomes

$$\|\rho_k(\bar{\mathbf{z}}) - \rho_k(\bar{\mathbf{z}}')\|_{M_k}^2 \leq (\rho_k(\bar{\mathbf{z}}) - \rho_k(\bar{\mathbf{z}}'))^\top M_k(\bar{\mathbf{z}} - \bar{\mathbf{z}}') \tag{3.70}$$
$$= (\rho_k(\bar{\mathbf{z}}) - \rho_k(\bar{\mathbf{z}}'))^\top M_k^{1/2} M_k^{1/2}(\bar{\mathbf{z}} - \bar{\mathbf{z}}').$$

We obtain in view of the Cauchy-Schwarz inequality

$$\left\|\rho_k(\bar{\mathbf{z}}) - \rho_k(\bar{\mathbf{z}}')\right\|_{M_k}^2 \leq \left\|M_k^{1/2}(\rho_k(\bar{\mathbf{z}}) - \rho_k(\bar{\mathbf{z}}'))\right\| \left\|M_k^{1/2}(\bar{\mathbf{z}} - \bar{\mathbf{z}}')\right\|. \tag{3.71}$$

We take the square of (3.71) and get

$$\left\|\rho_k(\bar{\mathbf{z}}) - \rho_k(\bar{\mathbf{z}}')\right\|_{M_k}^4 \leq \left\|M_k^{1/2}(\rho_k(\bar{\mathbf{z}}) - \rho_k(\bar{\mathbf{z}}'))\right\|^2 \left\|M_k^{1/2}(\bar{\mathbf{z}} - \bar{\mathbf{z}}')\right\|^2$$
$$= \|\rho_k(\bar{\mathbf{z}}) - \rho_k(\bar{\mathbf{z}}')\|_{M_k}^2 \|\bar{\mathbf{z}} - \bar{\mathbf{z}}'\|_{M_k}^2. \tag{3.72}$$

If $\rho_k(\bar{\mathbf{z}}) \neq \rho_k(\bar{\mathbf{z}}')$, we can divide by $\|\rho_k(\bar{\mathbf{z}}) - \rho_k(\bar{\mathbf{z}}')\|_{M_k}^2$ to get the desired inequality (3.62). Otherwise, we get $0 \leq 0$ in (3.72). $\qquad\square$

In the absence of disturbances, evaluating (3.62) in Lemma 3.4 at $\bar{\mathbf{z}} = \mathbf{z}_k$ and $\bar{\mathbf{z}}' = \bar{\mathbf{z}}_k$, i.e., at the true system state and at the pMHE a priori estimate, respectively, yields

$$\|\mathbf{z}_k - \hat{\mathbf{z}}_k^\star\|_{M_k}^2 \leq \|\mathbf{z}_k - \bar{\mathbf{z}}_k\|_{M_k}^2, \tag{3.73}$$

which is due to the fact that, by Lemma 3.3, $\rho_k(\mathbf{z}_k) = \mathbf{z}_k$. In this disturbance-free case, the above inequality is exactly (3.21) in Lemma 3.1 when the quadratic Bregman distance $D_{\psi_k}(\mathbf{z}, \bar{\mathbf{z}}) = \frac{1}{2}\|\mathbf{z} - \bar{\mathbf{z}}\|_{M_k}^2$ is employed. Notice however that the proof techniques used in Lemma 3.4 and Lemma 3.1 are different, in the sense that we resorted in the former to proximal operator theory and in the latter to central properties of Bregman distances.

In order to be able arrive to a result similar to (3.21) that holds in the presence of disturbances, we need to evaluate the modified nonexpansivenss property (3.62) at two suitable vectors $\bar{\mathbf{z}}, \bar{\mathbf{z}}' \in \mathbb{R}^{(N+1)n}$. Their choice should allow for an upper bound on the pMHE error $\mathbf{z}_k - \hat{\mathbf{z}}_k^\star$ in terms of the error between the true system state $\mathbf{z}_k$ and the pMHE a priori estimate $\bar{\mathbf{z}}_k$, which can be then directly used in the main ISS analysis. This is achieved in the subsequent lemma due to the following additional assumptions on the stage cost.

Assumption 3.7 (Strongly smooth stage cost). *For any $i \in \mathbb{N}$, the functions r_i and q_i are continuously differentiable and the gradients of r_i and q_i are Lipschitz continuous with the uniform Lipschitz constants $L_r \in \mathbb{R}_{++}$ and $L_q \in \mathbb{R}_{++}$, respectively, i.e.,*

$$\|\nabla r_i(x) - \nabla r_i(y)\| \leq L_r \|x - y\| \quad \forall i, \ \forall x, y \in \mathbb{R}^p, \tag{3.74}$$
$$\|\nabla q_i(x) - \nabla q_i(y)\| \leq L_q \|x - y\| \quad \forall i, \ \forall x, y \in \mathbb{R}^n. \tag{3.75}$$

Lemma 3.5. *Consider system* (3.18) *and the pMHE problem* (3.57). *Let Assumptions 3.3 and 3.7 hold and suppose that the set $\mathcal{X}_k$ contains the true state in its interior, i.e., $x_k \in \mathrm{int}(\mathcal{X}_k)$, $k \in \mathbb{N}$. Then,*

$$\|\mathbf{z}_k - \hat{\mathbf{z}}_k^\star\|_{M_k}^2 \leq \|\mathbf{z}_k - \bar{\mathbf{z}}_k + M_k^{-1}\nabla F_k(\mathbf{z}_k)\|_{M_k}^2 \tag{3.76}$$

for any $k \in \mathbb{N}$, where $\mathbf{z}_k$ denotes the true system state, $\hat{\mathbf{z}}_k^\star$ the pMHE solution, and $\bar{\mathbf{z}}_k$ the a priori estimate based on the Kalman filter.

Proof. We invoke the nonexpansivenss property (3.62) derived in Lemma 3.4. We choose $\bar{\mathbf{z}}'$ as the Kalman filter a priori estimate, i.e., we set $\bar{\mathbf{z}}' = \bar{\mathbf{z}}_k$ with $\rho_k(\bar{\mathbf{z}}_k) = \hat{\mathbf{z}}_k^*$. As for the choice of $\bar{\mathbf{z}}$, we require it to fulfill $\rho_k(\bar{\mathbf{z}}) = \mathbf{z}_k$ such that we can obtain in (3.62)

$$\|\mathbf{z}_k - \hat{\mathbf{z}}_k^*\|_{M_k}^2 \leq \|\underbrace{\bar{\mathbf{z}}}_{?} - \bar{\mathbf{z}}_k\|_{M_k}^2. \tag{3.77}$$

Consider again the proof of Lemma 3.4. We evaluate the KKT conditions derived in (3.64) for problem (3.59) at the minimizer $\rho(\bar{\mathbf{z}}) = \mathbf{z}_k$. Since the true state x_{k-N} of system (3.18) strictly satisfies the inequality constraints, and given that $\mathcal{W}_k = \mathbb{R}^n$, it holds that $\mathbf{z}_k \in \text{int}(\mathcal{S}_k)$. Hence, the Lagrange multiplier associated with $\mathbf{z}_k$ is zero and (3.64a) at $\rho(\bar{\mathbf{z}}) = \mathbf{z}_k$ becomes

$$0 = M_k(\mathbf{z}_k - \bar{\mathbf{z}}) + \nabla F_k(\mathbf{z}_k), \tag{3.78}$$

due to the fact that F_k is differentiable by Assumption 3.7. Since M_k is invertible,

$$\bar{\mathbf{z}} = \mathbf{z}_k + M_k^{-1}\nabla F_k(\mathbf{z}_k). \tag{3.79}$$

Plugging (3.79) in (3.77) yields the desired result (3.76). □

We establish in the following theorem that the estimation error generated from the pMHE scheme (3.57) based on the Kalman filter is ISS with respect to additive process and measurement disturbances.

Theorem 3.3. *Consider system (3.18) and the pMHE problem (3.57). Let Assumptions 3.1, 3.3, 3.6 and 3.7 hold and suppose that the set $\mathcal{X}_k$ contains the true state in its interior, i.e., $x_k \in \text{int}(\mathcal{X}_k)$, $k \in \mathbb{N}$. Then, the estimation error (3.17) is ISS with respect to the process and measurement disturbances, i.e., there exist $\beta_{\mathrm{e}} \in (0,1)$ and $c_{\mathrm{e}}, c_{\mathrm{w}}, c_{\mathrm{v}} \in \mathbb{R}_{++}$ such that*

$$\|x_k - \hat{x}_k\| \leq c_{\mathrm{e}}\beta_{\mathrm{e}}^k \|x_0 - \bar{x}_0\| + c_{\mathrm{w}}\|\mathbf{w}_{[0:k-1]}\| + c_{\mathrm{v}}\|\mathbf{v}_{[0:k-1]}\| \tag{3.80}$$

holds for any $k \in \mathbb{N}_+$ and $x_0, \bar{x}_0 \in \mathbb{R}^n$, where $\|\mathbf{w}_{[i:k]}\| := \sup_{i \leq j \leq k}\{\|w_j\|\}$.

Proof. We prove this result by showing that there exists for the pMHE error (3.16) an ISS-Lyapunov function V_k satisfying conditions (A.9a) and (A.9b) in Definition A.9 in Appendix A. As a candidate Lyapunov function, we choose the quadratic Bregman distance (3.43), i.e.,

$$V_k(\mathbf{z}_k, \hat{\mathbf{z}}_k^*) = D_{\psi_k}(\mathbf{z}_k, \hat{\mathbf{z}}_k^*) = \frac{1}{2}\|\mathbf{z}_k - \hat{\mathbf{z}}_k^*\|_{M_k}^2 \tag{3.81}$$

with $M_k = \text{diag}(\Pi_{k-N}^-, W, W, \ldots, W) \in \mathbb{S}_{++}^{(N+1)n}$. Recall in the proof of Proposition 3.1 that $P_k^- = (\Pi_k^-)^{-1}$ is uniformly lower and upper bounded by $\underline{p}I \preceq P_k^- \preceq \bar{p}I$ due to Assumption 3.6. Hence, there exist $\mathcal{K}_\infty$-functions α_1, α_2 such that condition (A.9a) is fulfilled. In particular, $\alpha_1(\|\mathbf{z}_k - \hat{\mathbf{z}}_k^*\|) = \sigma\|\mathbf{z}_k - \hat{\mathbf{z}}_k^*\|^2$ and $\alpha_2(\|\mathbf{z}_k - \hat{\mathbf{z}}_k^*\|) = \gamma\|\mathbf{z}_k - \hat{\mathbf{z}}_k^*\|^2$, where $\sigma = \min(\frac{1}{\bar{p}}, \lambda_{\min}(\bar{W}))$ and $\gamma = \max(\frac{1}{\underline{p}}, \lambda_{\max}(\bar{W}))$. Moreover, we have by (3.76) in

Lemma 3.5 that

$$\Delta V_k := V_k(\mathbf{z}_k, \hat{\mathbf{z}}_k^*) - V_{k-1}(\mathbf{z}_{k-1}, \hat{\mathbf{z}}_{k-1}^*) \tag{3.82}$$

$$= \frac{1}{2}\left\|\mathbf{z}_k - \hat{\mathbf{z}}_k^*\right\|_{M_k}^2 - \frac{1}{2}\left\|\mathbf{z}_{k-1} - \hat{\mathbf{z}}_{k-1}^*\right\|_{M_{k-1}}^2$$

$$\leq \frac{1}{2}\left\|\mathbf{z}_k - \bar{\mathbf{z}}_k + M_k^{-1}\nabla F_k(\mathbf{z}_k)\right\|_{M_k}^2 - \frac{1}{2}\left\|\mathbf{z}_{k-1} - \hat{\mathbf{z}}_{k-1}^*\right\|_{M_{k-1}}^2$$

$$= \frac{1}{2}\left\|\mathbf{z}_k - \bar{\mathbf{z}}_k\right\|_{M_k}^2 + \nabla F_k(\mathbf{z}_k)^\top (\mathbf{z}_k - \bar{\mathbf{z}}_k) + \frac{1}{2}\left\|\nabla F_k(\mathbf{z}_k)\right\|_{M_k^{-1}}^2 - \frac{1}{2}\left\|\mathbf{z}_{k-1} - \hat{\mathbf{z}}_{k-1}^*\right\|_{M_{k-1}}^2.$$

By introducing the following matrices

$$H_k := \begin{bmatrix} G_k & E_k \end{bmatrix} \in \mathbb{R}^{Np \times (N+1)n}, \tag{3.83}$$

$$\tilde{H} := \begin{bmatrix} 0_{[Nn \times n]} & I_{[Nn]} \end{bmatrix} \in \mathbb{R}^{Nn \times (N+1)n}, \tag{3.84}$$

where

$$G_k := \begin{bmatrix} C_{k-N} \\ C_{k-N+1} A_{k-N} \\ C_{k-N+2} \Phi(k-N+2, k-N) \\ \vdots \\ C_{k-1} \Phi(k-1, k-N) \end{bmatrix} \in \mathbb{R}^{Np \times n}, \tag{3.85}$$

$$E_k := \begin{bmatrix} 0 & 0 & \dots & 0 \\ C_{k-N+1} & 0 & \dots & 0 \\ C_{k-N+2}A_{k-N+1} & C_{k-N+2} & \dots & 0 \\ \vdots & \vdots & \dots & \vdots \\ C_{k-1}\Phi(k-1, k-N+1) & C_{k-1}\Phi(k-1, k-N+2) & \dots & C_{k-1} \end{bmatrix} \in \mathbb{R}^{Np \times Nn},$$

as well as the vectors defined based on the true process and measurement disturbances

$$D_{\mathrm{r},k} := \begin{bmatrix} \nabla r_{k-N}^\top(v_{k-N}) & \dots & \nabla r_{k-1}^\top(v_{k-1}) \end{bmatrix}^\top \in \mathbb{R}^{Np}, \tag{3.86a}$$

$$D_{\mathrm{q},k} := \begin{bmatrix} \nabla q_{k-N}^\top(w_{k-N}) & \dots & \nabla q_{k-1}^\top(w_{k-1}) \end{bmatrix}^\top \in \mathbb{R}^{Nn}, \tag{3.86b}$$

we can compute

$$\nabla F_k(\mathbf{z}_k) = -H_k^\top D_{\mathrm{r},k} + \tilde{H}^\top D_{\mathrm{q},k} = \begin{bmatrix} -G_k^\top D_{\mathrm{r},k} \\ -E_k^\top D_{\mathrm{r},k} + D_{\mathrm{q},k} \end{bmatrix} \in \mathbb{R}^{(N+1)n}. \tag{3.87}$$

Recalling the notation (3.14) and using (3.87) and (3.43) in (3.82) yields

$$\Delta V_k \leq \frac{1}{2}\|x_{k-N} - \bar{x}_{k-N}\|_{\Pi_{k-N}^-}^2 - \frac{1}{2}\left\|x_{k-N-1} - \hat{x}_{k-N-1}^*\right\|_{\Pi_{k-N-1}^-}^2 + \frac{1}{2}\|\mathbf{w}_k\|_W^2 - \frac{1}{2}\|\mathbf{w}_{k-1} - \hat{\mathbf{w}}_{k-1}^*\|_W^2$$

$$- D_{\mathrm{r},k}^\top G_k(x_{k-N} - \bar{x}_{k-N}) + (D_{\mathrm{q},k}^\top - D_{\mathrm{r},k}^\top E_k)\mathbf{w}_k + \frac{1}{2}\|\nabla F_k(\mathbf{z}_k)\|_{M_k^{-1}}^2, \tag{3.88}$$

since we have zero a priori process disturbances. Given system (3.18), the a priori estimate (3.42b) based on the Kalman filter, and the definition of the pMHE error (3.16), we can compute

$$x_{k-N} - \bar{x}_{k-N} = A_{k-N-1}x_{k-N-1} + w_{k-N-1} - A_{k-N-1}\hat{x}^*_{k-N-1} \qquad (3.89a)$$
$$- A_{k-N-1}K_{k-N-1}\left(C_{k-N-1}x_{k-N-1} + v_{k-N-1} - C_{k-N-1}\hat{x}^*_{k-N-1}\right)$$
$$= \tilde{A}_{k-N-1}e_{k-N-1} + d_{k-N-1}$$

since the input terms cancel out. Here, we let $\tilde{A}_{k-N-1} := A_{k-N-1}(I - K_{k-N-1}C_{k-N-1})$ and

$$d_{k-N-1} := w_{k-N-1} - A_{k-N-1}K_{k-N-1}v_{k-N-1}. \qquad (3.89b)$$

Plugging (3.89) in (3.88) yields

$$\Delta V_k \leq \frac{1}{2}\left\|\tilde{A}_{k-N-1}e_{k-N-1}\right\|^2_{\Pi^-_{k-N}} - \frac{1}{2}\left\|e_{k-N-1}\right\|^2_{\Pi^-_{k-N-1}} - \frac{1}{2}\left\|\mathbf{w}_{k-1} - \hat{\mathbf{w}}^*_{k-1}\right\|^2_{\tilde{W}} \qquad (3.90)$$
$$+ d^\top_{k-N-1}\Pi^-_{k-N}\tilde{A}_{k-N-1}e_{k-N-1} + \frac{1}{2}\left\|d_{k-N-1}\right\|^2_{\Pi^-_{k-N}} + \frac{1}{2}\left\|\mathbf{w}_k\right\|^2_{\tilde{W}}$$
$$- D^\top_{\mathrm{r},k}G_k(\tilde{A}_{k-N-1}e_{k-N-1} + d_{k-N-1}) + (D^\top_{\mathrm{q},k} - D^\top_{\mathrm{r},k}E_k)\mathbf{w}_k + \frac{1}{2}\left\|\nabla F_k(\mathbf{z}_k)\right\|^2_{M^{-1}_k}.$$

We rearrange (3.90) and obtain

$$\Delta V_k \leq \frac{1}{2}\left\|\tilde{A}_{k-N-1}e_{k-N-1}\right\|^2_{\Pi^-_{k-N}} - \frac{1}{2}\left\|e_{k-N-1}\right\|^2_{\Pi^-_{k-N-1}} - \frac{1}{2}\left\|\mathbf{w}_{k-1} - \hat{\mathbf{w}}^*_{k-1}\right\|^2_{\tilde{W}} \qquad (3.91)$$
$$+ (d^\top_{k-N-1}\Pi^-_{k-N} - D^\top_{\mathrm{r},k}G_k)\tilde{A}_{k-N-1}e_{k-N-1} + \frac{1}{2}\left\|d_{k-N-1}\right\|^2_{\Pi^-_{k-N}} - D^\top_{\mathrm{r},k}G_k d_{k-N-1}$$
$$+ \frac{1}{2}\left\|\mathbf{w}_k\right\|^2_{\tilde{W}} + (D^\top_{\mathrm{q},k} - D^\top_{\mathrm{r},k}E_k)\mathbf{w}_k + \frac{1}{2}\left\|\nabla F_k(\mathbf{z}_k)\right\|^2_{M^{-1}_k}.$$

In the following, we derive an upper bound on the right-hand side of (3.91). Consider again the proof of Proposition 3.1. Setting $x = x_{k-N-1}$ and $\hat{x} = \hat{x}^*_{k-N-1}$ in (3.54) yields for the first two terms in the right-hand side of (3.91) that

$$\frac{1}{2}\left\|\tilde{A}_{k-N-1}e_{k-N-1}\right\|^2_{\Pi^-_{k-N}} - \frac{1}{2}\left\|e_{k-N-1}\right\|^2_{\Pi^-_{k-N-1}} \leq -\frac{\tilde{u}}{2}\left\|e_{k-N-1}\right\|^2, \qquad (3.92)$$

where $\tilde{u}$ is defined in (3.53). By Assumption 3.1, the function h_k is Lipschitz continuous with the uniform Lipschitz constant $c_\mathrm{h} \in \mathbb{R}_{++}$. Hence, C_k is uniformly bounded by c_h, i.e., $\|C_k\| \leq c_\mathrm{h}$ for all $k \in \mathbb{N}$. Furthermore, since $R \in \mathbb{S}^p_{++}$ and given the established bounds on $\|P^-_k\|$ and $\|C_k\|$, (3.44a) implies that there exists $\bar{k} \in \mathbb{R}_{++}$ such that the Kalman gain satisfies $\|K_k\| \leq \bar{k}$. By using Young's inequality, we therefore obtain for the fourth term in the right-hand side of (3.91) that

$$d^\top_{k-N-1}\Pi^-_{k-N}\tilde{A}_{k-N-1}e_{k-N-1} \leq s_1\left\|e_{k-N-1}\right\|^2 + \frac{c^2_\mathrm{f}(1 + \bar{k}c_\mathrm{h})^2}{s_1\,\underline{p}'^2}\left\|d_{k-N-1}\right\|^2, \qquad (3.93a)$$

$$- D^\top_{\mathrm{r},k}G_k\tilde{A}_{k-N-1}e_{k-N-1} \leq s_2\left\|e_{k-N-1}\right\|^2 + \frac{c^2_\mathrm{f}(1 + \bar{k}c_\mathrm{h})^2\,\bar{g}^2}{s_2}\left\|D_{\mathrm{r},k}\right\|^2, \qquad (3.93b)$$

where $s_1, s_2 \in \mathbb{R}_{++}$ are arbitrary and $\bar{g} \in \mathbb{R}_{++}$ is such that $\|G_k\| \leq \bar{g}$ for all $k \in \mathbb{N}$. Note that the uniform upper bound $\bar{g}$ exists since every element of G_k (as well as of E_k) defined in (3.85) can be upper bounded in terms of c_f and c_h due to Assumption 3.1. We compute for the fifth and sixth terms in the right-hand side of (3.91)

$$\frac{1}{2}\|d_{k-N-1}\|^2_{\Pi_{k-N}^{-1}} \le \frac{1}{2\underline{p}'}\|d_{k-N-1}\|^2,\tag{3.94a}$$

$$-D_{r,k}^\top G_k d_{k-N-1} \le s_3\|d_{k-N-1}\|^2 + \frac{\bar{g}^2}{s_3}\|D_{r,k}\|^2,\tag{3.94b}$$

where $s_3 \in \mathbb{R}_{++}$ is arbitrary. By exploiting (3.87) and the fact that the weight matrix M_k in the Bregman distance satisfies $\|M_k\| \le \gamma$ with $\gamma = \max(\frac{1}{\underline{p}'}, \lambda_{\max}(\bar{W}))$, we obtain for the last three terms of (3.91) that

$$\frac{1}{2}\|\mathbf{w}_k\|^2_{\bar{W}} \le \frac{\lambda_{\max}(\bar{W})}{2}\|\mathbf{w}_k\|^2,\tag{3.95a}$$

$$(D_{q,k}^\top - D_{r,k}^\top E_k)\mathbf{w}_k \le s_4\|\mathbf{w}_k\|^2 + \frac{1}{s_4}\|D_{q,k}\|^2 + s_5\|\mathbf{w}_k\|^2 + \frac{\bar{e}^2}{s_5}\|D_{r,k}\|^2,\tag{3.95b}$$

$$\frac{1}{2}\|\nabla F_k(\mathbf{z}_k)\|^2_{M_k^{-1}} \le \frac{\gamma}{2}(\|G_k^\top D_{r,k}\|^2 + \|D_{q,k} - E_k^\top D_{r,k}\|^2)\tag{3.95c}$$

$$\le \frac{\gamma\bar{g}^2}{2}\|D_{r,k}\|^2 + \frac{\gamma\bar{e}^2}{2}\|D_{r,k}\|^2 + \frac{\gamma}{2}\|D_{q,k}\|^2 + s_6\|D_{r,k}\|^2 + \frac{\bar{e}^2}{s_6}\|D_{q,k}\|^2,$$

where $s_4, s_5, s_6 \in \mathbb{R}_{++}$ are arbitrary and $\bar{e} \in \mathbb{R}_{++}$ is such that $\|E_k\| \le \bar{e}$ for all $k \in \mathbb{N}$. According to Assumption 3.7, the gradients ∇r_i and ∇q_i are Lipschitz continuous with the uniform Lipschitz constants $L_r, L_q \in \mathbb{R}_{++}$. Moreover, by Assumption 3.3, $\nabla r_i(0) = 0$ and $\nabla q_i(0) = 0$. Thus, it holds for the vectors defined in (3.86) that

$$\|D_{r,k}\|^2 = \|\nabla r_{k-N}(v_{k-N}) - \nabla r_{k-N}(0)\|^2 + \cdots + \|\nabla r_{k-1}(v_{k-1}) - \nabla r_{k-1}(0)\|^2\tag{3.96}$$

$$\le L_r^2\|v_{k-N}\|^2 + \cdots + L_r^2\|v_{k-1}\|^2 \le L_r^2\|\mathbf{v}_k\|^2,$$

$$\|D_{q,k}\|^2 \le L_q^2\|w_{k-N}\|^2 + \cdots + L_q^2\|w_{k-1}\|^2 \le L_q^2\|\mathbf{w}_k\|^2.\tag{3.97}$$

Substituting all the derived upper bounds (starting from (3.92)) in (3.91) finally yields

$$\Delta V_k \le -c_1\|e_{k-N-1}\|^2 - \frac{1}{2}\|\mathbf{w}_{k-1} - \hat{\mathbf{w}}^*_{k-1}\|^2_{\bar{W}} + c_d\|d_{k-N-1}\|^2 + \frac{\tilde{c}_w}{2}\|\mathbf{w}_k\|^2 + \frac{\tilde{c}_v}{2}\|\mathbf{v}_k\|^2,\tag{3.98a}$$

where

$$c_1 := \frac{\tilde{u}}{2} - s_1 - s_2\tag{3.98b}$$

$$c_d := \frac{c_f^2(1 + \bar{k}c_h)^2}{s_1\,\underline{p}'^2} + \frac{1}{2\underline{p}'} + s_3,\tag{3.98c}$$

$$\tilde{c}_w := \left(\frac{1}{s_4} + \frac{\gamma}{2} + \frac{\bar{e}^2}{s_6}\right)L_q^2 + \frac{\lambda_{\max}(\bar{W})}{2} + s_4 + s_5,\tag{3.98d}$$

$$\tilde{c}_v := \left(\frac{c_f^2(1 + \bar{k}c_h)^2\,\bar{g}^2}{s_2} + \frac{\bar{g}^2}{s_3} + \frac{\bar{e}^2}{s_5} + \frac{\gamma(\bar{g}^2 + \bar{e}^2)}{2} + s_6\right)L_r^2,\tag{3.98e}$$

are strictly positive if we choose $s_1, s_2 \in \mathbb{R}_{++}$ such that $s_1 + s_2 < \tilde{u}/2$. This implies that condition (A.9b) is fulfilled and V_k is a uniform ISS-Lyapunov function for the pMHE error

(3.16). Based on (3.98), we can derive an explicit characterization of the ISS property of the pMHE error in the form of (A.8) by simple, albeit tedious, algebraic manipulations. Following the last step in the proof of Theorem 3.2, we can then use Lipschitz arguments and the triangle inequality to compute the ISS gains c_e, β_e, c_w, c_v in (3.80). Since the aforementioned calculations are rather cumbersome, we refer to (Gharbi and Ebenbauer, 2020, Appendix C) for a thorough derivation of these gains. □

Theorem 3.3 implies that the pMHE scheme (3.57) based on the Kalman filter is inherently robust to arbitrary process and measurement disturbances, for which specific bounds need not be known a priori. More specifically, if the disturbances are bounded, then the estimation error remains bounded with a well-defined gain between the size of the error bound and the magnitude of disturbances, and if the disturbances decay, so does the estimation error. Furthermore, setting the disturbances in (3.80) to be zero yields that the estimation error is GUES, and we recover the stability result of Corollary 3.1.

Establishing robust stability for MHE under bounded disturbances can be a much more complicated task than proving nominal stability. In (Rawlings et al. (2017)), mainly convergent disturbances are considered in the underlying robustness analysis. More specifically, the estimation error generated by the linear constrained MHE scheme with quadratic stage cost (2.12) and prior weighting (2.15) is guaranteed to be ISS with respect to convergent disturbances satisfying condition (2.7). The matrix P_k in the prior weighting (2.15) denotes the covariance matrix of the Kalman filter and is computed via (2.14b) similar to the pMHE scheme with quadratic Bregman distance (3.43). Even though both schemes share this similarity in the design, in pMHE with convex stage cost, ISS is established for the more general case of bounded disturbances. As discussed in the previous chapter, various works either add max-terms to the objective function to achieve robust stability under bounded disturbances or require a large enough horizon length (cf. Müller (2017); Raković and Levine (2018)). In (Alessandri et al. (2010); Sui and Johansen (2014)) where LTI systems are considered, ISS of the estimation error generated by the associated MHE schemes with quadratic cost is achieved by incorporating the observability properties of the system in the design of the prior weighting. Compared to these methods, the simple design of the Bregman distance in our pMHE scheme guarantees robust stability in the ISS sense for any horizon length N and a convex stage cost satisfying Assumptions 3.3 and 3.7. Moreover, in the special case of LTI systems, we will discuss in Section 3.2.4 how robust stability of pMHE can be ensured under milder assumptions, where detectability rather than observability is assumed.

Remark 3.5. In principle, the ISS result derived in this section can be easily extended to the case where additional convex polytopic constraints $\hat{w}_i \in \mathcal{W}_k \subseteq \mathbb{R}^n$ and $\hat{v}_i \in \mathcal{V}_k \subseteq \mathbb{R}^p$ are imposed in the pMHE problem (3.57). Nevertheless, we only impose constraints on the state to reflect the fact that robustness holds for arbitrary and unknown additive disturbances.

3.2.3 Bayesian interpretation of proximity MHE

In this section, we consider the state estimation problem of system (3.18) from a probabilistic perspective, with which we can give some insight into the design of the components of the pMHE problem (3.19). For simplicity of analysis, we set the input matrices B_k in system (3.18) to be zero and consider the unconstrained case where $\mathcal{X}_k = \mathbb{R}^n$, $\mathcal{W}_k = \mathbb{R}^n$, and $\mathcal{V}_k = \mathbb{R}^p$ for all $k \in \mathbb{N}$. In the pMHE problem (3.19), we decompose D_{ψ_k} into a

Bregman distance $D_{k-N} : \mathbb{R}^n \times \mathbb{R}^n \to \mathbb{R}_+$ for the first estimated state in the horizon window $\hat{x}_{k-N}$, and quadratic Bregman distances for the estimated process disturbances $\hat{w}_i$. More specifically, we define $\mathbf{z} = \begin{bmatrix} x^\top & \mathbf{w}^\top \end{bmatrix}^\top$, $\hat{\mathbf{z}} = \begin{bmatrix} \hat{x}^\top & \hat{\mathbf{w}}^\top \end{bmatrix}^\top$, $x, \hat{x} \in \mathbb{R}^n$, $\mathbf{w}, \hat{\mathbf{w}} \in \mathbb{R}^{Nn}$ and let

$$D_{\psi_k}(\mathbf{z}, \hat{\mathbf{z}}) = D_{k-N}(x, \hat{x}) + \frac{1}{2}\|\mathbf{w} - \hat{\mathbf{w}}\|_{\bar{W}}^2, \tag{3.99}$$

where $\bar{W} = \mathrm{diag}(W, \ldots, W) \in \mathbb{S}_{++}^{Nn}$ with $W \in \mathbb{S}_{++}^n$ arbitrary. Moreover, we choose zero a priori disturbances in the a priori estimate $\bar{\mathbf{z}}_k$ defined in (3.14), i.e., we set $\bar{\mathbf{z}}_k = \begin{bmatrix} \bar{x}_{k-N}^\top & 0 & \ldots & 0 \end{bmatrix}^\top$, where $\bar{x}_{k-N} \in \mathbb{R}^n$ refers to some a priori estimate for the state $\hat{x}_{k-N}$. The resulting Bregman distance D_{ψ_k} centered around $\bar{\mathbf{z}}_k$ in the pMHE cost function (3.19a) is then

$$D_{\psi_k}(\hat{\mathbf{z}}_k, \bar{\mathbf{z}}_k) = D_{k-N}(\hat{x}_{k-N}, \bar{x}_{k-N}) + \sum_{i=k-N}^{k-1} \frac{1}{2}\|\hat{w}_i\|_W^2. \tag{3.100}$$

In essence, we focus on a Bayesian interpretation of the linear pMHE scheme based on the following unconstrained optimization problem

$$\min_{\hat{\mathbf{z}}_k} \quad \sum_{i=k-N}^{k-1} r_i(\hat{v}_i) + \tilde{q}_i(\hat{w}_i) + D_{k-N}(\hat{x}_{k-N}, \bar{x}_{k-N}) \tag{3.101a}$$

$$\text{s. t.} \quad \hat{x}_{i+1} = A_i\,\hat{x}_i + \hat{w}_i, \qquad i = k - N, \ldots, k - 1 \tag{3.101b}$$

$$\qquad y_i = C_i\,\hat{x}_i + \hat{v}_i, \qquad i = k - N, \ldots, k - 1, \tag{3.101c}$$

where the function $\tilde{q}_i(\hat{w}_i) := q_i(\hat{w}_i) + \frac{1}{2}\|\hat{w}_i\|_W^2$ with $i = k - N, \ldots, k - 1$ is strongly convex. In system (3.18), we assume that the disturbances w_k and v_k are realizations of continuous independent random variables W_k and V_k with probability densities ρ_{W_k} and ρ_{V_k}, respectively. The idea of Bayes filtering is to recursively estimate the conditional probability density of the state x_k given online measurements $\{y_0, \cdots, y_{k-1}\}$. Following (Rao (2000)), it can be shown using Bayes' theorem and the Markov property of the state space model (3.18) that the posterior conditional probability density of the state sequence $\{x_{k-N}, \ldots, x_k\}$ given the measurements $\{y_0, \ldots, y_{k-1}\}$ is

$$\rho(x_{k-N}, \ldots, x_k \mid y_0, \ldots, y_{k-1}) \propto \rho(x_{k-N} \mid y_0, \ldots, y_{k-N-1}) \prod_{i=k-N}^{k-1} \rho(y_i \mid x_i)\,\rho(x_{i+1} \mid x_i). \tag{3.102}$$

By abuse of notation, let $\rho(x_{k-N} \mid y_0, \ldots, y_{k-N-1})$ refer to the prior conditional probability density of x_{k-N} given $\{y_0, \ldots, y_{k-N-1}\}$. Note that we use the symbol $\propto$ to indicate proportionality. Under the assumptions on W_k and V_k, it holds that

$$\rho(x_{k-N}, \ldots, x_k \mid y_0, \ldots, y_{k-1}) \propto \rho(x_{k-N} \mid y_0, \ldots, y_{k-N-1})$$
$$\times \prod_{i=k-N}^{k-1} \rho_v(y_i - C_i\,x_i)\,\rho_w(x_{i+1} - A_i\,x_i). \tag{3.103}$$

In order to establish a Bayesian view on pMHE based on problem (3.101), we associate the Bregman distance D_{k-N} centered around the a priori estimate $\bar{x}_{k-N}$ to the prior conditional density, i.e.,

$$\rho(x_{k-N} \mid y_0, \ldots, y_{k-N-1}) \propto e^{-D_{k-N}(x_{k-N}, \bar{x}_{k-N})}. \tag{3.104}$$

Note that given the established connections between Bregman distances (also called Bregman divergences) and well-known exponential families of probability distributions (Banerjee et al. (2005); Dhillon and Tropp (2008)), $e^{-D_{k-N}(x_{k-N}, \bar{x}_{k-N})}$ is a meaningful choice for the prior density function. Let us further associate r_k and $\tilde{q}_k$ to the densities

$$\rho_{V_k}(v_k) \propto e^{-r_k(v_k)}, \qquad \rho_{W_k}(w_k) \propto e^{-\tilde{q}_k(w_k)}. \tag{3.105}$$

Substituting (3.104) and (3.105) in (3.103) yields

$$\rho\left(x_{k-N}, \ldots, x_k \,|\, y_0, \ldots, y_{k-1}\right) \propto e^{-D_{k-N}(x_{k-N}, \bar{x}_{k-N})} \prod_{i=k-N}^{k-1} e^{-(r_i(y_i - C_i x_i) + \tilde{q}_i(x_{i+1} - A_i x_i))}. \tag{3.106}$$

In view of the Maximum Likelihood method (Särkkä (2013)), we can compute estimates of $\{x_{k-N}, \ldots, x_k\}$ by maximizing the posterior conditional probability density (3.106) or by equivalently minimizing its negative logarithm, i.e.,

$$\min_{\hat{x}_{k-N}, \ldots, \hat{x}_k} \quad D_{k-N}\left(\hat{x}_{k-N}, \bar{x}_{k-N}\right) + \sum_{i=k-N}^{k-1} r_i\left(y_i - C_i \hat{x}_i\right) + \tilde{q}_i\left(\hat{x}_{i+1} - A_i \hat{x}_i\right). \tag{3.107}$$

Hence, we recover the pMHE problem (3.101) by using the system dynamics (3.101b) and (3.101c) and considering $\hat{\mathbf{z}}_k = \begin{bmatrix} \hat{x}_{k-N}^{\top} & \hat{w}_{k-N}^{\top} & \ldots & \hat{w}_{k-1}^{\top} \end{bmatrix}^{\top}$ as decision variable.

Thus, what we have shown is that the pMHE scheme based on problem (3.101) is equivalent to a Bayesian estimator with prior conditional probability density (3.104), densities (3.105) for the process and measurement disturbances, and with posterior conditional probability density (3.106). This is depicted in Figure 3.1.

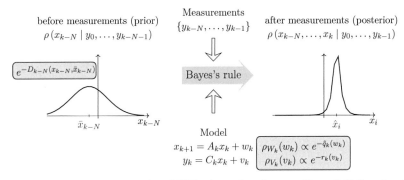

Figure 3.1: Bayesian view on the pMHE scheme based on problem (3.101) under the specified assumptions on W_k and V_k.

Design of the stage cost

In the following, we provide a probabilistic meaning and intuition to the design and tuning of the convex functions $\tilde{q}_i$, r_i, which we assumed rather arbitrary in the previous section, as well as of D_{k-N} and $\bar{x}_{k-N}$. More specifically, we guide the choice of the design components

$(\tilde{q}_i, r_i, D_{k-N}, \bar{x}_{k-N}, N)$ of the pMHE problem based on the statistics of the disturbances. If the process and measurement disturbances are assumed to be independent zero mean Gaussian random variables with $w_k \sim \mathcal{N}(0, Q_k)$ and $v_k \sim \mathcal{N}(0, R_k)$, we can select in the cost function (3.101) the following design components: i) the function $\tilde{q}_i(w_i) = \frac{1}{2}\|w_i\|^2_{Q_i^{-1}}$ for the process disturbances, ii) the function $r_i(v_i) = \frac{1}{2}\|v_i\|^2_{R_i^{-1}}$ for the output residuals, iii) the quadratic Bregman distance D_{k-N} as specified in (3.43), i.e., $D_{k-N}(\hat{x}_{k-N}, \bar{x}_{k-N}) = \frac{1}{2}\|\hat{x}_{k-N} - \bar{x}_{k-N}\|^2_{\Pi_{k-N}^-}$, and iv) the a priori estimate $\bar{x}_{k-N}$ given in (3.42b) and constructed based on the Kalman filter. In particular, setting the horizon length $N = 1$ leads in this case to the following optimization problem

$$\min_{\hat{x}_{k-1}, \hat{w}_{k-1}} \quad \frac{1}{2}\|\hat{v}_{k-1}\|^2_{R_{k-1}^{-1}} + \frac{1}{2}\|\hat{w}_{k-1}\|^2_{Q_{k-1}^{-1}} + \frac{1}{2}\|\hat{x}_{k-1} - \bar{x}_{k-1}\|^2_{\Pi_{k-1}^-} \tag{3.108a}$$

$$\text{s.\,t.} \quad \hat{x}_k = A_{k-1}\,\hat{x}_{k-1} + \hat{w}_{k-1}, \tag{3.108b}$$

$$y_{k-1} = C_{k-1}\,\hat{x}_{k-1} + \hat{v}_{k-1}, \tag{3.108c}$$

which can be solved analytically in order to obtain the pMHE solution $\hat{x}^*_{k-1}$ and consequently the state estimate $\hat{x}_k$. More specifically, substituting the measurement equation (3.108c) in (3.108a) yields

$$\min_{\hat{x}_{k-1}, \hat{w}_{k-1}} \quad \frac{1}{2}\|y_{k-1} - C_{k-1}\,\hat{x}_{k-1}\|^2_{R_{k-1}^{-1}} + \frac{1}{2}\|\hat{w}_{k-1}\|^2_{Q_{k-1}^{-1}} + \frac{1}{2}\|\hat{x}_{k-1} - \bar{x}_{k-1}\|^2_{\Pi_{k-1}^-}. \tag{3.109}$$

In view of the first-order necessary and sufficient condition for optimality of $(\hat{x}^*_{k-1}, \hat{w}^*_{k-1})$, evaluating the gradient of the above cost function at the solution $(\hat{x}^*_{k-1}, \hat{w}^*_{k-1})$ leads to

$$\hat{w}^*_{k-1} = 0, \tag{3.110a}$$

$$\hat{x}^*_{k-1} = \left(C_{k-1}^\top R_{k-1}^{-1} C_{k-1} + \Pi_{k-1}^-\right)^{-1}\left(C_{k-1}^\top R_{k-1}^{-1} y_{k-1} + \Pi_{k-1}^- \bar{x}_{k-1}\right), \tag{3.110b}$$

since both R_{k-1} an Π_{k-1} are positive definite matrices. Using the following matrix equalities (Jazwinski, 1970, Appendix 7B)

$$\left(C^\top R^{-1} C + P^{-1}\right)^{-1} P^{-1} = I - PC^\top\left(CPC^\top + R\right)^{-1} C, \tag{3.111a}$$

$$\left(C^\top R^{-1} C + P^{-1}\right)^{-1} C^\top R^{-1} = PC^\top\left(CPC^\top + R\right)^{-1}, \tag{3.111b}$$

where $P \succ 0$, $R \succ 0$ and C are matrices with suitable dimensions, in order to rearrange (3.110b) yields

$$\hat{x}^*_{k-1} = P_{k-1}^- C_{k-1}^\top\left(C_{k-1} P_{k-1}^- C_{k-1}^\top + R_{k-1}\right)^{-1} y_{k-1} \tag{3.112}$$

$$+ \left(I - P_{k-1}^- C_{k-1}^\top\left(C_{k-1} P_{k-1}^- C_{k-1}^\top + R_{k-1}\right)^{-1} C_{k-1}\right) \bar{x}_{k-1}$$

$$= \bar{x}_{k-1} + P_{k-1}^- C_{k-1}^\top(C_{k-1} P_{k-1}^- C_{k-1}^\top + R_{k-1})^{-1}(y_{k-1} - C_{k-1}\bar{x}_{k-1})$$

$$= \bar{x}_{k-1} + K_{k-1}(y_{k-1} - C_{k-1}\bar{x}_{k-1}),$$

where the Kalman gain K_{k-1} is defined in (3.44a). Given (3.108b), we finally get

$$\hat{x}_k = A_{k-1}\hat{x}^*_{k-1} \tag{3.113a}$$

$$\hat{x}^*_{k-1} = \bar{x}_{k-1} + K_{k-1}(y_{k-1} - C_{k-1}\bar{x}_{k-1}). \tag{3.113b}$$

Thus, similar to the Kalman filter, under Gaussian assumptions and with $N = 1$, the pMHE solution $\hat{x}_{k-1}^*$ uses the Kalman gain to update the a priori estimate, and the state estimate $\hat{x}_k$ predicts the pMHE solution based on the transition matrix A_{k-1}.

Consider now the case where the measurement disturbances have heavy tailed probability densities that tend to zero less rapidly than the Gaussian distribution. Suppose for instance that v_k has the Laplace distribution with probability density function $\rho_{V_k}(v_k) \propto e^{-\|v_k\|_1}$. In this case, we may choose in the pMHE scheme $r_i(v) = \|v\|_1$. Note that this density is related to the presence of outliers (Aravkin et al. (2017); Lauer et al. (2011)), which occur in many applications due to, for instance, sensor malfunction or failures of data transmitters. Robustness to outliers is therefore an important requirement for these applications which can be easily incorporated in our framework.

If v_k has a mixture of Gaussian and Laplace distributions, the Huber loss (Boyd and Vandenberghe (2004); Huber (1992)) which is defined as

$$\Phi_{\text{hub}}(v) = \begin{cases} v^2 & |v| \leq M \\ M\left(2|v| - M\right) & |v| > M \end{cases} \tag{3.114}$$

with parameter $M \in \mathbb{R}_{++}$ is a more suitable function that might improve the estimation quality since it behaves like the ℓ_2-norm near zero and like the ℓ_1-norm for larger disturbances. An illustration of the discussed choices of penalty functions can be found in Figure 3.2. Alternative convex penalty functions that reflect prior knowledge on the process and measurement disturbances can be found in (Aravkin et al. (2017)).

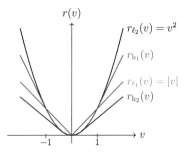

Figure 3.2: Illustration of the different convex penalty functions for the one-dimensional case. Here, $r_{\text{h}_1}(v) = \Phi_{\text{hub}}(v)$ with $M = 1$ and $r_{\text{h}_2}(v) = \Phi_{\text{hub}}(v)$ with $M = 0.4$.

Remark 3.6. It is worth mentioning that an outlier that takes place at time instant $k - N$ enters in the a priori estimate (3.42b) employed in the pMHE problem to be solved at time $k + 1$ via the output residual $y_{k-N} - C_{k-N}\hat{x}_{k-N}^*$. Obviously, this can lead to poor a priori estimates that might degenerate the overall pMHE performance. In order to avoid this, we can check for abnormal output residuals via a threshold test and skip the Kalman correction in the a priori estimate in these rare occurrences, i.e., (3.42b) becomes for such residuals $\bar{x}_{k-N+1} = A_{k-N}\hat{x}_{k-N}^* + B_{k-N}u_{k-N}$. Nevertheless, even without this outlier detection technique, pMHE should do better in the presence of outliers than the Kalman filter due to the more general cost.

In summary, the pMHE scheme based on problem (3.101) has the following properties. i) It yields GUES estimation errors if we employ the stabilizing a priori estimate (3.42b) and the Bregman distance (3.43) whose design is based on the Kalman filter. ii) In the absence of constraints and under the specified assumptions on W_k and V_k, it constitutes a Bayesian estimator with posterior conditional probability density (3.106). Here, the prior conditional probability density at time $k - N$ is approximated by D_{k-N} and the functions $\tilde{q}_i$ and r_i are used for defining the densities of the process and measurement disturbances (3.105). The estimate of the pMHE scheme is computed as the Maximum Likelihood estimator (3.107) and the stability of the estimator can be guaranteed by centering D_{k-N} around the a priori estimate $\bar{x}_{k-N}$. iii) Under Gaussian assumptions where quadratic functions $\tilde{q}_i$, r_i, and D_{k-N} are employed, setting $N = 1$ gives the pMHE problem (3.108), whose recursive solution given by (3.113) and (3.42b) is similar to that of the Kalman filter. In the case where the disturbances are non-Gaussian, the flexible formulation of the pMHE scheme allows to choose more adequate descriptions $\tilde{q}_i$, r_i and D_{k-N}. In addition, we may also improve performance by including constraints as prior information.

3.2.4 Proximity MHE for the special case of LTI systems

In the following, we present the pMHE problem for discrete-time LTI systems. In this case, we will see that a much simpler design of the associated Bregman distance and a priori estimate operator is sufficient for the global exponential stability (GES) of the underlying estimation error. More specifically, we consider a special case of system (3.18) of the form

$$x_{k+1} = A\,x_k + B\,u_k + w_k, \tag{3.115a}$$
$$y_k = C\,x_k + v_k, \tag{3.115b}$$

where it is known that the state and disturbances satisfy the following constraints

$$x_k \in \mathcal{X} \subseteq \mathbb{R}^n, \quad w_k \in \mathcal{W} \subseteq \mathbb{R}^n, \quad v_k \in \mathcal{V} \subseteq \mathbb{R}^p, \qquad k \in \mathbb{N}. \tag{3.116}$$

Here, the sets $\mathcal{X}$, $\mathcal{W}$, and $\mathcal{V}$ satisfy Assumption 3.1. In the pMHE problem (3.19), we set the stage costs $r_i(\cdot) := r(\cdot)$ and $q_i(\cdot) := q(\cdot)$ for all $i \in \mathbb{N}$, and assume they verify Assumption 3.3. Moreover, we fix the function $\psi_k(\cdot) := \psi(\cdot)$ for all time instants $k \in \mathbb{N}$. In addition, we let Assumption 3.4 hold and D_ψ refer to the associated Bregman distance constructed via (3.10). We present the resulting pMHE problem for the sake of completeness:

$$\min_{\hat{\mathbf{z}}_k} \quad \sum_{i=k-N}^{k-1} r(\hat{v}_i) + q(\hat{w}_i) + D_\psi\left(\hat{\mathbf{z}}_k, \bar{\mathbf{z}}_k\right) \tag{3.117a}$$

$$\text{s.t.} \quad \hat{x}_{i+1} = A\,\hat{x}_i + B\,u_i + \hat{w}_i, \qquad i = k - N, \ldots, k - 1 \tag{3.117b}$$

$$y_i = C\,\hat{x}_i + \hat{v}_i, \qquad i = k - N, \ldots, k - 1 \tag{3.117c}$$

$$\hat{x}_i \in \mathcal{X}, \qquad i = k - N, \ldots, k \tag{3.117d}$$

$$\hat{w}_i \in \mathcal{W}, \ \hat{v}_i \in \mathcal{V}, \qquad i = k - N, \ldots, k - 1. \tag{3.117e}$$

Based on the obtained pMHE solution $\hat{\mathbf{z}}_k^*$, the state estimate $\hat{x}_k$ is computed via a forward prediction of the dynamics (3.117b) as follows

$$\hat{x}_k = x\left(k; \hat{x}_{k-N}^*, k - N, \mathbf{u}_k, \hat{\mathbf{w}}_k^*\right) = A^N\,\hat{x}_{k-N}^* + \sum_{i=k-N}^{k-1} A^{k-1-i}\left(B\,u_i + \hat{w}_i^*\right). \tag{3.118}$$

As usual in pMHE, the stabilizing a priori estimate for the next time instant is calculated via the operator Φ_k as $\bar{\mathbf{z}}_{k+1} = \Phi_k(\hat{\mathbf{z}}_k^*)$.

Regrading the nominal stability properties of the estimation error generated by the pMHE problem (3.117), we state the following result which is a direct consequence of Theorem 3.2.

Corollary 3.2. *Consider system* (3.115) *with* $w_k = 0, v_k = 0, k \in \mathbb{N}$, *and the pMHE problem* (3.117). *Let Assumptions 3.1-3.5 hold. Suppose there exists a constant* $c \in \mathbb{R}_{++}$ *such that the Bregman distance* D_ψ *and the a priori estimate operator* Φ_k *satisfy*

$$D_\psi\left(\Phi_{k-1}(\mathbf{z}), \Phi_{k-1}(\hat{\mathbf{z}})\right) - D_\psi(\mathbf{z}, \hat{\mathbf{z}}) \leq -c\left\|\mathbf{z} - \hat{\mathbf{z}}\right\|^2 \qquad (3.119)$$

for all $k \in \mathbb{N}_+$ *and* $\mathbf{z}, \hat{\mathbf{z}} \in \mathbb{R}^{(N+1)n}$. *Then, the estimation error* (3.17) *is GES.*

In the following, we discuss specific design approaches for the a priori estimate operator and the associated Bregman distance which satisfy condition (3.119).

Φ_k and D_ψ based on the Luenberger observer

As stated previously, the basic idea of the pMHE framework is to use the Bregman distance D_ψ as a proximity measure to a stabilizing a priori estimate $\bar{\mathbf{z}}_k$ in order to inherit its stability properties. Since the Luenberger observer appears as a simple candidate in this time-invariant set up, we require that the operator Φ_k incorporates its dynamics as follows. Let $\mathbf{z} = \begin{bmatrix} x^\top & \mathbf{w}^\top \end{bmatrix}^\top$, $\hat{\mathbf{z}} = \begin{bmatrix} \hat{x}^\top & \hat{\mathbf{w}}^\top \end{bmatrix}^\top$, $x, \hat{x} \in \mathbb{R}^n$, $\mathbf{w}, \hat{\mathbf{w}} \in \mathbb{R}^{Nn}$. Given a time instant $k \in \mathbb{N}$ and a horizon length $N \in \mathbb{N}_+$, the a priori estimate operator based on the Luenberger observer is

$$\Phi_k(\mathbf{z}) = \begin{bmatrix} A\,x + B\,u_{k-N} + L(y_{k-N} - C x) \\ \mathbf{0} \end{bmatrix}, \qquad (3.120)$$

where $\mathbf{0} \in \mathbb{R}^{Nn}$ refers to the zero vector and the observer gain L is chosen such that all the eigenvalues of $A - LC$ are strictly within the unit circle. More specifically, given the pMHE solution $\hat{\mathbf{z}}_k^*$ of problem (3.117) at time instant k, the a priori estimate for the next time instant is computed as

$$\bar{\mathbf{z}}_{k+1} = \Phi_k(\hat{\mathbf{z}}_k^*) = \begin{bmatrix} \bar{x}_{k-N+1}^\top & 0 & \ldots & 0 \end{bmatrix}^\top. \qquad (3.121a)$$

As in the LTV case, we have here zero a priori process disturbances, i.e., $\bar{w}_i = 0$ for all $i = k - N + 1, \ldots, k$. Moreover,

$$\bar{x}_{k-N+1} = A\,\hat{x}_{k-N}^* + B\,u_{k-N} + L(y_{k-N} - C\hat{x}_{k-N}^*). \qquad (3.121b)$$

One simple way for the design of the Bregman distance is

$$D_\psi(\mathbf{z}, \hat{\mathbf{z}}) = \frac{1}{2}\left\|x - \hat{x}\right\|_P^2 + \frac{1}{2}\left\|\mathbf{w} - \hat{\mathbf{w}}\right\|_{\bar{W}}^2, \qquad (3.122)$$

where $P \in \mathbb{S}_{++}^n$ and the weight matrix $\bar{W} \in \mathbb{S}_{++}^{Nn}$ has the form $\bar{W} = \mathrm{diag}(W, \ldots, W)$ with $W \in \mathbb{S}_{++}^n$ arbitrary. The following proposition establishes a condition on the weight matrix P such that the proposed design of the a priori estimate operator and the Bregman distance satisfies the sufficient condition (3.119).

Proposition 3.2. *Consider system* (3.115) *with* $w_k = 0$, $v_k = 0$, $k \in \mathbb{N}$. *Then, the Bregman distance* (3.122) *and the a priori estimate operator* (3.120) *verify Assumptions 3.4 and 3.5. Moreover, if the weight matrix* $P \in \mathbb{S}_{++}^n$ *in the Bregman distance fulfills the linear matrix inequality (LMI)*

$$(A - LC)^\top P(A - LC) - P \preceq -Q \tag{3.123}$$

for some $Q \in \mathbb{S}_{++}^n$, *then there exists* $c \in \mathbb{R}_{++}$ *such that* Φ_k *and* D_ψ *satisfy condition* (3.119) *for all* $\mathbf{z}, \hat{\mathbf{z}} \in \mathbb{R}^{(N+1)n}$ *and* $k \in \mathbb{N}_+$.

Proof. The proof follows the same steps as in the proof of Proposition 3.1. Since $v_k = 0$, $k \in \mathbb{N}$, the true measurement satisfies $y_{k-N} = C\,x_{k-N}$. Evaluating the a priori estimate operator (3.120) at $\mathbf{z}_k$ therefore yields

$$\Phi_k(\mathbf{z}_k) = \begin{bmatrix} A\,x_{k-N} + B\,u_{k-N} + L\,(y_{k-N} - C\,x_{k-N}) \\ \mathbf{0} \end{bmatrix} = \begin{bmatrix} A\,x_{k-N} + B\,u_{k-N} \\ \mathbf{0} \end{bmatrix} = \mathbf{z}_{k+1}, \tag{3.124}$$

which implies that Assumption 3.5 is satisfied. Moreover, the Bregman distance (3.122) satisfies Assumption 3.4 with $\sigma = \min(\lambda_{\min}(P), \lambda_{\min}(\bar{W}))$ and $\gamma = \max(\lambda_{\max}(P), \lambda_{\max}(\bar{W}))$. Let $\mathbf{z} = \begin{bmatrix} x^\top & \mathbf{w}^\top \end{bmatrix}^\top$, $\hat{\mathbf{z}} = \begin{bmatrix} \hat{x}^\top & \hat{\mathbf{w}}^\top \end{bmatrix}^\top$, $x, \hat{x} \in \mathbb{R}^n$, $\mathbf{w}, \hat{\mathbf{w}} \in \mathbb{R}^{Nn}$. Based on (3.120) and (3.122), we compute

$$D_\psi\left(\Phi_{k-1}(\mathbf{z}), \Phi_{k-1}(\hat{\mathbf{z}})\right) = \frac{1}{2}\,\|(A - LC)\,(x - \hat{x})\|_P^2 \tag{3.125}$$

since we have zero a priori disturbances and the input and measurement terms cancel out. We obtain in view of the LMI (3.123) that

$$\begin{aligned} \Delta D &:= D_\psi\left(\Phi_{k-1}(\mathbf{z}), \Phi_{k-1}(\hat{\mathbf{z}})\right) - D_\psi(\mathbf{z}, \hat{\mathbf{z}}) \\ &= \frac{1}{2}\,\|(A - LC)\,(x - \hat{x})\|_P^2 - \frac{1}{2}\,\|x - \hat{x}\|_P^2 - \frac{1}{2}\|\mathbf{w} - \hat{\mathbf{w}}\|_{\bar{W}}^2 \\ &= \frac{1}{2}\,(x - \hat{x})^\top\left((A - LC)^\top P(A - LC) - P\right)(x - \hat{x}) - \frac{1}{2}\|\mathbf{w} - \hat{\mathbf{w}}\|_{\bar{W}}^2 \\ &\leq -\frac{1}{2}\,\|x - \hat{x}\|_Q^2 - \frac{1}{2}\|\mathbf{w} - \hat{\mathbf{w}}\|_{\bar{W}}^2, \end{aligned} \tag{3.126}$$

which establishes (3.119) with $c = \frac{1}{2}\min(\lambda_{\min}(Q), \lambda_{\min}(\bar{W}))$. $\qquad\square$

Observe that finding $P \in \mathbb{S}_{++}^n$ such that the LMI condition (3.123) is satisfied is equivalent to the detectability of the pair (A, C) in system (3.115).

Corollary 3.2 and Proposition 3.2 finally lead to the following (inherent) stability result for linear pMHE based on the Luenberger observer.

Corollary 3.3. *Consider system* (3.115) *with* $w_k = 0$, $v_k = 0$, $k \in \mathbb{N}$, *and the pMHE problem* (3.117) *with a priori estimate operator* (3.120) *and Bregman distance* (3.122) *based on the Luenberger observer. Let Assumptions 3.1-3.3 hold. Suppose that the weight matrix* $P \in \mathbb{S}_{++}^n$ *in the Bregman distance fulfills the LMI* (3.123). *Then, the estimation error* (3.17) *is GES.*

Notice that GES of the estimation error in the LTI case holds under very mild assumptions, where essentially only the detectability of the system matrices is needed instead of observability (cf. Alessandri et al. (2010); Sui and Johansen (2014)). Moreover, in contrast to the MHE formulations for LTI systems discussed in the previous chapter, this stability result holds despite the fact that the underlying optimization problem generalizes the classical least-squares formulation by considering a convex stage cost in the cost function. It is worth mentioning that this formulation of the pMHE problem is similar to the proposed MHE problem in (Sui and Johansen (2014)), where a quadratic prior weighting (2.19) is employed and a pre-stabilizing estimator constructed based on the Luenberger observer (2.20) is used in the a priori estimate. Even though our results are not limited to a quadratic cost function, we arrived to the same sufficient condition (2.21) on the weight matrix $P \in \mathbb{S}_{++}^n$. Furthermore, in (Sui and Johansen (2014)), since the correction term of the Luenberger observer enters also in the forward predictions of the MHE problem, an additional LMI (2.22) is imposed on the weight matrix $R \in \mathbb{S}_{++}^n$. In our setup, however, only the LMI (3.123) needs to be satisfied in order to guarantee GES of the estimation error, while R can be considered as a free tuning parameter in case we choose $r(v) = \|v\|_R^2$ in the stage cost.

While Corollary 3.3 is established by assuming that the system is disturbance-free, a robustness analysis similar to the analysis of the pMHE scheme based on the LTV Kalman filter in Section 3.2.2 can be easily adapted to the LTI case. More specifically, in order to show that the estimation error generated by the pMHE scheme (3.117) based on the Luenberger observer is ISS with respect to additive process and measurement disturbances, we would essentially replace the time-varying inverse of the error covariance matrix, i.e., Π_k^-, and the Kalman gain K_k by the weight matrix P and observer gain L satisfying the LMI condition (3.123) in the underlying calculations.

Remark 3.7. We can also resort to the steady-state Kalman filter in order to find suitable matrices L and P which fulfill the LMI (3.123). More specifically, $P \in \mathbb{S}_{++}^n$ can be chosen as the inverse of the steady-state value P_s of the error covariance matrix P_k^- in (3.44) which satisfies the algebraic Riccati equation

$$P_s = AP_sA^\top + Q - AP_sC^\top(CP_sC^\top + R)^{-1}CP_sA^\top, \qquad (3.127)$$

and the matrix L as the steady-state Kalman gain calculated as

$$L = AP_sC^\top(CP_sC^\top + R)^{-1}. \qquad (3.128)$$

This specific design of P and L can be shown to satisfy the LMI (3.123) by directly using (3.52), since its left-hand side applied to our case is exactly $(A - LC)^\top P (A - LC) - P$ and its right-hand side is negative definite. Note that a unique positive definite steady-state value P_s exists for LTI systems if (A, C) is detectable and $(A, \sqrt{Q})$ is controllable (De Souza et al. (1986)).

Remark 3.8. As discussed in Section 2.2, the choice of the Bregman distance can be guided by the geometry of the constraint set in the associated optimization problem such that it can act as a barrier function for the constraints. For instance, if we set $\mathcal{W} = \mathbb{R}^n$, $\mathcal{V} = \mathbb{R}^p$, and assume convex polytopic state constraints of the form

$$\mathcal{X} = \{x \in \mathbb{R}^n : C_x x \le d_x\}, \qquad (3.129)$$

where $C_x \in \mathbb{R}^{q_x \times n}$ and $d_x \in \mathbb{R}^{q_x}$, we can choose the Bregman distance in pMHE as

$$D_\psi (\mathbf{z}, \hat{\mathbf{z}}) = \frac{1}{2} \|x - \hat{x}\|_P^2 + \frac{1}{2} \|\mathbf{w} - \hat{\mathbf{w}}\|_W^2 + \varepsilon D_B (\mathbf{z}, \hat{\mathbf{z}}). \qquad (3.130)$$

Here, D_B refers to the so-called gradient recentered relaxed barrier function for the state constraints (3.117d) with (3.129) and the scalar $\varepsilon \in \mathbb{R}_{++}$ denotes a weighting parameter (Feller and Ebenbauer (2017a)). The Bregman distance D_B allows to incorporate the state constraints into the cost function and to formulate the pMHE problem (3.117) as an unconstrained optimization problem, which enables a simple and efficient numerical implementation. One further advantage of D_B is that it admits a quadratic upper bound which does not necessarily hold true for a general Bregman distance. More specifically,

$$D_B (\mathbf{z}, \hat{\mathbf{z}}) \leq \frac{1}{2} \|x - \hat{x}\|_M^2 + \frac{1}{2} \|\mathbf{w} - \hat{\mathbf{w}}\|_{M_1}^2 \qquad (3.131)$$

holds for all $\mathbf{z}, \hat{\mathbf{z}} \in \mathbb{R}^{(N+1)n}$, where $M \in \mathbb{S}_{++}^n$ and $M_1 \in \mathbb{S}_{++}^{Nn}$ can be computed using C_x in (3.129). If we consider the resulting pMHE problem (3.117) with a priori estimate operator based on the Luenberger observer (3.120) and Bregman distance (3.130), we can show that they verify condition (3.119) if the weight matrix $P \in \mathbb{S}_{++}^n$ fulfills the LMI

$$(A - LC)^\top (P + \varepsilon M)(A - LC) - P \preceq -Q \qquad (3.132)$$

for some $Q \in \mathbb{S}_{++}^n$. More detail about the relaxed barrier function based pMHE scheme can be found in (Gharbi and Ebenbauer (2019a)).

3.3 Proximity MHE for nonlinear systems

In this section, we consider two different classes of nonlinear systems and focus on the nominal stability properties of the associated pMHE schemes. Similar to the linear case, we use Lyapunov's direct method in order to prove in a deterministic setting exponential stability of the estimation errors under suitable assumptions. In Section 3.3.1, we investigate the stability properties of nonlinear pMHE designed such that the state estimate lies in proximity of a locally stable estimator. In Section 3.3.2, we focus on pMHE design approaches specific to the special case of nonlinear systems which can be transformed to systems that are affine in the output. The aforementioned sections are based on and taken in parts literally from (Gharbi et al. (2020a))[4] and (Gharbi and Ebenbauer (2020))[5].

3.3.1 Proximity MHE based on a locally stable estimator

In this section, we address the estimation problem of constrained discrete-time nonlinear systems of the form (3.1) which satisfy Assumption 3.1. For technical reasons which we discuss later, we only impose convex state constraints of the form $x_k \in \mathcal{X}$ on the initial state in the horizon window, and require that the set $\mathcal{X} \subseteq \mathbb{R}^n$ satisfies Assumption 3.2.

---

[4]M. Gharbi, F. Bayer and C. Ebenbauer. Proximity moving horizon estimation for discrete-time nonlinear systems. *IEEE Control Systems Letters*, 5(6):2090–2095, 2020a © 2020 IEEE.

[5]M. Gharbi and C. Ebenbauer. A proximity moving horizon estimator for a class of nonlinear systems. *International Journal of Adaptive Control and Signal Processing*, 34(6):721–742, 2020 © 2020 Wiley Online Library.

Concerning the constraints on the disturbances, we set $\mathcal{V}_k = \mathbb{R}^p$ and let $w_k \in \mathcal{W}_k$, where the set $\mathcal{W}_k \subseteq \mathbb{R}^{m_w}$ satisfies Assumption 3.2 for any $k \in \mathbb{N}$. In particular, the pMHE problem (3.8) becomes

$$\min_{\hat{\mathbf{z}}_k} \quad \sum_{i=k-N}^{k-1} r_i(\hat{v}_i) + q_i(\hat{w}_i) + D_{\psi_k}(\hat{\mathbf{z}}_k, \bar{\mathbf{z}}_k) \tag{3.133a}$$

$$\text{s.t.} \quad \hat{x}_{i+1} = f_i(\hat{x}_i, u_k, \hat{w}_i), \qquad i = k-N, \ldots, k-1 \tag{3.133b}$$

$$y_i = h_i(\hat{x}_i) + \hat{v}_i, \qquad i = k-N, \ldots, k-1 \tag{3.133c}$$

$$\hat{x}_i \in \mathcal{X}, \qquad i = k-N \tag{3.133d}$$

$$\hat{w}_i \in \mathcal{W}_i, \qquad i = k-N, \ldots, k-1. \tag{3.133e}$$

According to Theorem 3.1, a minimizer to problem (3.133) need not be unique. We therefore let the pMHE solution $\hat{\mathbf{z}}_k^*$ computed at time instant k refer to one element from the set of minimizers associated to (3.133). As usual, $\hat{\mathbf{z}}_k^*$ is decomposed into the estimated initial state $\hat{x}_{k-N}^*$ and process disturbances $\hat{\mathbf{w}}_k^*$. Based on $\hat{\mathbf{z}}_k^*$, the current state estimate $\hat{x}_k = x(k; \hat{x}_{k-N}^*, k-N, \mathbf{u}_k, \hat{\mathbf{w}}_k^*)$ can be obtained via a forward prediction of the system dynamics (3.133b). For the sake of completeness, we adapt (3.11) to this case and rewrite the pMHE problem (3.133) more compactly as

$$\min_{\hat{\mathbf{z}}_k \in \mathcal{S}_k} \quad J_k(\hat{\mathbf{z}}_k) = F_k(\hat{\mathbf{z}}_k) + D_{\psi_k}(\hat{\mathbf{z}}_k, \bar{\mathbf{z}}_k), \tag{3.134a}$$

where $F_k : \mathbb{R}^{N m_w + n} \to \mathbb{R}_+$ denotes the sum of stage cost

$$F_k(\hat{\mathbf{z}}_k) = \sum_{i=k-N}^{k-1} r_i\left(y_i - h_i\left(x\left(i; \hat{x}_{k-N}, k-N, \mathbf{u}, \hat{\mathbf{w}}\right)\right)\right) + q_i(\hat{w}_i) \tag{3.134b}$$

with $\mathbf{u} = \{u_{k-N}, \cdots, u_{i-1}\}$ and $\hat{\mathbf{w}} = \{\hat{w}_{k-N}, \cdots, \hat{w}_{i-1}\}$ for each $i = k-N, \cdots, k-1$ and the set $\mathcal{S}_k \subseteq \mathbb{R}^{N m_w + n}$ represents the *convex* feasible set of problem (3.133)

$$\mathcal{S}_k := \left\{ \hat{\mathbf{z}}_k = \begin{bmatrix} \hat{x}_{k-N}^\top & \hat{\mathbf{w}}_k^\top \end{bmatrix}^\top : \ \hat{x}_{k-N} \in \mathcal{X}, \ \hat{w}_i \in \mathcal{W}_i, \quad i = k-N, \ldots, k-1 \right\}. \tag{3.134c}$$

In the following, we investigate the nominal stability properties of the nonlinear pMHE scheme (3.133) applied to system (3.1) with $w_k = 0, v_k = 0, k \in \mathbb{N}$. More specifically, we establish sufficient conditions for the local uniform exponential stability (UES) of the resulting estimation error by making use of two preparatory lemmas.

Lemma 3.6. *Consider system (3.1) with $w_k = 0, v_k = 0, k \in \mathbb{N}$, and the pMHE problem (3.133) at a given time instant k. Let Assumptions 3.1-3.4 and 3.7 hold. Then, there exist constants $\bar{\epsilon}, \eta \in \mathbb{R}_{++}$ such that*

$$D_{\psi_k}\left(\mathbf{z}_k, \hat{\mathbf{z}}_k^*\right) \leq D_{\psi_k}\left(\mathbf{z}_k, \Phi_{k-1}\left(\hat{\mathbf{z}}_{k-1}^*\right)\right) + \eta\left\|\mathbf{z}_k - \hat{\mathbf{z}}_k^*\right\|^3 \tag{3.135}$$

holds for all $\mathbf{z}_k, \hat{\mathbf{z}}_k^*$ with $\|\mathbf{z}_k - \hat{\mathbf{z}}_k^*\| \leq \bar{\epsilon}$, where $\mathbf{z}_k$ denotes the true system state, $\hat{\mathbf{z}}_k^*$ the pMHE solution, and Φ_k the a priori estimate operator as introduced in (3.13).

Proof. Consider the nonlinear pMHE problem (3.134). As discussed in (Mine and Fukushima, 1981, Example 1.1.), this problem can be rewritten as

$$\min_{\hat{\mathbf{z}}_k} \quad J_k(\hat{\mathbf{z}}_k) + I_{\mathcal{S}_k}(\hat{\mathbf{z}}_k), \tag{3.136}$$

where $I_{\mathcal{S}_k}$ denotes the indicator function associated with the set $\mathcal{S}_k$ as defined in (2.24). Note that $I_{\mathcal{S}_k}$ is a convex function since $\mathcal{S}_k$ in (3.134c) is a convex set as implied by Assumption 3.2. Due to Assumptions 3.1 and 3.7, J_k is a continuously differentiable function and a solution $\hat{\mathbf{z}}_k^*$ of (3.136) must satisfy $-\nabla J_k(\hat{\mathbf{z}}_k^*) \in \partial I_{\mathcal{S}_k}(\hat{\mathbf{z}}_k^*)$, where $\partial I_{\mathcal{S}_k}(\hat{\mathbf{z}}_k^*)$ is the subdifferential of the function $I_{\mathcal{S}_k}$ at $\hat{\mathbf{z}}_k^*$ (see Definition 2.3). Hence, $-\nabla J_k(\hat{\mathbf{z}}_k^*)$ is a subgradient of the convex function $I_{\mathcal{S}_k}$ and it holds by the subgradient inequality (2.23) that

$$I_{\mathcal{S}_k}(\mathbf{z}) \geq I_{\mathcal{S}_k}(\hat{\mathbf{z}}_k^*) - \nabla J_k(\hat{\mathbf{z}}_k^*)^\top (\mathbf{z} - \hat{\mathbf{z}}_k^*) \qquad \forall \mathbf{z} \in \mathbb{R}^{Nm_{\mathrm{w}}+n}. \tag{3.137}$$

Since $\hat{\mathbf{z}}_k^* \in \mathcal{S}_k$, we have that $I_{\mathcal{S}_k}(\hat{\mathbf{z}}_k^*) = 0$ and therefore $\nabla J_k(\hat{\mathbf{z}}_k^*)^\top (\mathbf{z} - \hat{\mathbf{z}}_k^*) \geq 0$ for all $\mathbf{z} \in \mathcal{S}_k$. Hence, by (3.134a),

$$\left(\nabla F_k(\hat{\mathbf{z}}_k^*) + \nabla D_{\psi_k}(\hat{\mathbf{z}}_k^*, \bar{\mathbf{z}}_k)\right)^\top (\mathbf{z} - \hat{\mathbf{z}}_k^*) \geq 0 \qquad \forall \mathbf{z} \in \mathcal{S}_k. \tag{3.138}$$

Given that the true system state satisfies the inequality constraints, it holds that $\mathbf{z}_k \in \mathcal{S}_k$. Hence, we get by setting $\mathbf{z} = \mathbf{z}_k$ in (3.138) and using the definition of the Bregman distance in (3.10) that

$$\left(\nabla F_k(\hat{\mathbf{z}}_k^*) + \nabla \psi_k(\hat{\mathbf{z}}_k^*) - \nabla \psi_k(\bar{\mathbf{z}}_k)\right)^\top (\mathbf{z}_k - \hat{\mathbf{z}}_k^*) \geq 0. \tag{3.139}$$

Let us now introduce the matrix

$$\tilde{H} := \begin{bmatrix} 0_{[Nm_{\mathrm{w}} \times n]} & I_{[Nm_{\mathrm{w}}]} \end{bmatrix} \in \mathbb{R}^{Nm_{\mathrm{w}} \times (Nm_{\mathrm{w}}+n)} \tag{3.140}$$

and the following vectors defined based on the optimal output residuals and process disturbances

$$D_{\mathrm{r},k}^* := \begin{bmatrix} \nabla r_{k-N}^\top(\hat{v}_{k-N}^*) & \cdots & \nabla r_{k-1}^\top(\hat{v}_{k-1}^*) \end{bmatrix}^\top \in \mathbb{R}^{Np}, \tag{3.141}$$

$$D_{\mathrm{q},k}^* := \begin{bmatrix} \nabla q_{k-N}^\top(\hat{w}_{k-N}^*) & \cdots & \nabla q_{k-1}^\top(\hat{w}_{k-1}^*) \end{bmatrix}^\top \in \mathbb{R}^{Nm_{\mathrm{w}}}, \tag{3.142}$$

$$\hat{\mathbf{v}}_k^* := \begin{bmatrix} \hat{v}_{k-N}^{*\top} & \cdots & \hat{v}_{k-1}^{*\top} \end{bmatrix}^\top \in \mathbb{R}^{Np}. \tag{3.143}$$

Moreover, we introduce the function $H_k : \mathbb{R}^{Nm_{\mathrm{w}}+n} \rightarrow \mathbb{R}^{Np}$

$$H_k(\hat{\mathbf{z}}_k) := \begin{bmatrix} h_{k-N}(\hat{x}_{k-N}) \\ h_{k-N+1}(x(k-N+1; \hat{x}_{k-N}, k-N, u_{k-N}, \hat{w}_{k-N})) \\ \vdots \\ h_{k-1}(x(k-1; \hat{x}_{k-N}, k-N, \mathbf{u}, \hat{\mathbf{w}})) \end{bmatrix}. \tag{3.144}$$

By (3.134b), and since the true disturbances are $w_k = 0, v_k = 0, k \in \mathbb{N}$, it holds for the optimal output residual that

$$\hat{v}_i^* = y_i - h_i(x(i; \hat{x}_{k-N}^*, k-N, \mathbf{u}, \hat{\mathbf{w}})) \tag{3.145}$$
$$= h_i(x(i; x_{k-N}, k-N, \mathbf{u})) - h_i(x(i; \hat{x}_{k-N}^*, k-N, \mathbf{u}, \hat{\mathbf{w}}))$$

with $\mathbf{u} = \{u_{k-N}, \cdots, u_{i-1}\}$ and $\hat{\mathbf{w}} = \{\hat{w}_{k-N}, \cdots, \hat{w}_{i-1}\}$ for each $i = k-N, \cdots, k-1$. Hence, based on (3.143) and (3.144), we have that

$$\hat{\mathbf{v}}_k^* = H_k(\mathbf{z}_k) - H_k(\hat{\mathbf{z}}_k^*). \tag{3.146}$$

By Taylor's theorem (Nocedal and Wright (2006)), it holds for the j−th component of the vector H_k defined in (3.144) that

$$H_k^j(\mathbf{z}_k) = H_k^j(\hat{\mathbf{z}}_k^*) + \nabla H_k^j(\hat{\mathbf{z}}_k^*)^\top (\mathbf{z}_k - \hat{\mathbf{z}}_k^*) + d_k^j(\mathbf{z}_k, \hat{\mathbf{z}}_k^*), \tag{3.147a}$$

where

$$d_k^j(\mathbf{z}_k, \hat{\mathbf{z}}_k^*) = \frac{1}{2}(\mathbf{z}_k - \hat{\mathbf{z}}_k^*)^\top \nabla^2 H_k^j(\xi_j)(\mathbf{z}_k - \hat{\mathbf{z}}_k^*). \tag{3.147b}$$

Here, $\xi_j := \hat{\mathbf{z}}_k^* + t_j (\mathbf{z}_k - \hat{\mathbf{z}}_k^*)$ for some $t_j \in (0, 1)$. Thus,

$$H_k(\mathbf{z}_k) = H_k(\hat{\mathbf{z}}_k^*) + \nabla H_k(\hat{\mathbf{z}}_k^*)^\top (\mathbf{z}_k - \hat{\mathbf{z}}_k^*) + \mathbf{d}_k(\mathbf{z}_k, \hat{\mathbf{z}}_k^*), \tag{3.148}$$

where $\mathbf{d}_k$ is the column vector of the stacked d_k^j defined in (3.147b) with $j = 1, \cdots, Np$. In view of (3.146) and (3.148), we can therefore write

$$\hat{\mathbf{v}}_k^* = \nabla H_k(\hat{\mathbf{z}}_k^*)^\top (\mathbf{z}_k - \hat{\mathbf{z}}_k^*) + \mathbf{d}_k(\mathbf{z}_k, \hat{\mathbf{z}}_k^*). \tag{3.149}$$

Moreover, we have that $\hat{\mathbf{w}}_k^* = \tilde{H}\hat{\mathbf{z}}_k^* = \tilde{H}(\hat{\mathbf{z}}_k^* - \mathbf{z}_k)$. According to Assumption 3.3, the functions r_i and q_i are convex and achieve their minimum value at zero for all i. Furthermore, by Assumption 3.7, they are smooth. Hence, for any $i \in \mathbb{N}$,

$$0 \le r_i(\hat{v}_i^*) - r_i(0) \le \nabla r_i(\hat{v}_i^*)^\top (\hat{v}_i^* - 0), \tag{3.150a}$$
$$0 \le q_i(\hat{w}_i^*) - q_i(0) \le \nabla q_i(\hat{w}_i^*)^\top (\hat{w}_i^* - 0). \tag{3.150b}$$

Based on (3.150) and (3.149), we get

$$\begin{aligned}
0 &\le \sum_{i=k-N}^{k-1} \nabla r_i(\hat{v}_i^*)^\top \hat{v}_i^* + \nabla q_i(\hat{w}_i^*)^\top \hat{w}_i^* \\
&= D_{\mathrm{r},k}^{*\top} \hat{\mathbf{v}}_k^* + D_{\mathrm{q},k}^{*\top} \hat{\mathbf{w}}_k^* \\
&= D_{\mathrm{r},k}^{*\top} \left(\nabla H_k(\hat{\mathbf{z}}_k^*)^\top (\mathbf{z}_k - \hat{\mathbf{z}}_k^*) + \mathbf{d}_k(\mathbf{z}_k, \hat{\mathbf{z}}_k^*) \right) + D_{\mathrm{q},k}^{*\top} \tilde{H}(\hat{\mathbf{z}}_k^* - \mathbf{z}_k) \\
&= \left(-\nabla H_k(\hat{\mathbf{z}}_k^*) D_{\mathrm{r},k}^* + \tilde{H}^\top D_{\mathrm{q},k}^* \right)^\top (\hat{\mathbf{z}}_k^* - \mathbf{z}_k) + D_{\mathrm{r},k}^{*\top} \mathbf{d}_k(\mathbf{z}_k, \hat{\mathbf{z}}_k^*).
\end{aligned} \tag{3.151}$$

In view of (3.134b) and the introduced notations, the gradient of F_k at the solution $\hat{\mathbf{z}}_k^*$ can be expressed as

$$\nabla F_k(\hat{\mathbf{z}}_k^*) = -\nabla H_k(\hat{\mathbf{z}}_k^*) D_{\mathrm{r},k}^* + \tilde{H}^\top D_{\mathrm{q},k}^*, \tag{3.152}$$

which we plug in (3.151) to obtain

$$0 \le \nabla F_k(\hat{\mathbf{z}}_k^*)^\top (\hat{\mathbf{z}}_k^* - \mathbf{z}_k) + D_{\mathrm{r},k}^{*\top} \mathbf{d}_k(\mathbf{z}_k, \hat{\mathbf{z}}_k^*). \tag{3.153}$$

The last inequality and (3.139) therefore lead to

$$\begin{aligned}
\left(\nabla \psi_k(\hat{\mathbf{z}}_k^*) - \nabla \psi_k(\bar{\mathbf{z}}_k) \right)^\top (\mathbf{z}_k - \hat{\mathbf{z}}_k^*) &\ge \nabla F_k(\hat{\mathbf{z}}_k^*)^\top (\hat{\mathbf{z}}_k^* - \mathbf{z}_k) \\
&\ge -D_{\mathrm{r},k}^{*\top} \mathbf{d}_k(\mathbf{z}_k, \hat{\mathbf{z}}_k^*).
\end{aligned} \tag{3.154}$$

In view of the three-points identity of Bregman distances stated in Lemma 2.1 (with $a = \hat{\mathbf{z}}_k^*$, $b = \bar{\mathbf{z}}_k$, $c = \mathbf{z}_k$), it holds that

$$\left(\nabla \psi_k(\hat{\mathbf{z}}_k^*) - \nabla \psi_k(\bar{\mathbf{z}}_k)\right)^\top (\mathbf{z}_k - \hat{\mathbf{z}}_k^*) = D_{\psi_k}(\mathbf{z}_k, \bar{\mathbf{z}}_k) - D_{\psi_k}(\mathbf{z}_k, \hat{\mathbf{z}}_k^*) - D_{\psi_k}(\hat{\mathbf{z}}_k^*, \bar{\mathbf{z}}_k), \qquad (3.155)$$

which we substitute in (3.154) to get

$$D_{\psi_k}(\mathbf{z}_k, \bar{\mathbf{z}}_k) - D_{\psi_k}(\mathbf{z}_k, \hat{\mathbf{z}}_k^*) - D_{\psi_k}(\hat{\mathbf{z}}_k^*, \bar{\mathbf{z}}_k) \geq -D_{\mathrm{r},k}^{*\top} \mathbf{d}_k(\mathbf{z}_k, \hat{\mathbf{z}}_k^*). \qquad (3.156)$$

Using the fact that $D_{\psi_k}(\hat{\mathbf{z}}_k^*, \bar{\mathbf{z}}_k) \geq 0$ in (3.156) yields

$$D_{\psi_k}(\mathbf{z}_k, \bar{\mathbf{z}}_k) + D_{\mathrm{r},k}^{*\top} \mathbf{d}_k(\mathbf{z}_k, \hat{\mathbf{z}}_k^*) \geq D_{\psi_k}(\mathbf{z}_k, \hat{\mathbf{z}}_k^*). \qquad (3.157)$$

Let us now derive an upper bound on the second term in the left-hand side of (3.157). By the uniform Lipschitz continuity of ∇r_i in Assumption 3.7, we have

$$D_{\mathrm{r},k}^{*\top} \mathbf{d}_k(\mathbf{z}_k, \hat{\mathbf{z}}_k^*) \leq \|D_{\mathrm{r},k}^*\| \, \|\mathbf{d}_k(\mathbf{z}_k, \hat{\mathbf{z}}_k^*)\| \leq L_{\mathrm{r}} \|\hat{\mathbf{v}}_k^*\| \, \|\mathbf{d}_k(\mathbf{z}_k, \hat{\mathbf{z}}_k^*)\|. \qquad (3.158)$$

In view of Assumption 3.1, for any $k \in \mathbb{N}$, f_k, h_k are $\mathcal{C}^2$ functions uniformly over k. The Taylor approximation remainders of their compositions are therefore of second order. Hence, if we consider (3.148), there exist $\bar{\epsilon}, \kappa_{\mathrm{d}} \in \mathbb{R}_{++}$ such that

$$\|\mathbf{d}_k(\mathbf{z}_k, \hat{\mathbf{z}}_k^*)\| \leq \kappa_{\mathrm{d}} \|\mathbf{z}_k - \hat{\mathbf{z}}_k^*\|^2 \qquad (3.159)$$

holds for all $\mathbf{z}_k, \hat{\mathbf{z}}_k^*$ with $\|\mathbf{z}_k - \hat{\mathbf{z}}_k^*\| \leq \bar{\epsilon}$. Moreover, Assumption 3.1 states that the functions f_k, h_k are Lipschitz continuous uniformly over k, and hence their compositions in H_k defined in (3.144) is Lipschitz continuous. Let $c_{\mathrm{H}} \in \mathbb{R}_{++}$ refer to the uniform Lipschitz constant of the function H_k for all $k \in \mathbb{N}$. Thus, by (3.158) and (3.146), we obtain

$$\begin{aligned} D_{\mathrm{r},k}^{*\top} \mathbf{d}_k(\mathbf{z}_k, \hat{\mathbf{z}}_k^*) &\leq L_{\mathrm{r}} \|H_k(\mathbf{z}_k) - H_k(\hat{\mathbf{z}}_k^*)\| \, \|\mathbf{d}_k(\mathbf{z}_k, \hat{\mathbf{z}}_k^*)\| \\ &\leq L_{\mathrm{r}} \, c_{\mathrm{H}} \, \kappa_{\mathrm{d}} \|\mathbf{z}_k - \hat{\mathbf{z}}_k^*\|^3 \end{aligned} \qquad (3.160)$$

for all $\mathbf{z}_k, \hat{\mathbf{z}}_k^*$ with $\|\mathbf{z}_k - \hat{\mathbf{z}}_k^*\| \leq \bar{\epsilon}$. Applying the last inequality in (3.157) and setting $\bar{\mathbf{z}}_k = \Phi_{k-1}(\hat{\mathbf{z}}_{k-1}^*)$ as defined in (3.13) yields

$$D_{\psi_k}(\mathbf{z}_k, \hat{\mathbf{z}}_k^*) \leq D_{\psi_k}(\mathbf{z}_k, \Phi_{k-1}(\hat{\mathbf{z}}_{k-1}^*)) + L_{\mathrm{r}} \, c_{\mathrm{H}} \, \kappa_{\mathrm{d}} \|\mathbf{z}_k - \hat{\mathbf{z}}_k^*\|^3. \qquad (3.161)$$

This is exactly the desired result (3.135) with $\eta := L_{\mathrm{r}} \, c_{\mathrm{H}} \, \kappa_{\mathrm{d}}$. $\qquad \square$

Lemma 3.7. *Consider system* (3.1) *with* $w_k = 0$, $v_k = 0$, $k \in \mathbb{N}$, *and the pMHE problem* (3.133) *at time instant* k. *Let Assumptions 3.1-3.4 hold. Then, there exists a constant* $\tilde{c} \in \mathbb{R}_{++}$ *such that*

$$\|\mathbf{z}_k - \hat{\mathbf{z}}_k^*\| \leq \tilde{c} \, \|\mathbf{z}_k - \bar{\mathbf{z}}_k\|, \qquad (3.162)$$

where $\mathbf{z}_k$ *denotes the true system state,* $\hat{\mathbf{z}}_k^*$ *the pMHE solution, and* $\bar{\mathbf{z}}_k$ *the a priori estimate.*

Proof. Given the reformulation of the optimization problem (3.133) in (3.134) and the fact that the stage cost is nonnegative by Assumption 3.3, a minimizer $\hat{\mathbf{z}}_k^*$ satisfies

$$D_{\psi_k}(\hat{\mathbf{z}}_k^*, \bar{\mathbf{z}}_k) \leq J_k(\hat{\mathbf{z}}_k^*) \leq J_k(\mathbf{z}_k) = \sum_{i=k-N}^{k-1} r_i(0) + q_i(0) + D_{\psi_k}(\mathbf{z}_k, \bar{\mathbf{z}}_k). \qquad (3.163)$$

The second inequality holds by optimality of $\hat{\mathbf{z}}_k^*$ and due to the fact that, by Assumption 3.2, the true system state with zero disturbances satisfies the constraints, i.e., $\mathbf{z}_k \in \mathcal{S}_k$. In view of Assumption 3.3, $r_i(0) = 0$ and $q_i(0) = 0$ for any $i \in \mathbb{N}$. Hence, we obtain from (3.163) and the uniform bounds on the Bregman distance in Assumption 3.4 that

$$\frac{\sigma}{2}\|\hat{\mathbf{z}}_k^* - \bar{\mathbf{z}}_k\|^2 \leq D_{\psi_k}(\hat{\mathbf{z}}_k^*, \bar{\mathbf{z}}_k) \leq D_{\psi_k}(\mathbf{z}_k, \bar{\mathbf{z}}_k) \leq \frac{\gamma}{2}\|\mathbf{z}_k - \bar{\mathbf{z}}_k\|^2. \tag{3.164}$$

By the triangle inequality, we therefore get

$$\|\mathbf{z}_k - \hat{\mathbf{z}}_k^*\| \leq \|\mathbf{z}_k - \bar{\mathbf{z}}_k\| + \|\hat{\mathbf{z}}_k^* - \bar{\mathbf{z}}_k\| \tag{3.165}$$

$$\leq \|\mathbf{z}_k - \bar{\mathbf{z}}_k\| + \sqrt{\frac{\gamma}{\sigma}}\|\mathbf{z}_k - \bar{\mathbf{z}}_k\|,$$

which proves the lemma with $\tilde{c} := (1 + \sqrt{\frac{\gamma}{\sigma}})$. $\qquad\square$

Some comments on the lemmas stated above are in order. Lemma 3.6 can be interpreted as an extension of Lemma 3.1 to the nonlinear case. As in the linear case, we exploit central properties of Bregman distances such as Lemma 2.1 and tools from the theory of PMD algorithms introduced in Section 2.2.1. In contrast to (3.21), however, the Inequality (3.135) holds only locally and might be rather conservative due to the fact that the analysis is based on Taylor approximations of the functions f_k, h_k. Moreover, its proof is such that we can only impose the state constraints on the initial state $\hat{x}_{k-N}$ in the pMHE problem (3.133). More specifically, we can see that the tools we used therein (see the derivation of (3.137)) are suitable for a constrained minimization problem of a $\mathcal{C}^1$ function subject to convex constraints. If we were to set $\hat{x}_i \in \mathcal{X}$ in the pMHE problem, where $i = k - N, \ldots, k - 1$, the system nonlinearities would yield nonconvex constraints in terms of the decision variable. Nevertheless, as we will discuss later in Remark 3.10, we could project the obtained estimate on the set $\mathcal{X}$ and our theoretical guarantees would still hold. Furthermore, it is worth pointing out that in our simulations later, the formulation of the pMHE problem (3.133) turned out to be sufficient for guaranteeing that the state estimate satisfies the inequality constraints. Lemma 3.7 is rather technical and establishes an upper bound on the norm of the pMHE error $\mathbf{z}_k - \hat{\mathbf{z}}_k^*$ at a given time instant k in terms of the error norm between the true system state $\mathbf{z}_k$ and the a priori estimate $\bar{\mathbf{z}}_k$.

We state in the following theorem the stability properties of the estimation error generated from nonlinear pMHE based on problem (3.133).

Theorem 3.4. *Consider system* (3.1) *with* $w_k = 0, v_k = 0, k \in \mathbb{N}$, *and the pMHE problem* (3.133). *Let Assumptions 3.1-3.5 and 3.7 hold. Suppose there exist constants* $\epsilon, c \in \mathbb{R}_{++}$ *such that the Bregman distance* D_{ψ_k} *and the a priori estimate operator* Φ_k *satisfy*

$$D_{\psi_k}(\Phi_{k-1}(\mathbf{z}), \Phi_{k-1}(\hat{\mathbf{z}})) - D_{\psi_{k-1}}(\mathbf{z}, \hat{\mathbf{z}}) \leq -c\|\mathbf{z} - \hat{\mathbf{z}}\|^2 \tag{3.166}$$

for all $k \in \mathbb{N}_+$ *and* $\mathbf{z}, \hat{\mathbf{z}}$ *with* $\|\mathbf{z} - \hat{\mathbf{z}}\| \leq \epsilon$. *Then, the estimation error* (3.17) *is locally UES, i.e., there exist* $\epsilon', \tilde{\alpha} \in \mathbb{R}_{++}$ *and* $\beta \in (0, 1)$ *such that*

$$\|x_k - \hat{x}_k\| \leq \tilde{\alpha}\beta^k\|x_0 - \bar{x}_0\| \tag{3.167}$$

holds for any $k \in \mathbb{N}_+$ *and* $x_0, \bar{x}_0$ *with* $\|x_0 - \bar{x}_0\| \leq \epsilon'$.

Proof. As in the linear case, we first prove that the pMHE error (3.16) is locally UES by using $V_k(\mathbf{z}_k, \hat{\mathbf{z}}_k^*) = D_{\psi_k}(\mathbf{z}_k, \hat{\mathbf{z}}_k^*)$ as a continuous time-varying candidate Lyapunov function. In particular, the goal is to show that V_k satisfies locally the conditions (A.6a) and (A.6b) in Theorem A.2 in Appendix A. Due to the quadratic lower and upper bounds (3.12) in Assumption 3.4, condition (A.6a) is satisfied with $c_1 = \frac{\sigma}{2}$ and $c_2 = \frac{\gamma}{2}$.
In view of (3.135) in Lemma 3.6, we have

$$\begin{aligned} \Delta V_k &:= V_k(\mathbf{z}_k, \hat{\mathbf{z}}_k^*) - V_{k-1}(\mathbf{z}_{k-1}, \hat{\mathbf{z}}_{k-1}^*) \\ &\leq D_{\psi_k}(\mathbf{z}_k, \Phi_{k-1}(\hat{\mathbf{z}}_{k-1}^*)) - D_{\psi_{k-1}}(\mathbf{z}_{k-1}, \hat{\mathbf{z}}_{k-1}^*) + \eta \|\mathbf{z}_k - \hat{\mathbf{z}}_k^*\|^3 \end{aligned} \tag{3.168}$$

for all $\mathbf{z}_k, \hat{\mathbf{z}}_k^*$ with $\|\mathbf{z}_k - \hat{\mathbf{z}}_k^*\| \leq \bar{\epsilon}$. Since $\mathbf{z}_k = \Phi_{k-1}(\mathbf{z}_{k-1})$ by Assumption 3.5, it holds that

$$D_{\psi_k}(\mathbf{z}_k, \Phi_{k-1}(\hat{\mathbf{z}}_{k-1}^*)) = D_{\psi_k}(\Phi_{k-1}(\mathbf{z}_{k-1}), \Phi_{k-1}(\hat{\mathbf{z}}_{k-1}^*)). \tag{3.169}$$

In view of the sufficient condition (3.166) and the uniform bounds on the Bregman distance in Assumption 3.4, we can compute

$$\begin{aligned} 0 \leq \frac{\sigma}{2} \|\Phi_{k-1}(\mathbf{z}_{k-1}) - \Phi_{k-1}(\hat{\mathbf{z}}_{k-1}^*)\|^2 &\leq D_{\psi_k}(\Phi_{k-1}(\mathbf{z}_{k-1}), \Phi_{k-1}(\hat{\mathbf{z}}_{k-1}^*)) \\ &\leq D_{\psi_{k-1}}(\mathbf{z}_{k-1}, \hat{\mathbf{z}}_{k-1}^*) - c\|\mathbf{z}_{k-1} - \hat{\mathbf{z}}_{k-1}^*\|^2 \\ &\leq (\frac{\gamma}{2} - c)\|\mathbf{z}_{k-1} - \hat{\mathbf{z}}_{k-1}^*\|^2 \end{aligned} \tag{3.170}$$

for all $\mathbf{z}_{k-1}, \hat{\mathbf{z}}_{k-1}^*$ with $\|\mathbf{z}_{k-1} - \hat{\mathbf{z}}_{k-1}^*\| \leq \epsilon$. Moreover, by Lemma 3.7, we have

$$\begin{aligned} \|\mathbf{z}_k - \hat{\mathbf{z}}_k^*\| &\leq \tilde{c}\|\mathbf{z}_k - \bar{\mathbf{z}}_k\| \\ &= \tilde{c}\|\Phi_{k-1}(\mathbf{z}_{k-1}) - \Phi_{k-1}(\hat{\mathbf{z}}_{k-1}^*)\|. \end{aligned} \tag{3.171}$$

Applying (3.170) to (3.171) yields

$$\|\mathbf{z}_k - \hat{\mathbf{z}}_k^*\| \leq \tilde{c}\sqrt{(\gamma - 2c)/\sigma}\,\|\mathbf{z}_{k-1} - \hat{\mathbf{z}}_{k-1}^*\| \tag{3.172}$$

for all $\mathbf{z}_{k-1}, \hat{\mathbf{z}}_{k-1}^*$ with $\|\mathbf{z}_{k-1} - \hat{\mathbf{z}}_{k-1}^*\| \leq \tilde{\epsilon}$, where

$$\tilde{\epsilon} := \min\left(\epsilon, \frac{\bar{\epsilon}}{\tilde{c}\sqrt{(\gamma - 2c)/\sigma}}\right). \tag{3.173}$$

Note that his choice of $\tilde{\epsilon}$ ensures that $\|\mathbf{z}_k - \hat{\mathbf{z}}_k^*\| \leq \bar{\epsilon}$ in (3.172) and $\|\mathbf{z}_{k-1} - \hat{\mathbf{z}}_{k-1}^*\| \leq \epsilon$. Hence, we obtain in (3.168) by (3.166), (3.169) and (3.172) that

$$\begin{aligned} \Delta V_k &\leq D_{\psi_k}(\Phi_{k-1}(\mathbf{z}_{k-1}), \Phi_{k-1}(\hat{\mathbf{z}}_{k-1}^*)) - D_{\psi_{k-1}}(\mathbf{z}_{k-1}, \hat{\mathbf{z}}_{k-1}^*) + \eta\|\mathbf{z}_k - \hat{\mathbf{z}}_k^*\|^3 \\ &\leq -c\|\mathbf{z}_{k-1} - \hat{\mathbf{z}}_{k-1}^*\|^2 + \eta\|\mathbf{z}_k - \hat{\mathbf{z}}_k^*\|^3 \\ &\leq -c\|\mathbf{z}_{k-1} - \hat{\mathbf{z}}_{k-1}^*\|^2 + \eta\tilde{c}^3\sqrt{(\gamma - 2c)/\sigma}^3\|\mathbf{z}_{k-1} - \hat{\mathbf{z}}_{k-1}^*\|^3 \end{aligned} \tag{3.174}$$

for all $\mathbf{z}_{k-1}, \hat{\mathbf{z}}_{k-1}^*$ with $\|\mathbf{z}_{k-1} - \hat{\mathbf{z}}_{k-1}^*\| \leq \tilde{\epsilon}$. We get by defining $\tilde{\eta} := \eta\tilde{c}^3\sqrt{(\gamma - 2c)/\sigma}^3$ that

$$\Delta V_k \leq (-c + \tilde{\eta}\|\mathbf{z}_{k-1} - \hat{\mathbf{z}}_{k-1}^*\|)\|\mathbf{z}_{k-1} - \hat{\mathbf{z}}_{k-1}^*\|^2 \tag{3.175}$$

for all $\mathbf{z}_{k-1}, \hat{\mathbf{z}}_{k-1}^*$ with $\|\mathbf{z}_{k-1} - \hat{\mathbf{z}}_{k-1}^*\| \leq \tilde{\epsilon}$. By introducing $\epsilon' := \min\left(\tilde{\epsilon}, \frac{c}{2\bar{\eta}}\right)$, we obtain that ΔV_k is locally negative definite, i.e., it satisfies (A.6b) with $c_3 = \frac{c}{2}$ for all $\mathbf{z}_{k-1}, \hat{\mathbf{z}}_{k-1}^*$ with $\|\mathbf{z}_{k-1} - \hat{\mathbf{z}}_{k-1}^*\| \leq \epsilon'$. More specifically, in view of (3.175) and the uniform lower bound in Assumption 3.4, we have

$$0 \leq V_k(\mathbf{z}_k, \hat{\mathbf{z}}_k^*) \leq V_{k-1}(\mathbf{z}_{k-1}, \hat{\mathbf{z}}_{k-1}^*) - \frac{c}{2}\|\mathbf{z}_{k-1} - \hat{\mathbf{z}}_{k-1}^*\|^2 \tag{3.176}$$

$$\leq \tilde{\beta}\, V_{k-1}(\mathbf{z}_{k-1}, \hat{\mathbf{z}}_{k-1}^*)$$

for all $\mathbf{z}_{k-1}, \hat{\mathbf{z}}_{k-1}^*$ with $\|\mathbf{z}_{k-1} - \hat{\mathbf{z}}_{k-1}^*\| \leq \epsilon'$, where $\tilde{\beta} := 1 - c/\sigma$. Note that $\tilde{\beta} \in (0,1)$ since $V_k = D_{\psi_k}$ is strictly positive for all $\mathbf{z}_k \neq \hat{\mathbf{z}}_k^*$ and $c/\sigma > 0$. Hence, we get

$$\frac{\sigma}{2}\|\mathbf{z}_k - \hat{\mathbf{z}}_k^*\|^2 \leq V_k(\mathbf{z}_k, \hat{\mathbf{z}}_k^*) \leq \tilde{\beta}^k V_0(\mathbf{z}_0, \hat{\mathbf{z}}_0^*) \leq \tilde{\beta}^k \frac{\gamma}{2}\|\mathbf{z}_0 - \hat{\mathbf{z}}_0^*\|^2 \tag{3.177}$$

for all $\mathbf{z}_0, \hat{\mathbf{z}}_0^*$ with $\|\mathbf{z}_0 - \hat{\mathbf{z}}_0^*\| \leq \epsilon'$. Thus, it holds that

$$\|\mathbf{z}_k - \hat{\mathbf{z}}_k^*\| \leq \alpha \beta^k \|\mathbf{z}_0 - \hat{\mathbf{z}}_0^*\| \tag{3.178}$$

for all $\mathbf{z}_0, \hat{\mathbf{z}}_0^*$ with $\|\mathbf{z}_0 - \hat{\mathbf{z}}_0^*\| \leq \epsilon'$, where $\alpha := \sqrt{\gamma/\sigma}$ and $\beta := \sqrt{\tilde{\beta}} \in (0,1)$. Hence, the pMHE error (3.16) is locally UES.

Concerning the estimation error (3.17), we have by the triangle inequality

$$\left\|x_k - \hat{x}_k\right\| \leq \left\|x\left(k; x_{k-N}, k-N, \mathbf{u}_k\right) - x\left(k; \hat{x}_{k-N}^*, k-N, \mathbf{u}_k\right)\right\| \tag{3.179}$$

$$+ \left\|x\left(k; \hat{x}_{k-N}^*, k-N, \mathbf{u}_k\right) - x\left(k; \hat{x}_{k-N}^*, k-N, \mathbf{u}_k, \hat{\mathbf{w}}_k^*\right)\right\|.$$

Due to Assumption 3.1 and recalling (3.16), we can compute

$$\left\|x\left(k; x_{k-N}, k-N, \mathbf{u}_k\right) - x\left(k; \hat{x}_{k-N}^*, k-N, \mathbf{u}_k\right)\right\| \leq c_{\mathrm{f}}^N \left\|e_{k-N}\right\|, \tag{3.180a}$$

$$\left\|x\left(k; \hat{x}_{k-N}^*, k-N, \mathbf{u}_k\right) - x\left(k; \hat{x}_{k-N}^*, k-N, \mathbf{u}_k, \hat{\mathbf{w}}_k^*\right)\right\| \leq \sum_{i=k-N}^{k-1} c_{\mathrm{f}}^{k-i} \left\|\hat{w}_i^*\right\|. \tag{3.180b}$$

Substituting (3.180) in (3.179) yields

$$\|x_k - \hat{x}_k\| \leq \bar{c}\, \|\mathbf{z}_k - \hat{\mathbf{z}}_k^*\|, \tag{3.181}$$

where $\bar{c} := c_{\mathrm{f}}^N + \sum_{i=k-N}^{k-1} c_{\mathrm{f}}^{k-i}$. As discussed at the end of the proof of Theorem 3.2, at $k = 0$, it holds that $\hat{\mathbf{z}}_0^* = \Pi_{S_0}^{\psi_0}(\bar{\mathbf{z}}_0)$, based on which we can compute that $\|\mathbf{z}_0 - \hat{\mathbf{z}}_0^*\| \leq \sqrt{\gamma/\sigma}\|\mathbf{z}_0 - \bar{\mathbf{z}}_0\|$. By Algorithm 3.1, we have that $\bar{\mathbf{z}}_0 = \bar{x}_0$, where $\bar{x}_0$ denotes the initial guess. Moreover, $\mathbf{z}_0 = x_0$ is the true initial state. We get in (3.181) by using (3.178) and the above arguments

$$\|x_k - \hat{x}_k\| \leq \tilde{\alpha} \beta^k \|\mathbf{z}_0 - \bar{\mathbf{z}}_0\| = \tilde{\alpha} \beta^k \|x_0 - \bar{x}_0\|, \tag{3.182}$$

for all $x_0, \bar{x}_0$ with $\|x_0 - \bar{x}_0\| \leq \epsilon'$, where $\tilde{\alpha} := \bar{c}\,\alpha\sqrt{\gamma/\sigma}$, which proves local UES of the estimation error, i.e., (3.167). $\qquad\square$

Similar to Theorem 3.2 for the linear case, Theorem 3.4 and more precisely condition (3.166) state that any locally exponentially stable estimation strategy of nonlinear systems can be used in the design of the a priori estimate operator, if we know the associated

Lyapunov function, and use it as the Bregman distance in pMHE. In fact, the proof of Theorem 3.4 demonstrates that in turn choosing this Bregman distance D_{ψ_k} as the pMHE Lyapunov function allows to transfer the stability analysis of pMHE to that of the a priori estimate. Because of this, however, the size of the region in which the local stability property (3.167) is guaranteed to hold depends on the local stability properties of the employed estimator as well as on the size of the region where (3.135) in Lemma 3.6 holds, and can be therefore small. Moreover, its size which is characterized by ϵ' depends on many parameters (see the proof of the theorem) and is therefore rather hard to quantify. Nevertheless, we expect when considering specific applications that the size of the region of initial errors which yield convergent estimation errors should be larger than our estimate.

Remark 3.9. Although exponential stability of the estimation error holds independently of the choice of the horizon length N, it is rather nontrivial to establish from the analysis how N really affects the performance of the estimator. We therefore use simulation examples later to illustrate the pMHE performance with increasing values of N.

Remark 3.10. In order to make sure that the estimate of the state x_k satisfies the constraints, we can perform a projection of the obtained pMHE estimate $\hat{x}_k$ onto the convex set $\mathcal{X}$ as

$$\hat{x}_k^{\mathrm{proj}} = \arg\min_{x \in \mathcal{X}} \|x - \hat{x}_k\|. \qquad (3.183)$$

In fact, this does not jeopardize the stability properties of the resulting estimation error, which we briefly discuss in the following. Since the proximal operator $\mathrm{prox}_{I_{\mathcal{X}}}$ of the indicator function $I_{\mathcal{X}}$ is the Euclidean projection onto $\mathcal{X}$ (see (2.42)), i.e., $\hat{x}_k^{\mathrm{proj}} = \mathrm{prox}_{I_{\mathcal{X}}}(\hat{x}_k)$, and due to the fact that $\mathrm{prox}_{I_{\mathcal{X}}}$ is nonexpansive, it holds in view of (2.47) that

$$\|\mathrm{prox}_{I_{\mathcal{X}}}(x_k) - \mathrm{prox}_{I_{\mathcal{X}}}(\hat{x}_k)\| \leq \|x_k - \hat{x}_k\|. \qquad (3.184)$$

The true state x_k satisfies the constraints, hence $\mathrm{prox}_{I_{\mathcal{X}}}(x_k) = x_k$ and we obtain by the nonexpansiveness property that $\|x_k - \hat{x}_k^{\mathrm{proj}}\| \leq \|x_k - \hat{x}_k\|$. Substituting this inequality in (3.167) yields

$$\|x_k - \hat{x}_k^{\mathrm{proj}}\| \leq \tilde{\alpha}\beta^k \|x_0 - \bar{x}_0\| \qquad (3.185)$$

for any $k \in \mathbb{N}_+$ and $x_0, \bar{x}_0$ with $\|x_0 - \bar{x}_0\| \leq \epsilon'$. This proves local UES of the estimation error with respect to the projected estimate $\hat{x}_k^{\mathrm{proj}}$.

Φ_k and D_{ψ_k} based on the extended Kalman filter

In the following, we discuss a design approach for the a priori estimate operator and the associated Bregman distance, which will be demonstrated to satisfy condition (3.166). From the various exponential observer design methods in the literature (Sundarapandian (2002)), we choose to construct the a priori estimate based on the EKF as a direct extension of the linear pMHE scheme based on the Kalman filter. Moreover, we choose the Bregman distance D_{ψ_k} as a weighted quadratic function, where the weight matrix consists of the inverse of the EKF covariance matrix.

Let $\mathbf{z} = \begin{bmatrix} x^\top & \mathbf{w}^\top \end{bmatrix}^\top$, $\hat{\mathbf{z}} = \begin{bmatrix} \hat{x}^\top & \hat{\mathbf{w}}^\top \end{bmatrix}^\top$, $x, \hat{x} \in \mathbb{R}^n$, $\mathbf{w}, \hat{\mathbf{w}} \in \mathbb{R}^{Nm_w}$. Given a time instant $k \in \mathbb{N}$ and a horizon length $N \in \mathbb{N}_+$, the a priori estimate operator based on the EKF is

$$x^+ = x + K_{k-N}\left(y_{k-N} - h_{k-N}(x)\right) \tag{3.186a}$$

$$\Phi_k(\mathbf{z}) = \begin{bmatrix} f_{k-N}\left(x^+, u_{k-N}, 0\right) \\ 0 \end{bmatrix}, \tag{3.186b}$$

where $0 \in \mathbb{R}^{Nm_w}$ and the matrix $K_{k-N} \in \mathbb{R}^{n \times p}$ refers to the EKF gain computed as

$$K_{k-N} = P_{k-N}^- C_{k-N}^\top \left(C_{k-N} P_{k-N}^- C_{k-N}^\top + R\right)^{-1}, \tag{3.187a}$$

$$P_{k-N+1}^- = A_{k-N} P_{k-N}^+ A_{k-N}^\top + Q, \qquad \Pi_{k-N}^- := \left(P_{k-N}^-\right)^{-1}, \tag{3.187b}$$

$$P_{k-N}^+ = (I - K_{k-N} C_{k-N}) P_{k-N}^-, \qquad \Pi_{k-N}^+ := \left(P_{k-N}^+\right)^{-1}. \tag{3.187c}$$

Here, $P_0^- \in \mathbb{S}_{++}^n$, $Q \in \mathbb{S}_{++}^n$ and $R \in \mathbb{S}_{++}^p$ are given weight matrices and

$$A_{k-N} = \left.\frac{\partial f_{k-N}}{\partial x}\right|_{(x^+, u_{k-N}, 0)}, \qquad C_{k-N} = \left.\frac{\partial h_{k-N}}{\partial x}\right|_x. \tag{3.187d}$$

In particular, consider again the notation introduced in (3.14) and the pMHE solution $\hat{\mathbf{z}}_k^*$ of problem (3.133) at time k. In view of (3.13), the a priori estimate for the next time instant is calculated as

$$\bar{\mathbf{z}}_{k+1} = \Phi_k(\hat{\mathbf{z}}_k^*) = \begin{bmatrix} \bar{x}_{k-N+1}^\top & 0 & \dots & 0 \end{bmatrix}^\top, \tag{3.188a}$$

where, as in the linear case, we have zero a priori process disturbances, i.e., $\bar{w}_i = 0$ for $i = k - N + 1, \dots, k$, and

$$\bar{x}_{k-N}^+ = \hat{x}_{k-N}^* + K_{k-N}\left(y_{k-N} - h_{k-N}(\hat{x}_{k-N}^*)\right) \tag{3.188b}$$

$$\bar{x}_{k-N+1} = f_{k-N}\left(\bar{x}_{k-N}^+, u_{k-N}, 0\right). \tag{3.188c}$$

Similar to (3.43) in the LTV case, the associated Bregman distance employed in the pMHE problem at time instant k is

$$D_{\psi_k}(\mathbf{z}, \hat{\mathbf{z}}) = \frac{1}{2}\|x - \hat{x}\|_{\Pi_{k-N}^-}^2 + \frac{1}{2}\|\mathbf{w} - \hat{\mathbf{w}}\|_{\bar{W}}^2, \tag{3.189}$$

where $\bar{W} \in \mathbb{S}_{++}^{Nm_w}$ is a weight matrix of the form $\bar{W} = \operatorname{diag}(W, \dots, W)$ with $W \in \mathbb{S}_{++}^{m_w}$ arbitrary. Moreover, $\Pi_{k-N}^- \in \mathbb{S}_{++}^n$ denotes the inverse of the EKF covariance matrix as specified in (3.187b).

In order to be able to establish that the proposed design based on the EKF satisfies condition (3.166), we require the following assumption.

Assumption 3.8 (Nonlinear system properties). *For any $k \in \mathbb{N}$, A_k is nonsingular, the spectral norms of the Jacobian matrices A_k and C_k defined in (3.187d) are uniformly bounded by $\|A_k\| \leq \bar{a}$, $\|C_k\| \leq \bar{c}$ for some $\bar{a}, \bar{c} \in \mathbb{R}_{++}$, and the matrices P_k^-, P_k^+ are uniformly bounded as*

$$\underline{p}'I \preceq P_k^- \preceq \bar{p}'I, \quad \underline{p}I \preceq P_k^+ \preceq \bar{p}I, \tag{3.190}$$

where $\underline{p}', \bar{p}', \underline{p}, \bar{p} \in \mathbb{R}_{++}$.

Although Assumption 3.8 might be rather strong and difficult to verify a priori, it constitutes a common assumption for the local convergence analysis of the EKF (Reif and Unbehauen (1999); Reif et al. (1999)). As will be demonstrated later, it allows for a simple design of nonlinear constrained MHE with local stability guarantees for any horizon length N, and a superior performance compared to the widely-used EKF. Note that, in view of Lemma 3.2, the bounds (3.190) are satisfied if the pairs (A_k, C_k) and $(A_k, \sqrt{Q})$ of the linearized system fulfill the uniform observability and controllability conditions stated in Definitions 3.1 and 3.2, respectively (Reif and Unbehauen (1999)).

We present the following result which states that the design of the a priori estimate operator and the Bregman distance based on the EKF satisfies amongst others the sufficient condition (3.166) in Theorem 3.4 for the local UES of the estimation error generated by the pMHE problem (3.133).

Proposition 3.3. *Consider system (3.1) with $w_k = 0$, $v_k = 0$, $k \in \mathbb{N}$ and let Assumptions 3.1 and 3.8 hold. Then, the Bregman distance (3.189) and the a priori estimate operator (3.186) verify Assumptions 3.4 and 3.5. Moreover, there exist $c, \epsilon \in \mathbb{R}_{++}$ such that they satisfy condition (3.166) for all $k \in \mathbb{N}_+$ and $\mathbf{z}, \hat{\mathbf{z}}$ with $\|\mathbf{z} - \hat{\mathbf{z}}\| \leq \epsilon$.*

Proof. Evaluating the a priori estimate operator (3.186) at $\mathbf{z}_k = \begin{bmatrix} x_{k-N}^\top & 0 & \dots & 0 \end{bmatrix}^\top$, i.e., the true state x_{k-N} with zero process disturbances, yields

$$\Phi_k(\mathbf{z}_k) = \begin{bmatrix} f_{k-N}\Big(x_{k-N} + K_{k-N}\left(y_{k-N} - h_{k-N}(x_{k-N})\right), u_{k-N}, 0\Big) \\ \mathbf{0} \end{bmatrix} \tag{3.191}$$

$$= \begin{bmatrix} f_{k-N}\Big(x_{k-N}, u_{k-N}, 0\Big) \\ \mathbf{0} \end{bmatrix} = \begin{bmatrix} x_{k-N+1} \\ \mathbf{0} \end{bmatrix} = \mathbf{z}_{k+1},$$

and hence Assumption 3.5 is satisfied. By the uniform bounds (3.190) in Assumption 3.8, the Bregman distance (3.189) satisfies (3.48). Therefore, Assumption 3.4 is fulfilled with $\sigma = \min(\frac{1}{\bar{p}}, \lambda_{\min}(\bar{W}))$ and $\gamma = \max(\frac{1}{\underline{p}}, \lambda_{\max}(\bar{W}))$.

Concerning the last statement of the proposition, we proceed with similar steps as in the proofs of Proposition 3.1 as well as (Reif and Unbehauen, 1999, Lemma 4).

Let $\mathbf{z} = \begin{bmatrix} x^\top & \mathbf{w}^\top \end{bmatrix}^\top$, $\hat{\mathbf{z}} = \begin{bmatrix} \hat{x}^\top & \hat{\mathbf{w}}^\top \end{bmatrix}^\top$, $x, \hat{x} \in \mathbb{R}^n$, $\mathbf{w}, \hat{\mathbf{w}} \in \mathbb{R}^{Nm_w}$. Moreover, let

$$x^+ = x + K_{k-N-1}\left(y_{k-N-1} - h_{k-N-1}(x)\right) \tag{3.192a}$$

$$\hat{x}^+ = \hat{x} + K_{k-N-1}\left(y_{k-N-1} - h_{k-N-1}(\hat{x})\right). \tag{3.192b}$$

In view of the EKF-based a priori estimate (3.186) and Bregman distance (3.189), and given the above definitions, we compute

$$\Delta D_k := D_{\psi_k}(\Phi_{k-1}(\mathbf{z}), \Phi_{k-1}(\hat{\mathbf{z}})) - D_{\psi_{k-1}}(\mathbf{z}, \hat{\mathbf{z}}) \tag{3.193}$$

$$= \frac{1}{2}\|f_{k-N-1}(x^+, u_{k-N-1}, 0) - f_{k-N-1}(\hat{x}^+, u_{k-N-1}, 0)\|_{\Pi_{k-N}^-}^2 - \frac{1}{2}\|x - \hat{x}\|_{\Pi_{k-N-1}^-}^2$$

$$- \frac{1}{2}\|\mathbf{w} - \hat{\mathbf{w}}\|_{\bar{W}}^2$$

since we have zero a priori process disturbances. By Assumption 3.1, f_k and h_k are $\mathcal{C}^2$

functions uniformly over k. Hence, they can be expanded as follows

$$f_{k-N-1}(x^+, u_{k-N-1}, 0) = f_{k-N-1}(\hat{x}^+, u_{k-N-1}, 0) + A_{k-N-1}(x^+ - \hat{x}^+) \tag{3.194}$$
$$+ \varphi_{k-N-1}(x^+, \hat{x}^+),$$
$$h_{k-N-1}(x) = h_{k-N-1}(\hat{x}) + C_{k-N-1}(x - \hat{x}) + \chi_{k-N-1}(x, \hat{x}), \tag{3.195}$$

where A_k, C_k are obtained according to (3.187d) and φ_k and χ_k are higher order terms. In addition, since f_k and h_k are $\mathcal{C}^2$ functions uniformly over k, there exist uniform constants $\epsilon_\varphi, \epsilon_\chi, \kappa_\varphi, \kappa_\chi \in \mathbb{R}_{++}$ such that

$$\|\varphi_k(x^+, \hat{x}^+)\| \leq \kappa_\varphi \|x^+ - \hat{x}^+\|^2, \tag{3.196a}$$
$$\|\chi_k(x, \hat{x})\| \leq \kappa_\chi \|x - \hat{x}\|^2 \tag{3.196b}$$

hold for any $k \in \mathbb{N}$, $x^+, \hat{x}^+$ with $\|x^+ - \hat{x}^+\| \leq \epsilon_\varphi$, and $x, \hat{x}$ with $\|x - \hat{x}\| \leq \epsilon_\chi$. Given (3.195) and (3.192), we can compute

$$x^+ - x = K_{k-N-1} y_{k-N-1} - K_{k-N-1} h_{k-N-1}(x) \tag{3.197}$$
$$= \hat{x}^+ - \hat{x} + K_{k-N-1}(h_{k-N-1}(\hat{x}) - h_{k-N-1}(x))$$
$$= \hat{x}^+ - \hat{x} + K_{k-N-1}(C_{k-N-1}(\hat{x} - x) - \chi_{k-N-1}(x, \hat{x})).$$

Rearranging the above equality yields

$$x^+ - \hat{x}^+ = (I - K_{k-N-1} C_{k-N-1})(x - \hat{x}) - K_{k-N-1}\chi_{k-N-1}(x, \hat{x}). \tag{3.198}$$

By plugging (3.198) in (3.194), we therefore get

$$f_{k-N-1}(x^+, u_{k-N-1}, 0) - f_{k-N-1}(\hat{x}^+, u_{k-N-1}, 0) = \tilde{A}_{k-N-1}(x - \hat{x}) + d_{k-N-1}, \tag{3.199a}$$

with $\tilde{A}_{k-N-1} := A_{k-N-1}(I - K_{k-N-1} C_{k-N-1})$ and the higher-order term

$$d_{k-N-1} = \varphi_{k-N-1}(x^+, \hat{x}^+) - A_{k-N-1} K_{k-N-1}\chi_{k-N-1}(x, \hat{x}). \tag{3.199b}$$

Substituting (3.199) in (3.193) leads to

$$\Delta D_k = \frac{1}{2}\left\|\tilde{A}_{k-N-1}(x - \hat{x}) + d_{k-N-1}\right\|^2_{\Pi^-_{k-N}} - \frac{1}{2}\left\|x - \hat{x}\right\|^2_{\Pi^-_{k-N-1}} - \frac{1}{2}\left\|\mathbf{w} - \hat{\mathbf{w}}\right\|^2_{\bar{W}} \tag{3.200}$$
$$= \frac{1}{2}(x - \hat{x})^\top \left(\tilde{A}^\top_{k-N-1}\Pi^-_{k-N}\tilde{A}_{k-N-1} - \Pi^-_{k-N-1}\right)(x - \hat{x}) + T_{k-N-1} - \frac{1}{2}\|\mathbf{w} - \hat{\mathbf{w}}\|^2_{\bar{W}},$$

where

$$T_{k-N-1} := \frac{1}{2}\left\|d_{k-N-1}\right\|^2_{\Pi^-_{k-N}} + d^\top_{k-N-1}\Pi^-_{k-N}\tilde{A}_{k-N-1}(x - \hat{x}). \tag{3.201}$$

In the following, we compute an upper bound for each of the first two summands in (3.200). Assumption 3.8 implies that we can exploit (3.54) in order to establish that

$$\frac{1}{2}(x - \hat{x})^\top \left(\tilde{A}^\top_{k-N-1}\Pi^-_{k-N}\tilde{A}_{k-N-1} - \Pi^-_{k-N-1}\right)(x - \hat{x}) \leq -\frac{\tilde{u}}{2}\|x - \hat{x}\|^2, \tag{3.202}$$

where $\tilde{u} := \frac{1}{\bar{p}'^2(1/\bar{p} + \bar{a}^2/\lambda_{\min}(Q))}$. Applying the Cauchy-Schwarz inequality in (3.201) yields

$$T_{k-N-1} \leq \|\Pi^-_{k-N}\| \|\tilde{A}_{k-N-1}\| \|d_{k-N-1}\| \|x - \hat{x}\| + \frac{1}{2}\|\Pi^-_{k-N}\| \|d_{k-N-1}\|^2. \tag{3.203}$$

Let us first compute an upper bound for the higher-order term d_k. In view of Assumption 3.8 and (3.187a), there exists $\bar{k} \in \mathbb{R}_{++}$ such that the spectral norm of the EKF gain satisfies $\|K_k\| \leq \bar{k}$ for any $k \in \mathbb{N}$. Hence, by (3.199b) and (3.196), it holds that

$$\|d_k\| \leq \kappa_\varphi \|x^+ - \hat{x}^+\|^2 + \bar{a}\bar{k}\,\kappa_\chi \|x - \hat{x}\|^2 \tag{3.204}$$

for all $k \in \mathbb{N}$, $x^+, \hat{x}^+$ with $\|x^+ - \hat{x}^+\| \leq \epsilon_\varphi$ and $x, \hat{x}$ with $\|x - \hat{x}\| \leq \epsilon_\chi$. Furthermore, by Assumption 3.8, (3.198), (3.196) and the triangle inequality, we have

$$\|x^+ - \hat{x}^+\| \leq (1 + \bar{k}\bar{c} + \bar{k}\kappa_\chi\epsilon_\chi)\|x - \hat{x}\| \tag{3.205}$$

for all $x, \hat{x}$ with $\|x - \hat{x}\| \leq \epsilon'$, where

$$\epsilon' := \min\left(\epsilon_\chi, \frac{\epsilon_\varphi}{1 + \bar{k}\bar{c} + \bar{k}\kappa_\chi\epsilon_\chi}\right). \tag{3.206}$$

Note that his choice of ϵ' ensures that we satisfy $\|x^+ - \hat{x}^+\| \leq \epsilon_\varphi$ in (3.205) as well as $\|x - \hat{x}\| \leq \epsilon_\chi$. We therefore obtain in (3.204) that

$$\|d_k\| \leq \kappa \|x - \hat{x}\|^2 \tag{3.207}$$

for all $x, \hat{x}$ with $\|x - \hat{x}\| \leq \epsilon'$, where $\kappa := \kappa_\varphi(1 + \bar{k}\bar{c} + \bar{k}\kappa_\chi\epsilon_\chi)^2 + \bar{a}\bar{k}\,\kappa_\chi$. By substituting the last inequality in (3.203) and using the bounds in Assumption 3.8, we arrive at

$$\begin{aligned} T_{k-N-1} &\leq \frac{\kappa}{\bar{p}'}\bar{a}\left(1 + \bar{k}\bar{c}\right)\|x - \hat{x}\|^3 + \frac{\kappa^2}{2\bar{p}'}\|x - \hat{x}\|^4 \\ &\leq \frac{\kappa}{\bar{p}'}\left(\bar{a}\left(1 + \bar{k}\bar{c}\right) + \frac{\kappa}{2}\epsilon'\right)\|x - \hat{x}\|^3 \end{aligned} \tag{3.208}$$

for all $x, \hat{x}$ with $\|x - \hat{x}\| \leq \epsilon'$. In view of the derived upper bounds on the first two summands in (3.200), i.e., (3.202) and the last inequality, we get

$$\Delta D_k \leq \left(-\frac{\tilde{u}}{2} + \tilde{\kappa}\|x - \hat{x}\|\right)\|x - \hat{x}\|^2 - \frac{1}{2}\|\mathbf{w} - \hat{\mathbf{w}}\|^2_{\bar{W}} \tag{3.209}$$

for all $x, \hat{x}$ with $\|x - \hat{x}\| \leq \epsilon'$, where $\tilde{\kappa} := \frac{\kappa}{\bar{p}'}\left(\bar{a}\left(1 + \bar{k}\bar{c}\right) + \frac{\kappa}{2}\epsilon'\right)$. Introducing $\epsilon := \min(\epsilon', \frac{\tilde{u}}{4\tilde{\kappa}})$ and considering all $x, \hat{x}$ with $\|x - \hat{x}\| \leq \epsilon$ in (3.209) proves the last statement of the proposition with $c = \frac{1}{2}\min(\frac{\tilde{u}}{2}, \lambda_{\min}(\bar{W}))$. $\qquad\square$

Overall, we obtain the following (inherent) stability result for nonlinear pMHE based on the EKF, which is a direct consequence of Theorem 3.4 and Proposition 3.3.

Corollary 3.4. *Consider system* (3.1) *with* $w_k = 0$, $v_k = 0$, $k \in \mathbb{N}$, *and the pMHE scheme* (3.133) *with a priori estimate operator* (3.186) *and Bregman distance* (3.189) *based on the EKF. Let Assumptions 3.1-3.3 and 3.7-3.8 hold. Then, the estimation error* (3.17) *is locally UES.*

As we have seen in the previous chapter, establishing stability of MHE for nonlinear systems is not trivial. If we consider the MHE scheme for nonlinear i-IOSS (detectable) systems in (Rawlings et al. (2017)), the sufficient condition on the approximation of the arrival cost in Assumption 2.2 such that GAS can be ensured in Theorem 2.2 is hard to verify. As a simple design approach, Rao (2000) suggests to use as prior weighting the filtering update (2.15), where a quadratic approximation of the arrival cost is constructed based on the covariance matrix of the EKF (2.18). Even though this approach is rather standard in MHE applications, it is only guaranteed to fulfill this stability condition when linear systems, quadratic objective, and convex constraints are considered. In nonlinear pMHE, however, by using the same simple design based on the EKF covariance matrix in the Bregman distance, and due to the additional advantage of the stabilizing a priori estimate, local UES of the estimation error is guaranteed. Regarding the constructive design methods discussed in Section 2.1, recall that RGAS of MHE under bounded disturbances is shown by Müller (2017) with the advantage that the prior weighting can be designed offline. Nevertheless, the horizon length has to be chosen large enough and the required i-IOSS assumption is difficult to verify. Moreover, in (Alessandri et al. (2010)), the prior weighting (2.19) in nonlinear MHE with quadratic stage cost can be also designed offline. However, its design depends on global Lipschitz and observability constants of the system. Concerning pMHE, a disadvantage of the proximity-based formulation of the MHE problem is that, in order for it to be able to inherit the stability properties from the a priori estimate, the assumptions for ensuring stability of the estimator based on which this a priori estimate is constructed have to be inherited as well. Especially in the nonlinear case where the EKF is used, these assumptions can be rather strong. Nevertheless, they allow for a simple design of MHE for the considered class of nonlinear systems with stability guarantees and a superior performance compared to the widely-used EKF. More specifically, this guarantee, albeit local, is ensured for arbitrary values of the horizon length N and for any convex and not necessarily quadratic stage cost satisfying Assumptions 3.3 and 3.7, such as the Huber penalty function. Although stability is inherited from the EKF, a satisfactory performance of pMHE can be therefore achieved independently based on the stage cost designed specifically for the problem at hand.

Remark 3.11. Note that in (Baumgärtner et al. (2020); Polóni et al. (2013)), the EKF is also employed in the a priori estimate and prior weighting as in the considered nonlinear pMHE scheme. In contrast to our case, however, no stability guarantees are provided in these works.

Remark 3.12. While the previous analysis does not account for the presence of process and measurement disturbances, we conjecture that an extension of the robustness analysis carried out in Section 3.2.2 for the linear pMHE scheme (3.57) based on the LTV Kalman filter to the nonlinear pMHE scheme (3.133) based on the EKF should be possible without any particular conceptual difficulties. More specifically, as we managed to extend Lemma 3.1 to Lemma 3.6, we could investigate extending the nonexpansiveness property (3.62) of the pMHE operator to the nonlinear case and adapting the local Lyapunov analysis accordingly. In the underlying ISS analysis, we would follow similar steps as in the proof of (Huang et al., 2012, Theorem 1), in which the ISS property (3.80) is shown to hold locally for the EKF, i.e., for all $x_0, \bar{x}_0$ such that $\|x_0 - \bar{x}_0\| \leq \epsilon$, with $\epsilon \in \mathbb{R}_{++}$. However, due to the fact that the calculations for establishing Theorem 3.3 were rather cumbersome despite the linear setup, we forgo the analysis in the nonlinear case.

3.3.2 pMHE for systems with output nonlinearities

In this section, we present a pMHE scheme for a special class of constrained discrete-time nonlinear systems that can be transformed into systems that are affine in the unmeasured state. In particular, we consider a class of discrete-time nonlinear systems of the form

$$z_{k+1} = f(z_k, u_k), \tag{3.210a}$$
$$y_k = h(z_k), \tag{3.210b}$$

where $z_k \in \mathbb{R}^n$ denotes the state vector, $u_k \in \mathbb{R}^m$ the input vector and $y_k \in \mathbb{R}^p$ the output vector. We assume that (3.210) can be transformed by means of a change of coordinates $x_k = T(z_k)$ into the equivalent system

$$x_{k+1} = A(y_k)\, x_k + \phi(y_k) + B(y_k)\, u_k, \tag{3.211a}$$
$$y_k = Cx_k, \tag{3.211b}$$

where $\phi : \mathbb{R}^p \to \mathbb{R}^n$ denotes a nonlinearity that depends on the output y_k. More specifically, we say that system (3.210) is equivalent to system (3.211) if there exists a diffeomorphism T which transforms system (3.210) into system (3.211). Hence, if we construct an estimator for system (3.211) such that the estimation error $x_k - \hat{x}_k$ vanishes, by continuity of the inverse coordinate transformation T^{-1}, $\hat{z}_k = T^{-1}(\hat{x}_k)$ converges to $z_k = T^{-1}(x_k)$ for $k \to \infty$ (Lee and Nam (1991)). We assume that there are no disturbances affecting the original system and that the state x_k of the new system (3.211) satisfies the constraints $x_k \in \mathcal{X}$, where the set $\mathcal{X} \subseteq \mathbb{R}^n$ is convex as specified by Assumption 3.2.

In the context of nonlinear observer design, this class of nonlinear systems for which such transformations exist is extensively studied in the literature (Chung and Grizzle (1990); Krener and Isidori (1983); Lee and Nam (1991); Lin and Byrnes (1995); Moraal and Grizzle (1995)). For instance, the transformation into system (3.211) includes nonlinear state transformations in which system (3.210) is transformed into the so-called nonlinear observer form (Lee and Nam (1991); Lin and Byrnes (1995)). As a an example, consider the following discrete-time nonlinear system (Lin and Byrnes (1995))

$$z_{k+1} = \begin{bmatrix} z_{k+1}^{(1)} \\ z_{k+1}^{(2)} \\ z_{k+1}^{(3)} \\ z_{k+1}^{(4)} \end{bmatrix} = \begin{bmatrix} \left(z_k^{(3)}\right)^2 \left(z_k^{(4)} + \left(z_k^{(2)}\right)^2\right) \\ z_k^{(1)} \\ z_k^{(2)} \\ z_k^{(4)} - \left(z_k^{(1)}\right)^2 + \left(z_k^{(2)}\right)^2 \end{bmatrix} = f(z_k), \tag{3.212a}$$

$$y_k = \begin{bmatrix} y_k^{(1)} \\ y_k^{(2)} \end{bmatrix} = \begin{bmatrix} z_k^{(3)} \\ z_k^{(4)} + \left(z_k^{(2)}\right)^2 \end{bmatrix} = h(z_k). \tag{3.212b}$$

This system can be transformed via the state transformation

$$x_k = T(z_k) = \begin{bmatrix} z_k^{(1)} & z_k^{(2)} & z_k^{(3)} & z_k^{(4)} + \left(z_k^{(2)}\right)^2 \end{bmatrix}^\top \tag{3.213}$$

to a system in the nonlinear observer form given by (3.211), in which the system matrices A and B do not depend on the output y_k. More specifically, we have

$$
A = \begin{bmatrix} 0 & 0 & 0 & 0 \\ 1 & 0 & 0 & 0 \\ 0 & 1 & 0 & 0 \\ 0 & 0 & 0 & 0 \end{bmatrix}, \quad B = 0, \quad C = \begin{bmatrix} 0 & 0 & 1 & 0 \\ 0 & 0 & 0 & 1 \end{bmatrix}, \quad \phi(y_k) = \begin{bmatrix} \left(y_k^{(1)}\right)^2 y_k^{(2)} \\ 0 \\ 0 \\ y_k^{(2)} \end{bmatrix}. \quad (3.214)
$$

Moreover, our framework includes transformations to systems of the form

$$
x_{k+1} = A(y_k)\, x_k + B(y_k) u_k, \tag{3.215a}
$$
$$
y_k = C x_k, \tag{3.215b}
$$

where the nonlinearity ϕ is zero. The resulting system (3.215) can be therefore considered as a discrete-time LTV system in which $A(y_k), B(y_k)$ vary at each time instant k based on a given output trajectory.

Overall, the transformation of the nonlinear system allows to obtain system dynamics which are affine in the unmeasured state x_k and hence a pMHE formulation in which the underlying optimization problem is convex. As in the previous cases, the cost function in pMHE includes a convex stage cost satisfying Assumption 3.3 as well as a suitable Bregman distance for the a priori estimate. More specifically, we consider the following pMHE problem

$$
\min_{\hat{\mathbf{z}}_k} \quad \sum_{i=k-N}^{k-1} r(\hat{v}_i) + q(\hat{w}_i) + D_\psi\left(\hat{\mathbf{z}}_k, \bar{\mathbf{z}}_k\right) \tag{3.216a}
$$

$$
\text{s.t.} \quad \hat{x}_{i+1} = A(y_i)\,\hat{x}_i + \phi(y_i) + B(y_i)\, u_i + \hat{w}_i, \quad i = k-N, \ldots, k-1 \tag{3.216b}
$$
$$
y_i = C\,\hat{x}_i + \hat{v}_i, \quad i = k-N, \ldots, k-1 \tag{3.216c}
$$
$$
\hat{x}_i \in \mathcal{X}, \quad i = k-N, \ldots, k. \tag{3.216d}
$$

Let $\hat{\mathbf{z}}_k^*$ refer to the unique pMHE solution of (3.216) at time instant k which yields the state estimate

$$
\hat{x}_k = \prod_{i=k-N}^{k-1} A(y_i)\, \hat{x}_{k-N}^* + \sum_{i=k-N}^{k-1} \prod_{j=i+1}^{k-1} A(y_j)\left(\phi(y_i) + B(y_i)\, u_i + \hat{w}_i^*\right), \tag{3.217}
$$

where $\prod_{i=j}^{k} A(y_i) := A(y_k)\, A(y_{k-1}) \ldots A(y_j)$ denotes the left multiplication of the matrices $A(y_i)$ for $i = j, \ldots, k$ with $j \leq k$. Concerning the design of the Bregman distance D_ψ and the a priori estimate operator Φ_k as introduced in (3.13), recall that in previous cases, we started by first deriving a general sufficient condition which D_ψ and Φ_k have to satisfy in order to ensure nominal stability of the resulting estimation error. Subsequently, we discussed explicit design approaches like the Kalman filter, the Luenberger observer, or the EKF that are shown to verify this condition. In this section, however, we directly specify the design of the a priori estimate and the Bregman distance. More specifically, we center the Bregman distance around a stabilizing a priori estimate which is constructed based on the (time-varying) Luenberger observer for system (3.211) as follows.

Let $\mathbf{z} = \begin{bmatrix} x^\top & \mathbf{w}^\top \end{bmatrix}^\top$, $\hat{\mathbf{z}} = \begin{bmatrix} \hat{x}^\top & \hat{\mathbf{w}}^\top \end{bmatrix}^\top$, $x, \hat{x} \in \mathbb{R}^n$ and $\mathbf{w}, \hat{\mathbf{w}} \in \mathbb{R}^{Nn}$. Given the pMHE solution

$\hat{\mathbf{z}}_k^*$ of problem (3.216) at time instant k, the a priori estimate for the next time instant is constructed based on the Luenberger observer as

$$\bar{\mathbf{z}}_{k+1} = \begin{bmatrix} \bar{x}_{k-N+1}^\top & 0 & \dots & 0 \end{bmatrix}^\top, \tag{3.218a}$$

where we have zero a priori process disturbances and

$$\bar{x}_{k-N+1} = A\left(y_{k-N}\right)\hat{x}_{k-N}^* + \phi\left(y_{k-N}\right) + L\left(y_{k-N}\right)\left(y_{k-N} - C\,\hat{x}_{k-N}^*\right) + B\left(y_{k-N}\right)u_{k-N}. \tag{3.218b}$$

Note that we let the observer gain L depend on the output in order to reflect this specific feature of system (3.211). Similar to the Bregman distance (3.122) designed for linear pMHE based on the Luenberger observer, we let

$$D_\psi\left(\mathbf{z}, \hat{\mathbf{z}}\right) = \frac{1}{2}\left\|x - \hat{x}\right\|_P^2 + \frac{1}{2}\left\|\mathbf{w} - \hat{\mathbf{w}}\right\|_{\bar{W}}^2, \tag{3.219}$$

where $P \in \mathbb{S}_{++}^n$ and $\bar{W} = \mathrm{diag}(W, \dots, W) \in \mathbb{S}_{++}^{Nn}$ with $W \in \mathbb{S}_{++}^n$ arbitrary. It is worth mentioning that this quadratic (time-invariant) Bregman distance fulfills Assumption 3.4. Let us now investigate the stability properties of the estimation error generated by the pMHE scheme (3.216) with a priori estimate (3.218) and Bregman distance (3.219). We will show that, if the weight matrix P in the Bregman distance satisfies a suitable system of LMIs , then the underlying estimation error is guaranteed to be GUES. Similar to the previous sections, this will be proven by employing the Bregman distance D_ψ as a Lyapunov function and on the basis of the following lemma.

Lemma 3.8. *Consider system (3.211) and the pMHE problem (3.216) at a given time instant k. Let Assumptions 3.1-3.3 hold. Then,*

$$D_\psi\left(\mathbf{z}_k, \hat{\mathbf{z}}_k^*\right) \le D_\psi(\mathbf{z}_k, \bar{\mathbf{z}}_k), \tag{3.220}$$

where $\mathbf{z}_k$ denotes the true system state, $\hat{\mathbf{z}}_k^*$ the pMHE solution, and $\bar{\mathbf{z}}_k$ the a priori estimate.

Proof. The proof of this result is rather straightforward. Since the optimization problem of (3.216) is convex, it suffices to reformulate it more compactly as in (3.11) and then to apply the steps for showing (3.1) in Lemma 3.1 established for LTV systems. The only major difference to the LTV case is the additional nonlinearity $\phi(y_i)$ in (3.216b). Nevertheless, this can be treated as an extra input term. More specifically, we can eliminate the forward prediction equations (3.216b), (3.216c) in problem (3.216) by defining

$$\Gamma_i(y) := \prod_{j=k-N}^{i-1} A\left(y_j\right) \in \mathbb{R}^{n \times n}, \tag{3.221}$$

$$\alpha_i(y) := \sum_{j=k-N}^{i-1} \prod_{l=j+1}^{i-1} A\left(y_l\right)\left(\phi\left(y_j\right) + B\left(y_j\right)u_j\right) \in \mathbb{R}^n,$$

in order to write down each state in the horizon window as an affine function of the decision variable $\hat{\mathbf{z}}_k = \begin{bmatrix} \hat{x}_{k-N}^\top & \hat{w}_{k-N}^\top & \dots & \hat{w}_{k-1}^\top \end{bmatrix}^\top$ as follows

$$\hat{x}_i = \Gamma_i(y)\,\hat{x}_{k-N} + \alpha_i(y) + \sum_{j=k-N}^{i-1} \prod_{l=j+1}^{i-1} A\left(y_l\right)\hat{w}_j. \tag{3.222}$$

Hence, the resulting (compact) formulation of the estimation problem (3.216) given by

$$\min_{\hat{\mathbf{z}}_k \in \mathcal{S}_k} \quad J_k(\hat{\mathbf{z}}_k) = F_k(\hat{\mathbf{z}}_k) + D_\psi\left(\hat{\mathbf{z}}_k, \bar{\mathbf{z}}_k\right) \tag{3.223}$$

is a convex optimization problem in which the function F_k and the feasible set $\mathcal{S}_k$ are convex in $\hat{\mathbf{z}}_k$. From now on, we can use the steps of the proof of Lemma 3.1 to obtain the desired result. □

Theorem 3.5. *Consider system* (3.211) *and the pMHE scheme* (3.216) *with a priori estimate* (3.218) *and Bregman distance* (3.219). *Let Assumptions 3.1-3.3 hold. Suppose that the weight matrix* $P \in \mathbb{S}_{++}^n$ *in the Bregman distance fulfills the LMI*

$$\left(A(y_k) - L(y_k)C\right)^\top P \left(A(y_k) - L(y_k)C\right) - P \preceq -Q \tag{3.224}$$

for each output $y_k \in \mathbb{R}^p$ *and a given* $Q \in \mathbb{S}_{++}^n$. *Then, the estimation error* $x_k - \hat{x}_k$ *is GUES.*

Proof. As usual, we start by proving that the pMHE error (3.16) is GUES by employing the Bregman distance D_ψ as a candidate Lyapunov function, i.e., we set $V(\mathbf{z}_k, \hat{\mathbf{z}}_k^*) = D_\psi(\mathbf{z}_k, \hat{\mathbf{z}}_k^*)$. Due to the quadratic Bregman distance (3.219), condition (A.6a) in Theorem A.2 is satisfied with $c_1 = \frac{1}{2}\min(\lambda_{\min}(P), \lambda_{\min}(\bar{W}))$ and $c_2 = \frac{1}{2}\max(\lambda_{\max}(P), \lambda_{\max}(\bar{W}))$. In order to establish condition (A.6b), we employ (3.220) in Lemma 3.8 to get

$$\begin{aligned} \Delta V &:= V(\mathbf{z}_k, \hat{\mathbf{z}}_k^*) - V(\mathbf{z}_{k-1}, \hat{\mathbf{z}}_{k-1}^*) \\ &\leq D_\psi(\mathbf{z}_k, \bar{\mathbf{z}}_k) - D_\psi(\mathbf{z}_{k-1}, \hat{\mathbf{z}}_{k-1}^*) \\ &= \frac{1}{2}\|x_{k-N} - \bar{x}_{k-N}\|_P^2 - \frac{1}{2}\|x_{k-N-1} - \hat{x}_{k-N-1}^*\|_P^2 - \frac{1}{2}\|\hat{\mathbf{w}}_{k-1}^*\|_{\bar{W}}^2 \end{aligned} \tag{3.225}$$

since we have zero true and a priori process disturbances. By system (3.211) and (3.218b), and recalling that $e_{k-N-1} = x_{k-N-1} - \hat{x}_{k-N-1}^*$ as specified in (3.16), we get

$$x_{k-N} - \bar{x}_{k-N} = \left(A\left(y_{k-N-1}\right) - L\left(y_{k-N-1}\right)C\right)e_{k-N-1} \tag{3.226}$$

due to the fact that the term $B\left(y_{k-N-1}\right)u_{k-N-1}$ as well as the nonlinearity $\phi\left(y_{k-N-1}\right)$ cancel out. Since the weight matrix P in the Bregman distance (3.219) satisfies the system of LMIs (3.224), we obtain by substituting the last equation in (3.225) that

$$\begin{aligned} \Delta V &\leq \frac{1}{2}\left\|\tilde{A}_{k-N-1}\left(y_{k-N-1}\right)e_{k-N-1}\right\|_P^2 - \frac{1}{2}\left\|e_{k-N-1}\right\|_P^2 - \frac{1}{2}\left\|\hat{\mathbf{w}}_{k-1}^*\right\|_{\bar{W}}^2 \\ &= \frac{1}{2}e_{k-N-1}^\top\left(\tilde{A}_{k-N-1}^\top\left(y_{k-N-1}\right)P\,\tilde{A}_{k-N-1}\left(y_{k-N-1}\right) - P\right)e_{k-N-1} - \frac{1}{2}\|\hat{\mathbf{w}}_{k-1}^*\|_{\bar{W}}^2 \\ &\leq -\frac{1}{2}\left\|e_{k-N-1}\right\|_Q^2 - \frac{1}{2}\|\hat{\mathbf{w}}_{k-1}^*\|_{\bar{W}}^2, \end{aligned} \tag{3.227}$$

where $\tilde{A}_{k-N-1}\left(y_{k-N-1}\right) := A\left(y_{k-N-1}\right) - L\left(y_{k-N-1}\right)C$. This establishes condition (A.6b) with $c = \frac{1}{2}\min(\lambda_{\min}(Q), \lambda_{\min}(\bar{W}))$, which proves GUES of the pMHE error. By adapting the steps of the proof of Theorem 3.2 starting from (3.34) to our scenario, and using the fact that, by Assumption 3.1 on system (3.211), $\|A(y_k)\|$ is uniformly bounded by c_f, we can establish GUES of the estimation error. □

Observe that the LMI (3.224) implies that $(A(y_k), C)$ is uniformly detectable for any $y_k \in \mathbb{R}^p$. Moreover, the above result states that designing the matrix P such that the LMI (3.224) is satisfied for each output y_k is sufficient for guaranteeing stability. We mention in the following useful approaches for computing P in accordance with this condition. First, one could check if the observer gain $L(y_k)$ in (3.224) can be constructed such that $A(y_k) - L(y_k) C$ is constant for all time instants k and does not depend on the output y_k. Having this option is particularly convenient since it would reduce the problem to just one LMI that has to be fulfilled instead of a total of $T - N + 1$ LMIs, with T refering to the simulation time. Second, if $A(y_k)$ is polynomial in y_k and the output trajectory is available offline, one could solve the LMI (3.224) for the weight matrix P by using sum of squares techniques (Ebenbauer and Allgöwer (2006)). Third, we mention the case where the resulting system matrices in (3.211) do not depend on the output y_k. This corresponds for instance to transformations of system (3.210) into the nonlinear observer form discussed earlier. Note that in this case, a constant observer gain L can be chosen and the system of LMIs given by (3.224) becomes

$$(A - LC)^\top P(A - LC) - P \preceq -Q. \tag{3.228}$$

Moreover, in this case, the estimation error can be shown to be ISS with respect to additive disturbances via a robustness analysis analogous to that in Section 3.2.2.

Remark 3.13. Recall that in the pMHE framework, any stabilizing a priori estimate for which the Bregman distance constitutes a Lyapunov function can be used to ensure stability. For instance, one can construct the a priori estimate based on the LTV Kalman filter as done in Section 3.2, where the nonlinearity ϕ in system (3.211) is cast as an additional input. More specifically, instead of the time-invariant Bregman distance (3.219) in which the weight matrix $P \in \mathbb{S}^n_{++}$ is required to satisfy the LMI (3.224) for each output y_k, we can employ the time-varying Bregman distance

$$D_{\psi_k}(\mathbf{z}, \hat{\mathbf{z}}) = \frac{1}{2} \|x - \hat{x}\|^2_{\Pi^-_{k-N}} + \frac{1}{2} \|\mathbf{w} - \hat{\mathbf{w}}\|^2_{\hat{W}} \tag{3.229}$$

in the cost function (3.216a), where

$$K_k = P_k^- C^\top \left(CP_k^- C^\top + R\right)^{-1}, \tag{3.230a}$$

$$P_{k+1}^- = A(y_k) P_k^+ A(y_k)^\top + Q, \qquad \Pi_k^- := \left(P_k^-\right)^{-1}, \tag{3.230b}$$

$$P_k^+ = (I - K_k C) P_k^-, \qquad \Pi_k^+ := \left(P_k^+\right)^{-1}, \tag{3.230c}$$

with $Q \in \mathbb{S}^n_{++}$ and $R \in \mathbb{S}^p_{++}$. Furthermore, instead of (3.218b), we can compute $\bar{x}_{k-N+1}$ in the a priori estimate $\bar{\mathbf{z}}_k$ as

$$\bar{x}_{k-N+1} = A(y_{k-N})\left(\hat{x}^*_{k-N} + K_{k-N}\left(y_{k-N} - C\,\hat{x}^*_{k-N}\right)\right) + \phi\left(y_{k-N}\right) + B(y_{k-N})\,u_{k-N}. \tag{3.231}$$

In order to be able to establish GUES of the estimation error generated by solving the pMHE problem (3.216) designed based on the Kalman filter, we have to additionally impose and adapt Assumption 3.6, where uniform observability is required to hold for each output trajectory.

3.4 Numerical examples

In this section, we consider four numerical examples with which we illustrate some of the theoretical and practical properties of pMHE for linear and nonlinear systems. More specifically, the goal is to illustrate the obtained stability results as well as to demonstrate how to ensure a satisfactory performance of pMHE by choosing stage costs that are suitable for the problem under consideration. We first consider an example of an LTV system for which we employ the pMHE scheme (3.19) with Bregman distance and a priori estimate based on the Kalman filter. In the second example, we consider an LTI system for which we use the pMHE scheme (3.117) with Bregman distance and a priori estimate based on the Luenberger observer. Third, we consider the state estimation problem of a nonlinear system of the from (3.1) and employ the pMHE scheme (3.133) which is based on the EKF. Finally, we consider a nonlinear system which can be transformed into a system with output nonlinearities and design for the resulting system a convex pMHE scheme (3.216) which is based on the Luenberger observer. This section is based on and taken in parts literally from (Gharbi and Ebenbauer (2018))[6], (Gharbi and Ebenbauer (2019b))[7], (Gharbi et al. (2020a))[8] and (Gharbi and Ebenbauer (2020)).[9]

3.4.1 pMHE for LTV systems: Two-phase induction motor

We consider the example of a two-phase induction motor (Boutayeb and Aubry (1999)), for which the discretized nonlinear model with step size $h = 0.1$ ms is given by

$$x_{k+1} = x_k + h\, g(x_k), \qquad (3.232a)$$

where

$$x_k = \begin{bmatrix} x_k^{(1)} & x_k^{(2)} & x_k^{(3)} & x_k^{(4)} & x_k^{(5)} \end{bmatrix}^\top, \qquad (3.232b)$$

and

$$g(x_k) = \begin{bmatrix} -\gamma x_k^{(1)} + \frac{K}{T_r} x_k^{(3)} + K p\, x_k^{(5)}\, x_k^{(4)} + \frac{1}{\sigma L_s} u_k^{(1)} \\ -\gamma x_k^{(2)} + \frac{K}{T_r} x_k^{(4)} - K p\, x_k^{(5)}\, x_k^{(3)} + \frac{1}{\sigma L_s} u_k^{(2)} \\ \frac{M}{T_r} x_k^{(1)} - \frac{1}{T_r} x_k^{(3)} - p\, x_k^{(5)}\, x_k^{(4)} \\ \frac{M}{T_r} x_k^{(2)} - \frac{1}{T_r} x_k^{(4)} + p\, x_k^{(5)}\, x_k^{(3)} \\ \frac{p M}{J L_r} \left(x_k^{(3)}\, x_k^{(2)} - x_k^{(4)}\, x_k^{(1)} \right) - \frac{T_L}{J} \end{bmatrix}. \qquad (3.232c)$$

The states $x_k^{(1)}, x_k^{(2)}$ represent the stator currents, $x_k^{(3)}, x_k^{(4)}$ the rotor fluxes and $x_k^{(5)}$ the angular speed. The input signals which represent the stator voltages are given by

---

[6]M. Gharbi and C. Ebenbauer. A proximity approach to linear moving horizon estimation. In *Proc. 6th IFAC Conference on Nonlinear Model Predictive Control*, volume 51, pages 549–555, 2018 © 2018 Elsevier Ltd.

[7]M. Gharbi and C. Ebenbauer. Proximity moving horizon estimation for linear time-varying systems and a Bayesian filtering view. In *Proc. 58th Conference on Decision and Control (CDC)*, pages 3208–3213. IEEE, 2019 © 2019 IEEE.

[8]M. Gharbi, F. Bayer and C. Ebenbauer. Proximity moving horizon estimation for discrete-time nonlinear systems. *IEEE Control Systems Letters*, 5(6):2090–2095, 2020a © 2020 IEEE.

[9]M. Gharbi and C. Ebenbauer. A proximity moving horizon estimator for a class of nonlinear systems. In *International Journal of Adaptive Control and Signal Processing*, pages 721–742. IEEE, 2020 © 2020 Wiley Online Library.

$u_k^{(1)} = 350\cos(0.03k)$ and $u_k^{(2)} = 300\sin(0.03k)$. We assume to measure the stator currents and the angular velocity, i.e., we have

$$y_k := \begin{bmatrix} y_k^{(1)} & y_k^{(2)} & y_k^{(3)} \end{bmatrix}^\top = \begin{bmatrix} x_k^{(1)} & x_k^{(2)} & x_k^{(5)} \end{bmatrix}^\top. \tag{3.233}$$

The time constant T_r of the rotor and the parameters σ, K and γ are given by

$$T_r = \frac{L_r}{R_r}, \quad \sigma = 1 - \frac{M^2}{L_s L_r}, \quad K = \frac{M}{\sigma L_s L_r}, \quad \gamma = \frac{R_s}{\sigma L_s} + \frac{R_r M^2}{\sigma L_s L_r^2}. \tag{3.234}$$

The numerical values of the remaining parameters of the model are chosen as in (Boutayeb and Aubry (1999)), i.e., we set $R_s = 0.18\,\Omega$, $R_r = 0.15\,\Omega$, $M = 0.0068$ H, $L_s = 0.0699$ H, $L_r = 0.0699$ H, $J = 0.0586$ kgm^2, $T_L = 10$ Nm, and $p = 1$. The nonlinear system (3.232) can be transformed to the form (3.215), where $u_k = \begin{bmatrix} u_k^{(1)} & u_k^{(2)} & 1 \end{bmatrix}^\top$ and

$$A(y_k) = I_5 + h \begin{bmatrix} -\gamma & 0 & \frac{K}{T_r} & Kpy_k^{(3)} & 0 \\ 0 & -\gamma & -Kpy_k^{(3)} & \frac{K}{T_r} & 0 \\ \frac{M}{T_r} & 0 & -\frac{1}{T_r} & -py_k^{(3)} & 0 \\ 0 & \frac{M}{T_r} & py_k^{(3)} & -\frac{1}{T_r} & 0 \\ 0 & 0 & \frac{pM}{JL_r}y_k^{(2)} & \frac{pM}{JL_r}y_k^{(1)} & 0 \end{bmatrix}, \quad B = h \begin{bmatrix} \frac{1}{\sigma L_s} & 0 & 0 \\ 0 & \frac{1}{\sigma L_s} & 0 \\ 0 & 0 & 0 \\ 0 & 0 & 0 \\ 0 & 0 & -\frac{T_L}{J} \end{bmatrix}, \tag{3.235}$$

$$C = \begin{bmatrix} 1 & 0 & 0 & 0 & 0 \\ 0 & 1 & 0 & 0 & 0 \\ 0 & 0 & 0 & 0 & 1 \end{bmatrix}.$$

The resulting system can be considered as a discrete-time LTV system in which $A(y_k)$ varies at each time instant k based on a given output trajectory. We now assume that the initial condition x_0 of the resulting system is zero and that this system is subject to additive process and measurement disturbances as introduced in (3.18), where the process disturbances are independent zero mean Gaussian random variables with $w_k \sim \mathcal{N}(0, Q)$ and $Q = 0.05\,I_5$. In order to estimate the state x_k, we implement the pMHE scheme (3.19) with a priori estimate (3.42) and Bregman distance (3.43) based on the Kalman filter. Note that we do not consider inequality constraints on the state and disturbances. For the sake of comparison, we also employ a time-varying Kalman filter designed accordingly in order to illustrate the relationship of pMHE to this classical filter. We initialize the Kalman filter with $P_0^- = 100\,I_5$ and the quadratic Bregman distance (3.43) in pMHE with $\Pi_0^- = 0.01 I_5$. Moreover, we choose in the stage cost $q_i(w) = \frac{1}{2}\|w\|_{Q^{-1}}^2$ for all $i \in \mathbb{N}$. We set the horizon length $N = 1$ and the initial guess for the estimators as $\bar{x}_0 = [80\ 80\ 50\ 50\ 80]^\top$.

In the following, we demonstrate the benefits and potential of the pMHE approach by taking advantage of the freedom in its design as discussed in Section 3.2.3. In particular, we will illustrate that we can ensure a performance similar to that of the Kalman filter under Gaussian assumptions and otherwise obtain a better performance by customizing the function r_i in the cost function (3.19a) based on prior knowledge on the structure of the measurement disturbances. For this reason, we consider the following two scenarios.

Gaussian noise We assume that the measurement disturbances are independent zero mean Gaussian random variables with $v_k \sim \mathcal{N}(0, R)$ and $R = I_3$. We therefore choose the ℓ_2-norm as penalty function in the pMHE problem (3.19) and set $r_i(v) = \frac{1}{2}\|v\|_{R^{-1}}^2$ for all

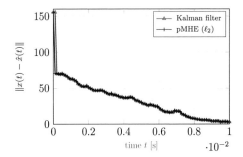

Figure 3.3: Estimation errors of the Kalman filter and the pMHE scheme with quadratic cost in the case of Gaussian process and measurement disturbances. © 2019 IEEE.

$i \in \mathbb{N}$. The resulting estimation errors are plotted in Figure 3.3.

We can see that under Gaussian assumptions, the unconstrained pMHE scheme (3.19) with quadratic stage cost, a priori estimate (3.42) and Bregman distance (3.43) yields estimates which coincide with those of the Kalman filter designed accordingly. Observe also that Theorem 3.3 is validated since pMHE exhibits ISS of the underlying estimation errors with respect to the disturbances.

Outliers In order to simulate outliers, we assume that a mixture of two Gaussian distributions take place in $y_k^{(2)}$. More specifically, $v_k^{(2)} \sim 0.95\,\mathcal{N}(0, \sigma^2) + 0.05\,\mathcal{N}(0, (100\sigma)^2)$ where $\sigma = 0.1$, such that the outliers are realized from a distribution with standard deviation 100σ and with probability 0.05 (Aravkin et al. (2017)). In this case, we set the function $r_i(v) = 10\|v\|_1$ in the pMHE stage cost. The estimation results are depicted in Figure 3.4.

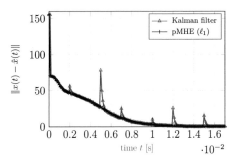

Figure 3.4: Estimation errors of the Kalman filter and the pMHE scheme with $r_i(v) = 10\|v\|_1$ in the case where outliers are present in the measurements of the stator current $x_k^{(2)}$. © 2019 IEEE.

Figure 3.4 shows that the Kalman filter is very vulnerable to the presence of outliers while the pMHE scheme designed with the ℓ_1-norm performs much better and is rather insensitive to the outliers. Note that there exists a number of methods aimed at improving the robustness of the Kalman filter against outliers and non-Gaussian noise (Gandhi and Mili (2009)). In pMHE, however, prior knowledge about non-Gaussian noise can be incorporated easily and more naturally into the structure of the estimator.

3.4.2 pMHE for LTI systems: Constant volume batch reactor

We consider the example of a well-mixed, constant volume, isothermal batch reactor where the associated nonlinear model is given in (Haseltine and Rawlings (2005)). In (Sui and Johansen (2014)), the system is linearized and discretized with a sampling time of $T_s = 0.25$ s, which yields a linear system of the form (3.115) with

$$A = \begin{bmatrix} 0.8831 & 0.0078 & 0.0022 \\ 0.1150 & 0.9563 & 0.0028 \\ 0.1178 & 0.0102 & 0.9954 \end{bmatrix}, \qquad B = \begin{bmatrix} 0 \\ 0 \\ 0 \end{bmatrix},$$
$$C = \begin{bmatrix} 32.84 & 32.84 & 32.84 \end{bmatrix}, \tag{3.236}$$

and (A, C) observable. Since the states represent concentrations, the physical state constraints are given by $x_k \geq 0$. Moreover, we assume zero process disturbances w_k in (3.115). We employ the pMHE scheme (3.117) with a priori estimate (3.121) based on the Luenberger observer and quadratic Bregman distance (3.122). Following (Sui and Johansen (2014)), we choose $L = [0.003\ 0.009\ 0.00043]^\top$ such that all the eigenvalues of $A - LC$ are strictly within the unit circle. Moreover, we design the weight matrix $P \in \mathbb{S}_{++}^3$ such that the LMI (3.123) is satisfied. In the cost function (3.117a), we set $q = 0$ such that we ignore the process disturbances and choose the horizon length $N = 11$. The true initial state and the initial guess are given by $x_0 = [0.5\ 0.05\ 0]^\top$ and $\bar{x}_0 = [0\ 0\ 4]^\top$, respectively.

Similar to the previous example, we illustrate how to ensure an improved performance of the pMHE scheme based on problem (3.117) by choosing a suitable stage cost depending on the nature of the measurement disturbances v_k. More specifically, we consider in the cost function (3.117a) different convex functions r given by $r_{\ell_2}(v) := R\,\|v\|^2$, $r_{\ell_1}(v) := R\,\|v\|_1$ and $r_{\mathrm{h}}(v) := R\,\Phi_{\mathrm{hub}}(v)$, where $v \in \mathbb{R}$ and $R \in \mathbb{R}_{++}$. The latter function refers to the Huber penalty function defined in (3.114). Here, the Huber penalty is designed with two different values of the parameter M, where $M \in \{0.5, 0.001\}$.

Outliers In this scenario, the measurement disturbances v_k are described by outliers taking very large values for precisely three time instants. The resulting measurements are depicted in Figure 3.5. In pMHE, we choose the tuning parameter $R = 10$ for all the considered functions and plot the estimation errors in Figure 3.5.

We can see that r_{ℓ_2} yields in this case the worst results with a very high sensitivity to the large values of the outliers. However, using r_{ℓ_1} instead yields estimation errors which are much less affected by the outliers. As for the Huber function, the performance of the associated estimator depends on the choice of the parameter M, where a smaller value of M yields an improved performance. It is worth mentioning that nominal GES of the estimation error, which is stated in Corollary 3.3, can be also observed in Figure 3.5. If we consider for instance the poor performance of pMHE due to using the function r_{ℓ_2}, we can see that it only occurs in the time interval where the outliers take place. In fact, if v_k were to be zero everywhere, we would observe exponential stability instead. Note that the constrained linear MHE scheme with filtering update (Rao et al. (2001)), i.e.,

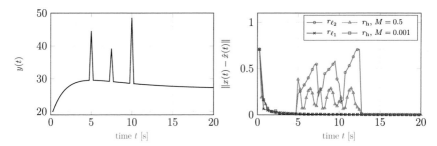

Figure 3.5: *Left:* True measurements corrupted by outliers $v(t)$, which are given by $v(5) = 15$, $v(7.5) = 10$ and $v(10) = 20$. *Right:* Estimation errors generated by the pMHE problem (3.117) with r_{ℓ_2}, r_{ℓ_1} and r_{h}. © 2018 Elsevier Ltd.

MHE with stage cost (2.12) and prior weighting (2.15), exhibits in this case a very similar performance to the pMHE estimator with r_{ℓ_2}, which is clearly due to the quadratic cost in both estimators.

Outliers and Gaussian noise In this scenario, the measurements are subject to the disturbances $v_k = z_k + e_k$, where z_k are independent zero mean Gaussian random variables with standard deviation $\sigma_z = 0.5$ and e_k consists of persistent severe outliers affecting y_k for a time interval of length 1.25 s. The measurements are depicted in Figure 3.6. After tuning, we choose in the pMHE scheme (3.117) the parameters $R = 0.001$ in all the employed functions r and $M = 0.5$ in the Huber function defined in (3.114). The resulting estimation errors are also depicted in Figure 3.6.

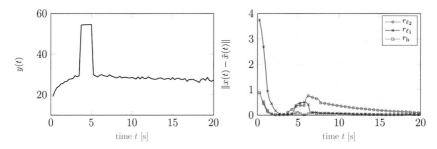

Figure 3.6: *Left:* True measurements corrupted by severe outliers and zero mean Gaussian noise with the standard deviation of 0.5. The outliers maintain the value of 25 between 3.75 s and 5 s. *Right:* Estimation errors corresponding to the pMHE strategy (3.117) with r_{ℓ_2}, r_{ℓ_1} and r_{h}. © 2018 Elsevier Ltd.

We can see that pMHE with r_{ℓ_1}, which remains a better choice than r_{ℓ_2}, fails to carry out the robust performance observed in the first scenario where consecutive jumps occur. Nevertheless, an improved robustness can still be attained in this scenario if we employ the Huber penalty function instead.

In summary, we demonstrated that the flexibility in the design of the pMHE scheme (3.117) allows us to obtain improved performance and robustness with respect to v_k if we choose the stage cost according to the given noise characteristics without jeopardizing the nominal stability of the estimation error.

3.4.3 pMHE for nonlinear systems: Constant volume batch reactor

In order to demonstrate the viability of the nonlinear pMHE approach (3.133), we consider the example of a constant volume batch reactor in which the reaction $2A \rightleftarrows B$ takes place (Rawlings et al. (2017); Tenny and Rawlings (2002)). The state consists of the partial pressures $x = \begin{bmatrix} p_A & p_B \end{bmatrix}^\top$ and the system is modeled by

$$\dot{x}_1 = -2k_1 x_1^2 + 2k_2 x_2, \tag{3.237a}$$

$$\dot{x}_2 = k_1 x_1^2 - k_2 x_2, \tag{3.237b}$$

$$y = x_1 + x_2, \tag{3.237c}$$

where $k_1 = 0.16\,\mathrm{min}^{-1}\,\mathrm{atm}^{-1}$ and $k_2 = 0.0064\,\mathrm{min}^{-1}$. Note that only the total pressure can be measured. The system is discretized using the Euler method with sample time $h = 0.1\,\mathrm{min}$, which yields a system of the form (3.1) without inputs u_k. Since the states represent pressures, the physical state constraints are given by $x_k \geq 0$. The initial condition of the system is $x_0 = \begin{bmatrix} 3 & 1 \end{bmatrix}^\top$.

We first consider the aforementioned system with zero process and measurement disturbances. In order to estimate the state x_k, we employ the pMHE scheme (3.133) with Bregman distance and a priori estimate based on the EKF. In the underlying cost function, we choose the functions $r_i(v) = \frac{1}{2}\|v\|_{R^{-1}}^2$ and $q_i(w) = \frac{1}{2}\|w\|_{Q^{-1}}^2$ for all $i \in \mathbb{N}$ as well as the Bregman distance defined in (3.189), where $R = 0.01$, $Q = 10^{-4}I_2$ and $P_0^- = I_2$. Given the fact that the function q_i is strongly convex, we can set the weight matrix for the a priori process disturbances as $\bar{W} = 0$ in (3.189). We compare the results with the EKF designed accordingly as well as with the MHE scheme with EKF filtering update (Rao et al. (2003)). In Chapter 2, this is the nonlinear MHE scheme (2.8) with quadratic prior weighting (2.15), where the weight P_k^{-1} is recursively updated via (2.18) and the a priori estimate is chosen as the state estimate computed N steps in the past. All estimators are initialized with a poor initial guess $\bar{x}_0 = [0.1\ 4.5]^\top$. We compare the estimated and true state trajectories in Figure 3.7 and plot the corresponding estimation errors as well.

In contrast to both MHE schemes, the EKF generates negative pressure estimates for this specific choice of the initial guess. Moreover, for $N = 3$, pMHE yields smaller estimation errors than MHE with filtering update. For $N = 10$, the corresponding state estimates become quite indistinguishable from each other. However, if we focus on the estimation errors, we can observe the periodic behavior of the MHE errors, where the period length consists with the horizon length. As discussed in (Tenny and Rawlings (2002)), this phenomenon can be mitigated by using the superior but more involved smoothing update, which uses the computed MHE solution from the previous time instant as discussed in Section 2.1. In the following, we compare the performance of pMHE with the MHE schemes with filtering update (MHE-F) and smoothing update (MHE-S) as well as with different values of the horizon length N. We choose as a performance index the root mean square

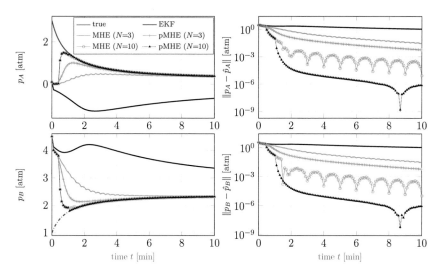

Figure 3.7: Resulting state estimates and estimation errors generated by the EKF, MHE with filtering update, and pMHE in the nominal case. © 2020 IEEE.

error (RMSE) defined as

$$RMSE = \sqrt{\sum_{k=N}^{T_{\text{sim}}} \frac{\|x_k - \hat{x}_k\|^2}{T_{\text{sim}} - N + 1}},$$ (3.238)

where $T_{\text{sim}} = 100$ denotes the simulation time. We get Table 3.1.

Table 3.1: The resulting RMSE of pMHE and MHE with filtering (MHE-F) and smoothing (MHE-S) update for different values of the horizon length N.

RMSE	$N = 3$	$N = 5$	$N = 10$	$N = 20$
MHE-F	0.9090	0.2156	0.0275	0.0063
MHE-S	0.7252	0.1533	0.0206	0.0051
pMHE	0.6493	0.1450	0.0193	0.0047

We can see that pMHE exhibits the best performance in terms of the calculated RMSE. More specifically, the RMSE of pMHE decreases with an increasing value of the horizon length N since a larger set of data is used in the optimization problem. Moreover, the RMSE of pMHE is smaller than the RMSE of MHE with smoothing update for every considered value of N, which in turn exhibits a better performance than MHE with filtering update. Since the only difference between the three moving horizon estimators lies in the design of the prior weighing, and due to the fact that pMHE and MHE with filtering update share the same weight matrix, a possible explanation to why pMHE performs the best in this scenario might be the use of the stabilizing a priori estimate. Note that although it is not established in this example whether the considered discrete-time system globally fulfills

Assumption 3.8, which is inherited form the EKF, we observed via simulations local UES stability of pMHE, at least for the considered scenarios.

For other choices of the initial guess, we expect the pMHE scheme to perform at least as good as the EKF and even exhibit a far superior performance due to the batch of N recent measurements and in terms of the ability to handle the nonlinearity of the system as well as constraints. This can be validated by plotting in grey the initial guesses $\hat{p}_A(0) \in [0, 10]$ and $\hat{p}_B(0) \in [0, 10]$ for which the EKF converges and in blue those for which it does not converge to the true state. We obtain Figure 3.8. In contrast, the pMHE scheme yields convergent state estimates for all the initial guesses considered in this region.

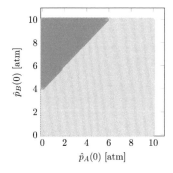

Figure 3.8: Convergence analysis of the EKF for given initial guesses $\hat{p}_A(0) \in [0, 10]$ and $\hat{p}_B(0) \in [0, 10]$. In grey, the initial guesses yield convergent state estimates and in blue non-convergent estimates.

Gaussian noise and outliers   In this scenario, the nonlinear system is subject to process and measurement disturbances, which are independent Gaussian random variables with zero mean and standard deviations $\sigma_w = 0.01$ and $\sigma_v = 0.1$, respectively. In addition, the measurements are contaminated by an outlier at the time instant $k = 10$ (or at $t = 1$ min) which takes the value $v_{10} = 20$. Analogous to previous case studies of linear systems, our goal is to illustrate that, in the nonlinear setup also, the design of the pMHE stage costs plays an important role in the performance of the estimator. We consider in the stage cost the Huber penalty function, i.e., we choose $r_i(v) = \frac{R^{-1}}{2} \Phi_{\text{hub}}(v)$, where Φ_{hub} is defined in (3.114). Here, the Huber penalty is designed with different values of the parameter M, where $M \in \{0.1, 5, 10\}$. The resulting pMHE scheme is therefore a Huber penalty estimation scheme with stability guarantees. We set $N = 3$ and compare the results of pMHE with the EKF, as well as with the MHE scheme with quadratic cost and filtering update. Obviously, one could design the latter estimator based on the Huber penalty function, however, there is no guarantee for its asymptotic stability (Rao et al. (2003)). Since our focus lies in demonstrating the effect of the measurement disturbances on the behavior of the estimators, we choose a better initial guess $\bar{x}_0 = [0.1 \ 0.1]^\top$ such that the EKF would yield positive estimates in the absence of the outlier.

In Figure 3.9, the smaller the parameter M is, the less sensitive pMHE to the outlier and the smaller the estimation error. This behavior is rather expected since with small values of

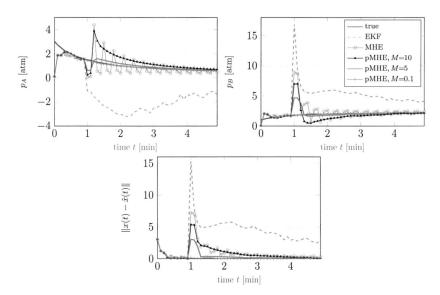

Figure 3.9: Resulting state estimates and estimation errors generated by the EKF, MHE with filtering update, and pMHE with Huber penalty function (3.114) with different values of the parameter M. Here, the nonlinear system is subject to Gaussian disturbances and an outlier. © 2020 IEEE.

M, the Huber penalty mimics the ℓ_1-norm, which is known to be robust to outliers. Note that with $M = 10$, we recover in this example the quadratic stage cost $r_i(v) = \frac{1}{2}\|v\|^2_{R^{-1}}$. We can also take advantage of the time-varying stage cost in pMHE if the interval in which the outliers are more likely to take place is known a priori. More specifically, we can set $r_i(v) = \frac{R^{-1}}{2}\Phi_{\mathrm{hub},i}(v)$ and adjust the values of the parameter M_i accordingly. For example, if we know that outliers take place in the time interval $[k_1, k_2]$ with $k_2 \geq k_1$, we can set $M_i = 0.1$ for all $k \in [k_1, k_2]$, and $M_i = 10$ otherwise.

3.4.4 pMHE for systems with output nonlinearities: Flexible joint robot link

In this section, we focus on an example of nonlinear systems in which every nonlinearity can be expressed in terms of the output. This yields a system of the form (3.211) with a constant matrix A, where all nonlinear terms are stacked into the nonlinearity $\phi(y_k)$. More specifically, we consider the example of a flexible joint robot link where joint flexibility is modeled by a stiffening torsional spring (Alessandri (2004); Zemouche et al. (2005)). The corresponding continuous-time nonlinear model is given by

$$\dot{\theta}_{\mathrm{m}} = \omega_{\mathrm{m}}$$
$$\dot{\omega}_{\mathrm{m}} = \frac{1}{J_{\mathrm{m}}}\tau - \frac{b}{J_{\mathrm{m}}}\omega_{\mathrm{m}} + \frac{K_{\tau}}{J_{\mathrm{m}}}u \tag{3.239}$$
$$\dot{\theta}_{\mathrm{l}} = \omega_{\mathrm{l}}$$
$$\dot{\omega}_{\mathrm{l}} = -\frac{1}{J_{\mathrm{l}}}\tau - \frac{Mgh}{J_{\mathrm{l}}}\sin\left(\theta_{\mathrm{l}}\right),$$

where θ_{m}, ω_{m}, θ_{l} and ω_{l} denote the motor and link positions and velocities. Moreover, the torque τ is computed as

$$\tau = (\theta_{\mathrm{l}} - \theta_{\mathrm{m}}) + (\theta_{\mathrm{l}} - \theta_{\mathrm{m}})^3. \tag{3.240}$$

The input is given by $u(t) = \sin(t)$ and we assume to measure the motor and link positions, i.e., $y = \begin{bmatrix} y^{(1)} & y^{(2)} \end{bmatrix}^{\mathsf{T}} = [\theta_{\mathrm{m}} \ \theta_{\mathrm{l}}]^{\mathsf{T}}$. The parameters of the model can be found in Table 3.2. We define the state vector $x = [\theta_{\mathrm{m}} \ \omega_{\mathrm{m}} \ \theta_{\mathrm{l}} \ \omega_{\mathrm{l}}]^{\mathsf{T}}$ and transform system (3.239) to

$$\dot{x} = A_{\mathrm{c}}\, x + \phi_{\mathrm{c}}(y) + B_{\mathrm{c}}\, u \tag{3.241a}$$
$$y = Cx \tag{3.241b}$$

where

$$A_{\mathrm{c}} = \begin{bmatrix} 0 & 1 & 0 & 0 \\ 0 & -\frac{b}{J_{\mathrm{m}}} & 0 & 0 \\ 0 & 0 & 0 & 1 \\ 0 & 0 & 0 & 0 \end{bmatrix}, \quad B_{\mathrm{c}} = \begin{bmatrix} 0 \\ \frac{K_{\tau}}{J_{\mathrm{m}}} \\ 0 \\ 0 \end{bmatrix}, \quad C = \begin{bmatrix} 1 & 0 & 0 & 0 \\ 0 & 0 & 1 & 0 \end{bmatrix}, \tag{3.241c}$$

$$\phi_{\mathrm{c}}(y) = \begin{bmatrix} 0 \\ \frac{\kappa_1}{J_{\mathrm{m}}}\left(y^{(2)} - y^{(1)}\right) + \frac{\kappa_2}{J_{\mathrm{m}}}\left(y^{(2)} - y^{(1)}\right)^3 \\ 0 \\ -\frac{\kappa_1}{J_{\mathrm{l}}}\left(y^{(2)} - y^{(1)}\right) - \frac{\kappa_2}{J_{\mathrm{l}}}\left(y^{(2)} - y^{(1)}\right)^3 - \frac{Mgh}{J_{\mathrm{l}}}\sin\left(y^{(2)}\right) \end{bmatrix}.$$

We discretize system (3.241) using the Euler method with a sampling time of $h = 0.01$ s and obtain a discrete-time system of the form (3.211) where $A = I_4 + h\, A_{\mathrm{c}}$, $B = h\, B_{\mathrm{c}}$ and $\phi(y_k) = h\, \phi_{\mathrm{c}}(y_k)$. Note that the pair (A, C) is observable. We set the initial condition to $x_0 = [0.25 \ 0.3 \ 0.2 \ 0.15]^{\mathsf{T}}$.

Table 3.2: Parameters of the model of a flexible joint robot link (Zemouche et al. (2005))

Symbol	Description	Value	Units
J_{m}	Inertia of the motor	3.7×10^{-3}	kgm^2
J_{l}	Inertia of the link	9.3×10^{-3}	kgm^2
$2h$	Length of the link	0.15	m
M	Mass of the link	0.21	kg
b	Viscous friction	0.046	m
K_{τ}	Amplifier gain	0.08	NmV^{-1}

We design for the resulting system two pMHE schemes based on problem (3.216), where we use different functions $r_{\ell_2}(v) = \|v\|^2$ and $r_{\ell_1}(v) = \|v\|_1$ in the stage cost. For both estimators, we set the horizon length $N = 5$ and the initial guess $\bar{x}_0 = [0.1 \ \ 0.1 \ \ 0.1 \ \ 0.1]^\top$. Due to the fact that the system matrix A does not depend on the output, the observer gain L in the a priori estimate (3.218) is chosen as a constant matrix that yields the following eigenvalues $\lambda_i(A - LC) = [0.9 \ \ 0.4 \ \ 0.8 \ \ 0.5]^\top$. In the Bregman distance (3.219), we design the weight matrix P such that the LMI (3.228) is satisfied.

As usual, we investigate the performance and robustness of the estimators by assuming that the measurements are corrupted by some additive noise, i.e., $y = Cx + v$. In particular, we consider the two following scenarios.

Outliers The measurement disturbances v encompass outliers, more specifically v is zero everywhere except for $v(t_1) = 5$ and $v(t_2) = 7$ where $t_1 = 0.2$ s and $t_2 = 0.5$ s. The resulting estimation errors are depicted in Figure 3.10.

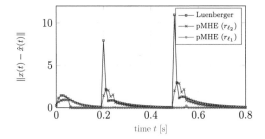

Figure 3.10: Resulting estimation errors for two pMHE schemes (3.216) with different stage costs designed for the discretized system of (3.241) and the case where the measurements are corrupted by two outliers. The estimation errors of the Luenberger observer with the same observer gain L are also depicted. © 2020 Wiley Online Library.

Figure 3.10 shows that both pMHE schemes exhibit a much better performance than the Luenberger observer. Nevertheless, we can see that the pMHE scheme with r_{ℓ_2} is more affected by the outliers and shows rather large estimation errors. In contrast to this behavior, observe that the estimator with r_{ℓ_1} is insensitive to the outliers and that the corresponding estimation errors are GES. Moreover, an interesting observation is that for this estimator, we can still indirectly see a very small effect of the outlier N times steps later, due to the fact that it enters in the a priori estimate $\bar{x}_k$ to be used in the pMHE problem at the time instant $k + N$ (see Remark 3.6). The above observations are validated by the computed RMSE defined in (3.238) associated to each pMHE scheme, where $T_{\text{sim}} = 100$ denotes the simulation time for system (3.241). These are reported in Table 3.3, where we also explored the effect of different values of the horizon length N on the pMHE performances.

Table 3.3: RMSE of the employed pMHE schemes for the first scenario (outliers) and different values of N.

N	1	2	5	10	20	30
pMHE (r_{ℓ_2})	1.5718	1.0259	0.5720	0.7592	0.4362	0.2327
pMHE (r_{ℓ_1})	1.6647	0.3031	0.1300	0.0048	$2.8916 \cdot 10^{-4}$	$1.1006 \cdot 10^{-4}$

We can observe that the performance of the pMHE scheme with the ℓ_1-norm improves monotonically with an increasing value of the horizon length N, as depicted in Table 3.3. In fact, for $N = 30$, its RMSE is much smaller than the one corresponding to the pMHE scheme with the ℓ_2-norm.

We also compare the estimation results with those provided from the EKF. The EKF estimates the state of the nonlinear system obtained after discretizing system (3.239) using the Euler method with a sampling time of $h = 0.01$ s. Note that the tuning of the process and measurement noise matrices Q_{ekf} and R_{ekf} is by no means trivial. More specifically, R_{ekf} has to be chosen large enough such that the influence of the outliers can be attenuated, but also not so large such that the information provided from the measurements becomes neglected. After tuning, we obtain the results depicted in Figure 3.11, in which we can observe that the EKF is highly affected by the presence of outliers and that both pMHE schemes exhibit a much better performance.

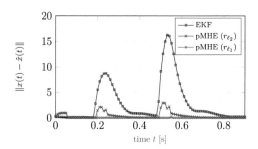

Figure 3.11: Resulting estimation errors for the pMHE schemes (3.216) compared to those of the EKF, where the measurements are corrupted by outliers. © 2020 Wiley Online Library.

Gaussian noise and outliers   The measurement disturbances take the form $v = z + e$, where z is a zero mean Gaussian noise with standard deviation $\sigma_z = 0.01[1, 1]^\top$ and $e = 3$ is an outlier that occurs at $t_2 = 0.5$ s. The resulting estimation errors are illustrated in Figure 3.12, which confirms observations from previous simulations. More specifically, we can see that the estimation error associated to each scheme is ISS with respect to the disturbances. Moreover, in the presence of Gaussian noise, the pMHE scheme with the ℓ_2-norm performs in this example much better than the pMHE scheme with the ℓ_1-norm. However, when the outlier takes place, the estimation errors of the quadratic estimator become larger than those of the pMHE with the ℓ_1-norm. We also compute in this scenario the resulting RMSE for each pMHE scheme in Table 3.4 with different values of the horizon length N. We can see in Table 3.4 that both schemes achieve a better performance with a larger horizon length. Moreover, their performances are more comparable, which could be

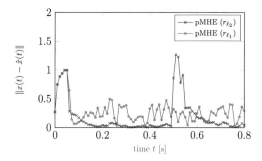

Figure 3.12: Resulting estimation errors for two pMHE schemes (3.216) with different functions r_{ℓ_2} and r_{ℓ_1} in the stage cost designed for the discretized system of (3.241) and the case where the measurements are corrupted by Gaussian noise and an outlier. © 2020 Wiley Online Library.

explained by the fact that, even though the ℓ_1-norm is more robust against the presence of the outlier, the ℓ_2-norm is more suitable when white noise is present.

Table 3.4: RMSE of the employed pMHE schemes for the second scenario (Gaussian noise and outliers) and different values of N.

N	1	2	5	10	20	30
pMHE (r_{ℓ_2})	0.6212	0.4572	0.3348	0.3260	0.1808	0.1086
pMHE (r_{ℓ_1})	0.6665	0.3330	0.2556	0.1089	0.0341	0.0222

It is important to mention that, if we set the weight matrix P as a free tuning parameter and choose it as the identity matrix for instance, the estimation error would diverge in this example. In fact, this specific choice of P does not satisfy the LMI (3.228). This demonstrates that, even though the employed pMHE scheme is based on a stabilizing Luenberger observer, the Bregman distance has to be chosen based on the corresponding Lyapunov function in order to guarantee stability.

3.5 Summary

In this chapter, we contributed to the existing research on constrained MHE for linear and nonlinear systems by employing a proximity-based formulation of the underlying optimization problem. In particular, the cost function in the pMHE formulation encompasses a Bregman distance centered around an a priori estimate obtained from an estimator which is used to guarantee stability as well as a convex and nonnegative stage cost which is used to improve performance based on the most recent measurements. By linking the MHE problem to proximal minimization algorithms with Bregman distances, we were able to carry out rigorous Lyapunov analyses of the resulting estimation errors and establish their stability properties. As a consequence, we have developed ℓ_1 and Huber penalty estimation schemes with stability guarantees.

In the first main part of the chapter, we established for LTV systems sufficient conditions on the Bregman distance and a priori estimate operator which ensure global uniform exponential stability of the estimation error. This condition reflects a key concept in pMHE, which is to construct the a priori estimate operator from an estimator with exponentially stable error dynamics, *and* to employ in the pMHE cost function the Bregman distance with which the stability of this a priori estimator can be verified. This gave rise to specific design strategies based on the Kalman filter or the Luenberger observer, from which the pMHE scheme is provably shown to inherit their stability properties. For the pMHE problem based on the LTV Kalman filter, we proved input-to-state stability of the estimation error with respect to additive process and measurement disturbances. Moreover, we demonstrated that the pMHE scheme based on this problem is a Bayesian estimator which exhibits inherent stability guarantees. In the LTI case, global exponential stability of the pMHE scheme based on the Luenberger observer is shown to hold under very mild assumptions, where only detectability of the system matrices is needed.

In the second main part of the chapter, we analyzed the stability properties of the pMHE problem for more general classes of nonlinear systems. First, we presented a nonlinear version of the linear pMHE scheme based on the Kalman filter, where we used the EKF as a stabilizing a priori estimate and designed the Bregman distance accordingly. We showed under suitable assumptions on the nonlinear system that the estimator yields uniformly exponentially stable estimation errors. Although the analysis was performed locally, this theoretical guarantee constitutes an important step in the framework of pMHE since it enables the estimator to be applied to a larger range of applications. Second, we considered a class of nonlinear systems that are globally equivalent to systems which are affine in the unmeasured state. This enabled us to formulate the associated pMHE problem as a convex optimization problem. By using a suitable Bregman distance as a proximity measure to a stabilizing a priori estimate constructed based on the Luenberger observer, we established global uniform exponential stability of the underlying estimation error.

For both linear and nonlinear systems, we reinforced the main advantages of pMHE through numerical examples, namely the nominal and robust stability guarantees, and the flexibility of its design that yields a satisfactory performance based on the problem at hand. From a practical point of view, these examples illustrated that the presented pMHE framework is well-suited for handling a rather wide range of imperfect measurements.

Chapter 4

Anytime proximity-based MHE algorithms

In this chapter, we develop an efficient algorithmic implementation of MHE for the constrained state estimation problem of discrete-time systems. More specifically, we present a novel proximity MHE (pMHE) iteration scheme, where only a finite number of optimization algorithm iterations are performed at each time instant and a potentially suboptimal state estimate is computed in real-time. In particular, a simple convex optimization problem is solved at each iteration, where a suitable Bregman distance is used as a proximity regularizing term to the previous iterate. The optimization algorithm is warm-started based on a stabilizing a priori estimate that incorporates the dynamics of a recursive and model-based estimator, from which stability can be implicitly inherited. As in the previous chapter, the primary goal is to establish exponential stability of the resulting estimation error. While the stability analysis in Chapter 3 was carried out under the assumption that a solution to the underlying pMHE optimization problem is available at each time instant, in this chapter, we take the internal iterations of the employed optimization algorithm into account in our theoretical studies. As a key result, we show that the proposed pMHE iteration scheme can be considered as an *anytime MHE algorithm*, where desirable stability properties of the resulting estimation error are guaranteed after any number of optimization algorithm iterations. This allows for example for a trade-off between estimation accuracy and computational cost through the choice of the number of iterations and based on the available computational resources. In addition, we answer the question of how good the proposed pMHE algorithm is by choosing the regret as a performance metric, based on which we can for instance characterize the performance of the iteration scheme relative to an estimator that generates the optimal solutions.

The chapter is structured as follows. In Section 4.1, we present the problem setup and formulate the constrained MHE problem that we aim to solve. In Section 4.2, we provide a detailed description of the proximity-based formulation of the MHE iteration scheme. In Section 4.3, we focus on linear systems, investigate the nominal stability properties of the pMHE iteration scheme, and derive regret upper bounds which reflect the performance properties of the algorithm. In Section 4.4, we consider nonlinear systems and establish local exponential stability of the underlying estimation error. We illustrate our theoretical results using numerical examples in Section 4.5 and conclude the chapter in Section 4.6. The results of this chapter are based on (Gharbi and Ebenbauer (2021); Gharbi et al. (2020b, 2021)).

4.1 Problem setup

Similar to Section 3.1, we consider in the following the state estimation problem of discrete-time nonlinear systems of the form

$$x_{k+1} = f_k(x_k, u_k, w_k), \tag{4.1a}$$
$$y_k = h_k(x_k) + v_k, \tag{4.1b}$$

where $x_k \in \mathbb{R}^n$ refers to the state vector, $u_k \in \mathbb{R}^m$ to the input vector, and $y_k \in \mathbb{R}^p$ to the measurement vector with $k \in \mathbb{N}$. The vectors $w_k \in \mathbb{R}^{m_w}$ and $v_k \in \mathbb{R}^p$ describe unknown process and measurement disturbances. Moreover, the initial condition $x_0 \in \mathbb{R}^n$ of system (4.1) is unknown. The state and disturbances are known to satisfy the following constraints

$$x_k \in \mathcal{X}_k \subseteq \mathbb{R}^n, \quad w_k \in \mathcal{W}_k \subseteq \mathbb{R}^{m_w}, \quad v_k \in \mathcal{V}_k \subseteq \mathbb{R}^p, \quad k \in \mathbb{N}. \tag{4.2}$$

We present again the required assumptions on the system and constraints for completeness.

Assumption 4.1 (System functions). *For any $k \in \mathbb{N}$, the functions $f_k : \mathbb{R}^n \times \mathbb{R}^m \times \mathbb{R}^{m_w} \to \mathbb{R}^n$ and $h_k : \mathbb{R}^n \to \mathbb{R}^p$ are twice continuously differentiable and Lipschitz continuous with Lipschitz constants $c_f \in \mathbb{R}_{++}$ and $c_h \in \mathbb{R}_{++}$, respectively, in all of their arguments and uniformly over $k \in \mathbb{N}$.*

Assumption 4.2 (Constraint sets). *For any $k \in \mathbb{N}$, the sets $\mathcal{X}_k$, $\mathcal{W}_k$, and $\mathcal{V}_k$ are closed and convex, $0 \in \mathcal{W}_k$, and $0 \in \mathcal{V}_k$.*

With $x(k; x_i, i, \mathbf{u}_k, \mathbf{w}_k)$, we denote the solution of system (4.1) at time instant k with initial state x_i at time i and input and disturbance sequences $\mathbf{u}_k = \{u_i, \cdots, u_{k-1}\}$ and $\mathbf{w}_k = \{w_i, \cdots, w_{k-1}\}$, respectively. This notation is simplified to $x(k; x_i, i, \mathbf{u}_k)$ in the disturbance-free case where $w_k = 0$ for all $k \in \mathbb{N}$.

We aim to compute an estimate of the state x_k in a moving horizon fashion, given the model (4.1), the constraints (4.2), and the last $N \in \mathbb{N}_+$ inputs $\{u_{k-N}, \cdots, u_{k-1}\}$ and measurements $\{y_{k-N}, \cdots, y_{k-1}\}$. More specifically, our goal is to find at each time instant k a solution to the following nonlinear MHE problem

$$\min_{\hat{\mathbf{z}}_k} \quad \sum_{i=k-N}^{k-1} r_i(\hat{v}_i) + q_i(\hat{w}_i) \tag{4.3a}$$

$$\text{s.t.} \quad \hat{x}_{i+1} = f_i(\hat{x}_i, u_k, \hat{w}_i), \quad i = k-N, \ldots, k-1 \tag{4.3b}$$
$$y_i = h_i(\hat{x}_i) + \hat{v}_i, \quad i = k-N, \ldots, k-1 \tag{4.3c}$$
$$\hat{x}_i \in \mathcal{X}_i, \quad i = k-N, \ldots, k \tag{4.3d}$$
$$\hat{w}_i \in \mathcal{W}_i, \ \hat{v}_i \in \mathcal{V}_i, \quad i = k-N, \ldots, k-1, \tag{4.3e}$$

where the decision variables $(\hat{x}_{k-N}, \hat{\mathbf{w}}_k)$ are collected in the vector $\hat{\mathbf{z}}_k$ as

$$\hat{\mathbf{z}}_k = \begin{bmatrix} \hat{x}_{k-N}^\top & \hat{w}_{k-N}^\top & \cdots & \hat{w}_{k-1}^\top \end{bmatrix}^\top \in \mathbb{R}^{Nm_w+n}. \tag{4.4}$$

In (4.3a), the stage cost consists of convex and nonnegative functions $r_i : \mathbb{R}^p \to \mathbb{R}_+$ and $q_i : \mathbb{R}^{m_w} \to \mathbb{R}_+$ which penalize the output residual $\hat{v}_i \in \mathbb{R}^p$ and the process disturbance

$\hat{w}_i \in \mathbb{R}^{m_w}$, respectively. By eliminating the system dynamics (4.3b) and (4.3c), we can reformulate the MHE problem (4.3) as

$$\min_{\hat{\mathbf{z}}_k \in \mathcal{S}_k} \quad F_k(\hat{\mathbf{z}}_k) = \sum_{i=k-N}^{k-1} r_i\left(y_i - h_i\left(x\left(i; \hat{x}_{k-N}, k-N, \mathbf{u}, \hat{\mathbf{w}}\right)\right)\right) + q_i\left(\hat{w}_i\right), \qquad (4.5a)$$

where $F_k : \mathbb{R}^{Nm_w+n} \to \mathbb{R}_+$ denotes the sum of stage cost with $\mathbf{u} = \{u_{k-N}, \cdots, u_{i-1}\}$ and $\hat{\mathbf{w}} = \{\hat{w}_{k-N}, \cdots, \hat{w}_{i-1}\}$ for each $i = k - N, \cdots, k - 1$, and $\mathcal{S}_k \subseteq \mathbb{R}^{Nm_w+n}$ represents the MHE feasible set defined as

$$\mathcal{S}_k := \left\{ \hat{\mathbf{z}}_k = \begin{bmatrix} \hat{x}_{k-N}^\top & \hat{\mathbf{w}}_k^\top \end{bmatrix}^\top : x\left(i; \hat{x}_{k-N}, k-N, \mathbf{u}, \hat{\mathbf{w}}\right) \in \mathcal{X}_i, \qquad i = k-N, \ldots, k \right.$$
$$\hat{w}_i \in \mathcal{W}_i, \qquad\qquad\qquad i = k-N, \ldots, k-1$$
$$\left. y_i - h(x\left(i; \hat{x}_{k-N}, k-N, \mathbf{u}, \hat{\mathbf{w}}\right)) \in \mathcal{V}_i, \quad i = k-N, \ldots, k-1 \right\}. \qquad (4.5b)$$

Since the true system satisfies the inequality constraints and therefore $(x_{k-N}, \mathbf{w}_k)$ lies in the set $\mathcal{S}_k$, the MHE problem is feasible at all times.

In the previous chapter, we proposed a proximity-based MHE scheme in which we solved a regularized form of (4.5). More specifically, the underlying optimization problem (3.11) is formulated by adding to the cost function (4.5a) a Bregman distance D_{ψ_k} as a generalized proximity measure to a stabilizing a priori estimate, which is referred to as $\bar{\mathbf{z}}_k$. In pMHE, the a priori estimate is constructed based on a simple, model-based, and recursive state estimator whose dynamics can be described by an operator Φ_k as specified in (3.13). A pseudo-code that summarizes the pMHE scheme can be found in Algorithm 3.1. While the performance of pMHE can be enforced with rather general convex functions r_i and q_i, stability can be ensured for any horizon length N with an appropriate choice of the a priori estimate operator Φ_k and the Bregman distance D_{ψ_k}. Despite the aforementioned advantages of pMHE, the computational requirements of solving the underlying constrained optimization problem at each time instant may render the estimator impractical for real-time applications. For instance, this includes applications where the computation power is limited or when the data sampling rates are high (Darby and Nikolaou (2007)).

In the following, we present a novel iteration scheme to the MHE problem (4.3), in which, rather than finding the solution at each time instant k, we reduce the computation time by executing only a finite number of optimization iterations of a first-order algorithm.

4.2 The proximity MHE iteration scheme

In this section, we propose a pMHE iteration scheme in which, at each time instant k, problem (4.5) is approximately solved by executing a limited number it$(k) \in \mathbb{N}_+$ of optimization algorithm iterations. In more details, at each time k, a suitable warm start solution $\hat{\mathbf{z}}_k^0 \in \mathbb{R}^{Nm_w+n}$ is generated from a stabilizing a priori estimate $\bar{\mathbf{z}}_k \in \mathbb{R}^{Nm_w+n}$ and an iterative optimization update is carried out, from which the sequence of (suboptimal) pMHE iterates $\left\{\hat{\mathbf{z}}_k^i\right\}$ with $i = 1, \cdots, \text{it}(k)$ is obtained. The iterative procedure is based on converting problem (4.5) into a sequence of simpler optimization problems with strictly convex objective functions obtained by adding a Bregman distance to a first-order approximation of F_k given in (4.5a). Before we explain the steps of the proposed algorithm in more detail, and for the sake of clarity, let us first introduce some notations. As before, the subscript k denotes the time instant in which we receive a new measurement. The superscript $i = 0, 1, \cdots, \mathrm{it}(k)$ refers to the i-th iteration of the optimization algorithm and $\mathrm{it}(k)$ to the total number of iterations between the time instants k and $k + 1$. Moreover, following the notation (3.14) used in the previous chapter, we introduce

$$\hat{\mathbf{z}}_k^i := \begin{bmatrix} \hat{x}_{k-N}^{i\top} & \hat{w}_{k-N}^{i\top} & \cdots & \hat{w}_{k-1}^{i\top} \end{bmatrix}^\top \in \mathbb{R}^{Nm_{\mathrm{w}}+n}, \tag{4.6a}$$

$$\bar{\mathbf{z}}_k := \begin{bmatrix} \bar{x}_{k-N}^\top & \bar{w}_{k-N}^\top & \cdots & \bar{w}_{k-1}^\top \end{bmatrix}^\top \in \mathbb{R}^{Nm_{\mathrm{w}}+n}, \tag{4.6b}$$

$$\mathbf{z}_k := \begin{bmatrix} x_{k-N}^\top & w_{k-N}^\top & \cdots & w_{k-1}^\top \end{bmatrix}^\top \in \mathbb{R}^{Nm_{\mathrm{w}}+n}. \tag{4.6c}$$

With $\hat{\mathbf{z}}_k^i$, we denote the i-th iterate of the optimization algorithm at time k. With $\bar{\mathbf{z}}_k$, we refer to the a priori estimate and with $\mathbf{z}_k$ to the true system state which encompasses the true state x_{k-N} and the true process disturbance sequence $\{w_{k-N}, \cdots, w_{k-1}\}$.

An iteration phase of the pMHE algorithm can be described as follows: At each time instant k, the sequence $\left\{ \hat{\mathbf{z}}_k^0, \hat{\mathbf{z}}_k^1, \cdots, \hat{\mathbf{z}}_k^{\mathrm{it}(k)} \right\}$ is generated by solving

$$\hat{\mathbf{z}}_k^0 = \arg\min_{\mathbf{z} \in \mathcal{S}_k} \quad D_{\psi_k}(\mathbf{z}, \bar{\mathbf{z}}_k) \tag{4.7a}$$

$$\hat{\mathbf{z}}_k^{i+1} = \arg\min_{\mathbf{z} \in \mathcal{S}_k} \left\{ \eta_k^i \nabla F_k \left(\hat{\mathbf{z}}_k^i \right)^\top \mathbf{z} + D_{\psi_k}(\mathbf{z}, \hat{\mathbf{z}}_k^i) \right\}, \tag{4.7b}$$

where $i = 0, 1, \cdots, \mathrm{it}(k) - 1$. More specifically, upon arrival of a new measurement at time k, the optimization algorithm is initialized based on the a priori estimate $\bar{\mathbf{z}}_k$. In particular, we compute the warm start $\hat{\mathbf{z}}_k^0$ in (4.7a) as the Bregman projection of $\bar{\mathbf{z}}_k$ onto the set $\mathcal{S}_k$ given in (4.5b) in order to make sure that the optimization algorithm is warm-started by an estimate that satisfies the constraints. Here, the Bregman distance $D_{\psi_k} : \mathbb{R}^{Nm_{\mathrm{w}}+n} \times \mathbb{R}^{Nm_{\mathrm{w}}+n} \to \mathbb{R}_+$ is constructed from the function $\psi_k : \mathbb{R}^{Nm_{\mathrm{w}}+n} \to \mathbb{R}$ as

$$D_{\psi_k}(\mathbf{z}_1, \mathbf{z}_2) = \psi_k(\mathbf{z}_1) - \psi_k(\mathbf{z}_2) - (\mathbf{z}_1 - \mathbf{z}_2)^\top \nabla \psi_k(\mathbf{z}_2). \tag{4.8}$$

Then, a simple iterative procedure is carried out. More specifically, a fixed number $\mathrm{it}(k)$ of optimization algorithm iterations is performed via (4.7b), in which the next iterate $\hat{\mathbf{z}}_k^{i+1}$ is obtained by minimizing a first-order approximation of the function F_k defined in (4.5a) at $\hat{\mathbf{z}}_k^i$ and the Bregman distance D_{ψ_k}. The optimizer update step (4.7b) is similar to the iteration step of the mirror descent algorithm (MDA) introduced in Section 2.2.2, in which D_{ψ_k} acts as a regularization that keeps $\mathbf{z}$ close to the previous iterate $\hat{\mathbf{z}}_k^i$, and not to the a priori estimate $\bar{\mathbf{z}}_k$ as in the pMHE problem (3.11). Here, $\eta_k^i \in \mathbb{R}_{++}$ denotes an appropriately chosen step size employed at the i-th iteration at time k. Notice that if the set $\mathcal{S}_k$ is convex, which directly holds if we consider for instance linear systems in Section 4.3, problem (4.7b) is a convex optimization problem. We impose the following assumptions on the cost function F_k and the Bregman distance D_{ψ_k}.

Assumption 4.3 (MHE stage cost). *For any $i \in \mathbb{N}$, the functions r_i and q_i in F_k defined in (4.5a) are continuously differentiable uniformly over i, convex and nonnegative, and attain their minimum zero at zero. Moreover, for any $k \in \mathbb{N}$, F_k is strongly smooth, i.e., its gradient ∇F_k is Lipschitz continuous with Lipschitz constant $L_k \in \mathbb{R}_{++}$, and there exists $L_{\mathrm{F}} \in \mathbb{R}_{++}$ such that $L_k \leq L_{\mathrm{F}}$ for all $k \in \mathbb{N}$.*

Assumption 4.4 (Bregman distance). *For any $k \in \mathbb{N}$, the function ψ_k is continuously differentiable, strongly convex with constant $\sigma_k \in \mathbb{R}_{++}$ and strongly smooth with constant $\gamma_k \in \mathbb{R}_{++}$, which implies the following for the associated Bregman distance*

$$\frac{\sigma_k}{2}\|\mathbf{z}_1 - \mathbf{z}_2\|^2 \leq D_{\psi_k}(\mathbf{z}_1, \mathbf{z}_2) \leq \frac{\gamma_k}{2}\|\mathbf{z}_1 - \mathbf{z}_2\|^2 \tag{4.9}$$

for all $\mathbf{z}_1, \mathbf{z}_2 \in \mathbb{R}^{Nm_{\mathbf{w}}+n}$. Moreover, these properties hold uniformly over k. In particular, there exist $\sigma \in \mathbb{R}_{++}$ and $\gamma \in \mathbb{R}_{++}$ such that $\sigma_k \geq \sigma$ and $\gamma_k \leq \gamma$ for all $k \in \mathbb{N}$.

Compared to Assumption 3.3, we additionally require the stage cost consisting of r_i, q_i to be continuously differentiable in Assumption 4.3. This enables us to employ a gradient-based optimizer update step (4.7b) for minimizing the cost function F_k. In particular, since f_k and h_k in system (4.1) are $\mathcal{C}^2$ functions by Assumption 4.1, F_k given in (4.5a) is continuously differentiable in the decision variable $\hat{\mathbf{z}}_k$. Note that assuming strong smoothness of F_k is rather customary in the analysis of first-order optimization algorithms (Mokhtari et al. (2016)) and will prove central in the subsequent theoretical studies of the pMHE iteration scheme. Assumption 4.4 is analogous to Assumption 3.4 and implies that a unique minimizer $\hat{\mathbf{z}}_k^{i+1}$ to problem (4.7b) exists if the set $\mathcal{S}_k$ is convex.

In the previous chapter, we used the Bregman distance D_{ψ_k} in the pMHE cost function (3.8a) as a proximity measure to a stabilizing a priori estimate $\bar{\mathbf{z}}_k$ in order to inherit its stability properties. Depending on the considered system class, we constructed the a priori estimate based on the computationally efficient Kalman filter, the Luenberger observer, or the extended Kalman filter (EKF). In this chapter, even though $\bar{\mathbf{z}}_k$ enters only in the warm start (4.7a), we can still exploit this key idea of the pMHE framework and generate $\bar{\mathbf{z}}_k$ from a stabilizing state estimator whose dynamics are incorporated into an operator $\Phi_k : \mathbb{R}^{Nm_{\mathbf{w}}+n} \to \mathbb{R}^{Nm_{\mathbf{w}}+n}$. In particular, given the last iterate $\hat{\mathbf{z}}_k^{\mathrm{it}(k)}$ obtained from (4.7b) at time k, the a priori estimate for the next time instant $k+1$ is calculated as

$$\bar{\mathbf{z}}_{k+1} = \Phi_k\left(\hat{\mathbf{z}}_k^{\mathrm{it}(k)}\right). \tag{4.10}$$

Analogous to Assumption 3.5, we require that the a priori estimate operator Φ_k evaluated at the true system state $\mathbf{z}_k$ defined in (4.6c) yields the next true system state in the absence of disturbances.

Assumption 4.5 (A priori estimate operator). *In the disturbance-free case where $\mathbf{z}_k = \begin{bmatrix} x_{k-N}^\top & 0 & \cdots & 0 \end{bmatrix}^\top$ and $\mathbf{z}_{k+1} = \begin{bmatrix} x_{k-N+1}^\top & 0 & \cdots & 0 \end{bmatrix}^\top$, it holds that $\Phi_k(\mathbf{z}_k) = \mathbf{z}_{k+1}$.*

Based on the last pMHE iterate $\hat{\mathbf{z}}_k^{\mathrm{it}(k)}$, the state estimate

$$\hat{x}_k = x\left(k; \hat{x}_{k-N}^{\mathrm{it}(k)}, k-N, \mathbf{u}_k, \hat{\mathbf{w}}_k^{\mathrm{it}(k)}\right) \tag{4.11}$$

is computed via the system dynamics (4.3b) given the initial condition $\hat{x}_{k-N}^{\mathrm{it}(k)}$ at initial time $k-N$, input sequence $\mathbf{u}_k = \{u_{k-N}, \cdots, u_{k-1}\}$, and estimated process disturbance sequence $\hat{\mathbf{w}}_k^{\mathrm{it}(k)} = \left\{\hat{w}_{k-N}^{\mathrm{it}(k)}, \cdots, \hat{w}_{k-1}^{\mathrm{it}(k)}\right\}$. In this chapter, we define the pMHE error at the time instant k as the difference between the true system state $\mathbf{z}_k$ and the last pMHE iterate $\hat{\mathbf{z}}_k^{\mathrm{it}(k)}$, i.e.,

$$\mathbf{z}_k - \hat{\mathbf{z}}_k^{\mathrm{it}(k)} = \begin{bmatrix} x_{k-N} \\ \mathbf{w}_k \end{bmatrix} - \begin{bmatrix} \hat{x}_{k-N}^{\mathrm{it}(k)} \\ \hat{\mathbf{w}}_k^{\mathrm{it}(k)} \end{bmatrix} = \begin{bmatrix} e_{k-N} \\ \mathbf{w}_k - \hat{\mathbf{w}}_k^{\mathrm{it}(k)} \end{bmatrix}, \tag{4.12}$$

where $e_{k-N} := x_{k-N} - \hat{x}_{k-N}^{\text{it}(k)}$, and the estimation error as

$$x_k - \hat{x}_k = x\left(k; x_{k-N}, k-N, \mathbf{u}_k, \mathbf{w}_k\right) - x\left(k; \hat{x}_{k-N}^{\text{it}(k)}, k-N, \mathbf{u}_k, \hat{\mathbf{w}}_k^{\text{it}(k)}\right). \qquad (4.13)$$

The steps of the pMHE iteration scheme are given in Algorithm 4.1 and illustrated in Figure 4.1.

Algorithm 4.1 General pMHE iteration scheme

Offline: Specify the functions r_i, q_i, the estimation horizon N, and design a suitable Bregman distance D_{ψ_k} and a priori estimate operator Φ_k. Choose an initial guess $\bar{x}_0 \in \mathbb{R}^n$ and set $\bar{\mathbf{z}}_0 = \bar{x}_0$.

Online:
1: **for** $k = 0, 1, 2, \cdots$ **do**
2:     get the inputs $\{u_{k-N}, \dots, u_{k-1}\}$ and measurements $\{y_{k-N}, \dots, y_{k-1}\}$
3:     warm start $\hat{\mathbf{z}}_k^0 = \arg\min\limits_{\mathbf{z} \in \mathcal{S}_k} D_{\psi_k}(\mathbf{z}, \bar{\mathbf{z}}_k)$
4:     **for** $i = 0, \dots, \text{it}(k) - 1$ **do** optimizer update

$$\hat{\mathbf{z}}_k^{i+1} = \arg\min_{\mathbf{z} \in \mathcal{S}_k} \left\{ \eta_k^i \, \nabla F_k\left(\hat{\mathbf{z}}_k^i\right)^\top \mathbf{z} + D_{\psi_k}(\mathbf{z}, \hat{\mathbf{z}}_k^i) \right\}$$

5:     **end for**
6:     obtain the state estimate $\hat{x}_k$ based on $\hat{\mathbf{z}}_k^{\text{it}(k)}$
7:     compute the a priori estimate $\bar{\mathbf{z}}_{k+1} = \Phi_k\left(\hat{\mathbf{z}}_k^{\text{it}(k)}\right)$ for the next time instant
8: **end for**

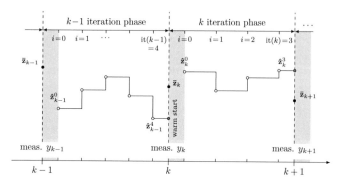

Figure 4.1: Illustration of the steps of the pMHE iteration scheme at the time instants $k-1$ and k, with the corresponding $\text{it}(k-1) = 4$ and $\text{it}(k) = 3$ optimization algorithm iterations, respectively.

Remark 4.1. Analogous to Remark 3.1, we can use Algorithm 4.1 for the time instants $k \leq N$ by setting $N = k$ in all of the underlying steps including the MHE problem (4.3). In other words, at each time $k \leq N$, we take all the available measurements into account in the estimation process. In particular, we have $\hat{\mathbf{z}}_k^i = \begin{bmatrix} \hat{x}_0^{i\top} & \hat{w}_0^{i\top} & \dots & \hat{w}_{k-1}^{i\top} \end{bmatrix}^\top$ and set $\bar{\mathbf{z}}_k = \begin{bmatrix} \bar{x}_0^\top & \bar{w}_0^\top & \dots & \bar{w}_{k-1}^\top \end{bmatrix}^\top$, where $\hat{\mathbf{z}}_k^i, \bar{\mathbf{z}}_k \in \mathbb{R}^{km_w + n}$ and $\bar{x}_0$ in the a priori estimate denotes the initial guess. Note that for $k = 0$, we have $\bar{\mathbf{z}}_0 = \bar{x}_0 \in \mathbb{R}^n$.

In the following, we compare the pMHE iteration scheme in Algorithm 4.1 with our earlier formulation of the pMHE scheme given in Algorithm 3.1. Observe that, while a solution of the pMHE optimization problem (3.11) is computed in line 3 of Algorithm 3.1, a suboptimal solution to the MHE problem (4.5) is obtained in line 4 of Algorithm 4.1. The underlying optimization algorithm consists of the MDA, which iterates (4.7b) until a given number of iterations $\mathrm{it}(k)$ is achieved. Notice that, for $\mathcal{S}_k = \mathbb{R}^{Nm_w + n}$ and a quadratic Bregman distance $D_{\psi_k}(\mathbf{z}_1, \mathbf{z}_2) = \frac{1}{2}\|\mathbf{z}_1 - \mathbf{z}_2\|_{P_k}^2$ induced from the function $\psi_k(\mathbf{z}) = \frac{1}{2}\|\mathbf{z}\|_{P_k}^2$ with $P_k \in \mathbb{S}_{++}^{Nm_w + n}$, we can obtain an analytical expression for the optimizer update step (4.7b). More specifically, taking the first-order necessary optimality condition of (4.7b) at the solution $\hat{\mathbf{z}}_k^{i+1}$ yields (cf. (2.50))

$$0 = \eta_k^i \, \nabla F_k(\hat{\mathbf{z}}_k^i) + P_k \left(\hat{\mathbf{z}}_k^{i+1} - \hat{\mathbf{z}}_k^i \right), \tag{4.14}$$

which we rearrange to obtain

$$\hat{\mathbf{z}}_k^{i+1} = \hat{\mathbf{z}}_k^i - \eta_k^i \, P_k^{-1} \, \nabla F_k(\hat{\mathbf{z}}_k^i). \tag{4.15}$$

As discussed in Section 2.2.2, if P_k is the identity matrix, we recover the iteration step of the classical gradient descent algorithm and hence (4.7b) can be executed very quickly. In more general situations, the MDA based on the iteration step (4.7b) can be regarded as a generalization of the projected gradient algorithm to the non-Euclidean setting (Beck and Teboulle (2003)). For this reason, we can view the pMHE iteration scheme in Algorithm 4.1 as a real-time version of the pMHE scheme given in Algorithm 3.1. An appealing feature of Algorithm 4.1 is that, depending on the available computation time between two subsequent time instants k and $k + 1$, the user can specify a maximum number of iterations $\mathrm{it}(k)$ after which the optimization algorithm at time k has to return a solution. Alternatively, the decision to terminate the iterative procedure (4.7) after a particular number $\mathrm{it}(k)$ of iterations can be guided by the first-order information $\nabla F_k(\hat{\mathbf{z}}_k^i)$ acquired at the time instant k.

Another key difference between the two algorithms is that Algorithm 3.1 is biased by the stabilizing a priori estimate $\bar{\mathbf{z}}_k$, while this bias is fading away in Algorithm 4.1. In other words, the stabilizing a priori estimate, which is generated from a model-based estimator through the operator Φ_k in (4.10), has less impact at each optimization iteration and hence an improved performance can be achieved. This is due to the fact that $\bar{\mathbf{z}}_k$ might degenerate performance in Algorithm 3.1, since the solution of the pMHE problem (3.11) lies in proximity to the a priori estimate. In Algorithm 4.1, however, the stabilizing estimator that generates $\bar{\mathbf{z}}_k$ enters only in the warm start (4.7a). From this point of view, it is quite surprising that, even though the effect of this stabilizing ingredient is fading away with each iteration, stability of the pMHE iteration scheme is provably preserved, as we will show in the subsequent sections. Thus, this "implicit stabilizing regularization" approach of the a priori estimate is in contrast to the explicit stabilizing regularization proposed in the previous chapter (see also (Belabbas (2020))). Moreover, we will see in the following sections that the proposed pMHE iteration scheme possesses the anytime property under suitable conditions on the Bregman distance D_{ψ_k} and the a priori estimate operator Φ_k. The anytime property refers to the fact that the algorithm will yield stable estimation errors after any number of optimization algorithm iterations. This is similar in spirit to anytime model predictive control (MPC) algorithms, which compute stabilizing control inputs after any optimization algorithm iteration (Feller and Ebenbauer (2017b, 2018)). In addition, since the internal iterations of the underlying optimization algorithm are taken into account in our theoretical studies, the natural question that arises is how to choose a suitable step size η_k^i. The subsequent stability analyses will uncover possible answers to this question.

4.3 Anytime proximity MHE for linear systems

As in Chapter 3, we first consider discrete-time LTV systems by setting in system (4.1) $f_k(x, u, w) = A_k\, x + B_k\, u + w$ and $h_k(x) = C_k\, x$, i.e., we have

$$x_{k+1} = A_k\, x_k + B_k\, u_k + w_k, \tag{4.16a}$$

$$y_k = C_k\, x_k + v_k, \tag{4.16b}$$

where $m_\mathrm{w} = n$ and the matrices A_k, B_k, C_k are of compatible dimensions. Since the functions r_i, q_i are convex by Assumption 4.3, and given the linear dynamics (4.16), the cost function F_k defined in (4.5a) is convex for any $k \in \mathbb{N}$. Moreover, the resulting feasible set $\mathcal{S}_k$ in (4.5b) is convex, which implies for the pMHE iteration scheme that a unique solution $\hat{\mathbf{z}}_k^{i+1}$ to the optimizer update step (4.7b) exists for any $i = 0, \cdots, \mathrm{it}(k) - 1$ and $k \in \mathbb{N}$. In Algorithm 4.1, based on the last iterate $\hat{\mathbf{z}}_k^{\mathrm{it}(k)}$ obtained from (4.7b), we can explicitly compute the state estimate $\hat{x}_k$ defined in (4.11) as

$$\hat{x}_k = x\left(k; \hat{x}_{k-N}^{\mathrm{it}(k)}, k - N, \mathbf{u}_k, \hat{\mathbf{w}}_k^{\mathrm{it}(k)}\right) \tag{4.17}$$

$$= \Phi(k, k - N)\, \hat{x}_{k-N}^{\mathrm{it}(k)} + \sum_{j=k-N}^{k-1} \Phi(k, j+1)\left(B_j\, u_j + \hat{w}_j^{\mathrm{it}(k)}\right),$$

where $\Phi(k, k_0) := A_{k-1} \ldots A_{k_0}$ with $\Phi(k+1, k) = A_k$ and $\Phi(k, k) = I_n$.
In Section 4.3.1, the nominal stability properties of the pMHE iteration scheme for system (4.16) without disturbances, i.e., $w_k = 0$, $v_k = 0$, $k \in \mathbb{N}$, are investigated. In Section 4.3.2, the focus is on the performance properties of the iteration scheme which are investigated in terms of a regret analysis. The aforementioned sections are based on and taken in parts literally from (Gharbi et al. (2020b))[1].

4.3.1 Nominal stability of the estimation error

In this section, we analyze the nominal stability properties of the proposed pMHE iteration scheme (Algorithm 4.1) used for the state estimation problem of the linear system (4.16). More specifically, we derive sufficient conditions on the Bregman distance D_{ψ_k}, the a priori estimate operator Φ_k, and the step size η_k^i, for the global uniform exponential stability (GUES) of the estimation error (4.13). Similar to the analysis in Chapter 3, we exploit

---

[1]M. Gharbi, B. Gharesifard, and C. Ebenbauer. Anytime proximity moving horizon estimation: Stability and regret. In *arXiv preprint arXiv:2006.14303*, 2020b.

central properties of Bregman distances such as Lemmas 2.1 and 2.2 as well as tools from the convergence analysis of the MDA introduced in Section 2.2.2.

The following lemma, which is the key ingredient for the proof of the subsequent theorem, results from applying the iterative procedure (4.7) at a given time instant k and holds independently of the specific design approaches of D_{ψ_k}, Φ_k, and η_k^i.

Lemma 4.1. *Consider system (4.16) with $w_k = 0, v_k = 0, k \in \mathbb{N}$, and let $\left\{ \hat{\mathbf{z}}_k^0, \hat{\mathbf{z}}_k^1, \cdots, \hat{\mathbf{z}}_k^{\mathrm{it}(k)} \right\}$ be the sequence of pMHE iterates generated by (4.7) at a given time instant k. Suppose Assumptions 4.1-4.4 hold. Then, for any $i \in \{0, ..., \mathrm{it}(k) - 1\}$,*

$$D_{\psi_k}\left(\mathbf{z}_k, \hat{\mathbf{z}}_k^{i+1}\right) \leq D_{\psi_k}\left(\mathbf{z}_k, \hat{\mathbf{z}}_k^i\right) + \frac{1}{2}\left(\eta_k^i L_k - \sigma_k\right)\left\|\hat{\mathbf{z}}_k^{i+1} - \hat{\mathbf{z}}_k^i\right\|^2. \tag{4.18}$$

Moreover, we have that

$$D_{\psi_k}\left(\mathbf{z}_k, \hat{\mathbf{z}}_k^{\mathrm{it}(k)}\right) \leq D_{\psi_k}\left(\mathbf{z}_k, \Phi_{k-1}\left(\hat{\mathbf{z}}_{k-1}^{\mathrm{it}(k-1)}\right)\right) + \frac{1}{2}\sum_{i=0}^{\mathrm{it}(k)-1}\left(\eta_k^i L_k - \sigma_k\right)\left\|\hat{\mathbf{z}}_k^{i+1} - \hat{\mathbf{z}}_k^i\right\|^2, \tag{4.19}$$

where $\mathbf{z}_k$ denotes the true system state, $\hat{\mathbf{z}}_k^{\mathrm{it}(k)}$ the last pMHE iterate, and Φ_k the a priori estimate operator as introduced in (4.10).

Proof. The proof generalizes and follows similar steps as in the proof of (Mokhtari et al., 2016, Proposition 2), in which the performance of the online gradient descent method is investigated. As discussed above, due to the linear setup, the cost function F_k and the set $\mathcal{S}_k$ are convex for any $k \in \mathbb{N}$. Convexity of F_k implies that for any $\mathbf{z} \in \mathbb{R}^{(N+1)n}$ and in particular for any $\mathbf{z} \in \mathcal{S}_k$

$$F_k\left(\mathbf{z}\right) \geq F_k\left(\hat{\mathbf{z}}_k^i\right) + \nabla F_k\left(\hat{\mathbf{z}}_k^i\right)^\top\left(\mathbf{z} - \hat{\mathbf{z}}_k^i\right). \tag{4.20}$$

Adding and subtracting $F_k\left(\hat{\mathbf{z}}_k^i\right)^\top \hat{\mathbf{z}}_k^{i+1}$ to the right-hand side of (4.20) yields

$$F_k\left(\mathbf{z}\right) \geq F_k\left(\hat{\mathbf{z}}_k^i\right) + \nabla F_k\left(\hat{\mathbf{z}}_k^i\right)^\top\left(\hat{\mathbf{z}}_k^{i+1} - \hat{\mathbf{z}}_k^i\right) + \nabla F_k\left(\hat{\mathbf{z}}_k^i\right)^\top\left(\mathbf{z} - \hat{\mathbf{z}}_k^{i+1}\right). \tag{4.21}$$

Given the definition of the Bregman distance D_{ψ_k} in (4.8) and the convex problem (4.7b), $\hat{\mathbf{z}}_k^{i+1}$ is optimal if and only if the following holds (Boyd and Vandenberghe (2004))

$$\left(\eta_k^i \nabla F_k\left(\hat{\mathbf{z}}_k^i\right) + \nabla \psi_k\left(\hat{\mathbf{z}}_k^{i+1}\right) - \nabla \psi_k\left(\hat{\mathbf{z}}_k^i\right)\right)^\top\left(\mathbf{z} - \hat{\mathbf{z}}_k^{i+1}\right) \geq 0 \tag{4.22}$$

for all $\mathbf{z} \in \mathcal{S}_k$. Using (4.22) in (4.21) yields

$$\eta_k^i F_k\left(\mathbf{z}\right) \geq \eta_k^i F_k\left(\hat{\mathbf{z}}_k^i\right) + \eta_k^i \nabla F_k\left(\hat{\mathbf{z}}_k^i\right)^\top\left(\hat{\mathbf{z}}_k^{i+1} - \hat{\mathbf{z}}_k^i\right) + \left(\nabla \psi_k\left(\hat{\mathbf{z}}_k^i\right) - \nabla \psi_k\left(\hat{\mathbf{z}}_k^{i+1}\right)\right)^\top\left(\mathbf{z} - \hat{\mathbf{z}}_k^{i+1}\right) \tag{4.23}$$

for all $\mathbf{z} \in \mathcal{S}_k$. By Assumption 4.3, the gradient of F_k is Lipschitz continuous for any $k \in \mathbb{N}$ with the Lipschitz constant $L_k \in \mathbb{R}_{++}$. This implies that

$$F_k\left(\hat{\mathbf{z}}_k^{i+1}\right) \leq F_k\left(\hat{\mathbf{z}}_k^i\right) + \nabla F_k\left(\hat{\mathbf{z}}_k^i\right)^\top\left(\hat{\mathbf{z}}_k^{i+1} - \hat{\mathbf{z}}_k^i\right) + \frac{L_k}{2}\left\|\hat{\mathbf{z}}_k^{i+1} - \hat{\mathbf{z}}_k^i\right\|^2, \tag{4.24}$$

which we substitute in (4.23) to get

$$\eta_k^i F_k\big(\mathbf{z}\big) \geq \eta_k^i F_k\left(\hat{\mathbf{z}}_k^{i+1}\right) - \frac{\eta_k^i L_k}{2}\left\|\hat{\mathbf{z}}_k^{i+1} - \hat{\mathbf{z}}_k^i\right\|^2 + \left(\nabla\psi_k\big(\hat{\mathbf{z}}_k^i\big) - \nabla\psi_k\big(\hat{\mathbf{z}}_k^{i+1}\big)\right)^\top\left(\mathbf{z} - \hat{\mathbf{z}}_k^{i+1}\right) \quad (4.25)$$

for all $\mathbf{z} \in \mathcal{S}_k$. In view of the three-points identity (2.33) of Bregman distances in Lemma 2.1 (with $a = \hat{\mathbf{z}}_k^{i+1}$, $b = \hat{\mathbf{z}}_k^i$, $c = \mathbf{z}$), and given the lower bound on D_{ψ_k} in (4.9) in Assumption 4.4, which is induced by assuming strong convexity of ψ_k, we have

$$\left(\nabla\psi_k\big(\hat{\mathbf{z}}_k^i\big) - \nabla\psi_k\big(\hat{\mathbf{z}}_k^{i+1}\big)\right)^\top\left(\mathbf{z} - \hat{\mathbf{z}}_k^{i+1}\right) = D_{\psi_k}\big(\mathbf{z}, \hat{\mathbf{z}}_k^{i+1}\big) + D_{\psi_k}\big(\hat{\mathbf{z}}_k^{i+1}, \hat{\mathbf{z}}_k^i\big) - D_{\psi_k}\big(\mathbf{z}, \hat{\mathbf{z}}_k^i\big)$$
$$\geq D_{\psi_k}\big(\mathbf{z}, \hat{\mathbf{z}}_k^{i+1}\big) + \frac{\sigma_k}{2}\left\|\hat{\mathbf{z}}_k^{i+1} - \hat{\mathbf{z}}_k^i\right\|^2 - D_{\psi_k}\big(\mathbf{z}, \hat{\mathbf{z}}_k^i\big).$$
$$(4.26)$$

Therefore, by using (4.26) in (4.25), we arrive at

$$\eta_k^i F_k\big(\mathbf{z}\big) \geq \eta_k^i F_k\left(\hat{\mathbf{z}}_k^{i+1}\right) - \frac{\eta_k^i L_k}{2}\left\|\hat{\mathbf{z}}_k^{i+1} - \hat{\mathbf{z}}_k^i\right\|^2 + D_{\psi_k}\big(\mathbf{z}, \hat{\mathbf{z}}_k^{i+1}\big) - D_{\psi_k}\big(\mathbf{z}, \hat{\mathbf{z}}_k^i\big) + \frac{\sigma_k}{2}\left\|\hat{\mathbf{z}}_k^{i+1} - \hat{\mathbf{z}}_k^i\right\|^2$$
$$= \eta_k^i F_k\left(\hat{\mathbf{z}}_k^{i+1}\right) + \frac{1}{2}\big(\sigma_k - \eta_k^i L_k\big)\left\|\hat{\mathbf{z}}_k^{i+1} - \hat{\mathbf{z}}_k^i\right\|^2 + D_{\psi_k}\big(\mathbf{z}, \hat{\mathbf{z}}_k^{i+1}\big) - D_{\psi_k}\big(\mathbf{z}, \hat{\mathbf{z}}_k^i\big) \quad (4.27)$$

for all $\mathbf{z} \in \mathcal{S}_k$. Due to Assumption 4.3 and given that $w_k = 0$, $v_k = 0$, $k \in \mathbb{N}$ in system (4.16), evaluating the cost function F_k at the true system state $\mathbf{z}_k$ yields $0 = F_k\big(\mathbf{z}_k\big) \leq F_k\left(\hat{\mathbf{z}}_k^{i+1}\right)$ as shown in (3.26). Moreover, $\mathbf{z}_k \in \mathcal{S}_k$. Hence, setting $\mathbf{z} = \mathbf{z}_k$ in (4.27) leads to

$$0 \geq \eta_k^i\Big(F_k\big(\mathbf{z}_k\big) - F_k\left(\hat{\mathbf{z}}_k^{i+1}\right)\Big) \quad (4.28)$$
$$\geq \frac{1}{2}\big(\sigma_k - \eta_k^i L_k\big)\left\|\hat{\mathbf{z}}_k^{i+1} - \hat{\mathbf{z}}_k^i\right\|^2 + D_{\psi_k}\big(\mathbf{z}_k, \hat{\mathbf{z}}_k^{i+1}\big) - D_{\psi_k}\big(\mathbf{z}_k, \hat{\mathbf{z}}_k^i\big).$$

Rearranging the above inequality establishes the first statement (4.18) in Lemma 4.1. Applying (4.18) for each two subsequent iterations i and $i+1$ (where $i = 0, \cdots, \mathrm{it}(k) - 1$) yields

$$D_{\psi_k}\left(\mathbf{z}_k, \hat{\mathbf{z}}_k^{\mathrm{it}(k)}\right) \leq D_{\psi_k}\left(\mathbf{z}_k, \hat{\mathbf{z}}_k^{\mathrm{it}(k)-1}\right) + \frac{1}{2}\Big(\eta_k^{\mathrm{it}(k)-1} L_k - \sigma_k\Big)\left\|\hat{\mathbf{z}}_k^{\mathrm{it}(k)} - \hat{\mathbf{z}}_k^{\mathrm{it}(k)-1}\right\|^2 \quad (4.29)$$
$$\leq D_{\psi_k}\left(\mathbf{z}_k, \hat{\mathbf{z}}_k^{\mathrm{it}(k)-2}\right) + \frac{1}{2}\Big(\eta_k^{\mathrm{it}(k)-2} L_k - \sigma_k\Big)\left\|\hat{\mathbf{z}}_k^{\mathrm{it}(k)-1} - \hat{\mathbf{z}}_k^{\mathrm{it}(k)-2}\right\|^2$$
$$\qquad + \frac{1}{2}\Big(\eta_k^{\mathrm{it}(k)-1} L_k - \sigma_k\Big)\left\|\hat{\mathbf{z}}_k^{\mathrm{it}(k)} - \hat{\mathbf{z}}_k^{\mathrm{it}(k)-1}\right\|^2$$
$$\leq \cdots$$
$$\leq D_{\psi_k}\left(\mathbf{z}_k, \hat{\mathbf{z}}_k^0\right) + \frac{1}{2}\sum_{i=0}^{\mathrm{it}(k)-1}\Big(\eta_k^i L_k - \sigma_k\Big)\left\|\hat{\mathbf{z}}_k^{i+1} - \hat{\mathbf{z}}_k^i\right\|^2.$$

In (4.7a), the warm start $\hat{\mathbf{z}}_k^0$ is computed as the Bregman projection of the a priori estimate $\bar{\mathbf{z}}_k$ onto the convex set $\mathcal{S}_k$, i.e., $\hat{\mathbf{z}}_k^0 = \Pi_{\mathcal{S}_k}^{\psi_k}(\bar{\mathbf{z}}_k)$ by Definition 2.5. Hence, given that $\mathbf{z}_k \in \mathcal{S}_k$, we can use (2.34) in Lemma 2.2 to get

$$0 \leq D_{\psi_k}\big(\hat{\mathbf{z}}_k^0, \bar{\mathbf{z}}_k\big) \leq D_{\psi_k}\big(\mathbf{z}_k, \bar{\mathbf{z}}_k\big) - D_{\psi_k}\big(\mathbf{z}_k, \hat{\mathbf{z}}_k^0\big). \quad (4.30)$$

Thus, $D_{\psi_k}\left(\mathbf{z}_k, \hat{\mathbf{z}}_k^0\right) \le D_{\psi_k}(\mathbf{z}_k, \bar{\mathbf{z}}_k)$ holds, which we substitute in (4.29) to obtain

$$D_{\psi_k}\left(\mathbf{z}_k, \hat{\mathbf{z}}_k^{\text{it}(k)}\right) \le D_{\psi_k}\left(\mathbf{z}_k, \bar{\mathbf{z}}_k\right) + \frac{1}{2} \sum_{i=0}^{\text{it}(k)-1} \left(\eta_k^i L_k - \sigma_k\right) \left\|\hat{\mathbf{z}}_k^{i+1} - \hat{\mathbf{z}}_k^i\right\|^2. \qquad (4.31)$$

The second statement of the lemma follows by setting $\bar{\mathbf{z}}_k = \Phi_{k-1}\left(\hat{\mathbf{z}}_{k-1}^{\text{it}(k-1)}\right)$ in the above inequality as specified by (4.10). $\qquad\square$

The first result (4.18) in Lemma 4.1 implies that by choosing the step size such that it satisfies $\eta_k^i \le \frac{\sigma_k}{L_k}$ for all $i = 0, \cdots, \text{it}(k) - 1$ and $k \in \mathbb{N}$, we can make sure that

$$D_{\psi_k}\left(\mathbf{z}_k, \hat{\mathbf{z}}_k^{i+1}\right) \le D_{\psi_k}\left(\mathbf{z}_k, \hat{\mathbf{z}}_k^i\right), \qquad (4.32)$$

i.e., that the Bregman distance to the true system state $\mathbf{z}_k$ decreases after each optimizer update step (4.7b). Moreover, in view of the second result (4.19) in the lemma, fulfilling this step size condition allows to bound the Bregman distance between $\mathbf{z}_k$ and the last pMHE iterate $\hat{\mathbf{z}}_k^{\text{it}(k)}$ used to generate the state estimate $\hat{x}_k$ in terms of the distance between $\mathbf{z}_k$ and the stabilizing a priori estimate $\bar{\mathbf{z}}_k$ as follows

$$D_{\psi_k}\left(\mathbf{z}_k, \hat{\mathbf{z}}_k^{\text{it}(k)}\right) \le D_{\psi_k}\left(\mathbf{z}_k, \bar{\mathbf{z}}_k\right). \qquad (4.33)$$

Note that a similar result which uses the optimal solution $\hat{\mathbf{z}}_k^*$ of the pMHE problem (3.19) is established in (3.21) in Lemma 3.1.

Remark 4.2. Observe that the proof of Lemma 4.1 indicates that, with the aforementioned choice of the step size η_k^i, we can also ensure that the sequence of function values of the pMHE iterates generated by (4.7), i.e., $\left\{F_k\left(\hat{\mathbf{z}}_k^0\right), \cdots, F_k\left(\hat{\mathbf{z}}_k^{\text{it}(k)}\right)\right\}$, is nonincreasing. More specifically, setting $\mathbf{z} = \hat{\mathbf{z}}_k^i \in \mathcal{S}_k$ in (4.27) yields

$$\eta_k^i F_k\left(\hat{\mathbf{z}}_k^i\right) \ge \eta_k^i F_k\left(\hat{\mathbf{z}}_k^{i+1}\right) + \frac{1}{2}\left(\sigma_k - \eta_k^i L_k\right) \left\|\hat{\mathbf{z}}_k^{i+1} - \hat{\mathbf{z}}_k^i\right\|^2 + D_{\psi_k}\left(\hat{\mathbf{z}}_k^i, \hat{\mathbf{z}}_k^{i+1}\right) - D_{\psi_k}\left(\hat{\mathbf{z}}_k^i, \hat{\mathbf{z}}_k^i\right)$$

$$\ge \eta_k^i F_k\left(\hat{\mathbf{z}}_k^{i+1}\right) + \frac{1}{2}\left(\sigma_k - \eta_k^i L_k\right) \left\|\hat{\mathbf{z}}_k^{i+1} - \hat{\mathbf{z}}_k^i\right\|^2 \qquad (4.34)$$

since $D_{\psi_k}\left(\hat{\mathbf{z}}_k^i, \hat{\mathbf{z}}_k^{i+1}\right) \ge 0$ and $D_{\psi_k}\left(\hat{\mathbf{z}}_k^i, \hat{\mathbf{z}}_k^i\right) = 0$. Hence, choosing the step size such that it satisfies $\eta_k^i \le \frac{\sigma_k}{L_k}$ for all $i = 0, \dots, \text{it}(k) - 1$ and $k \in \mathbb{N}$ yields

$$F_k\left(\hat{\mathbf{z}}_k^i\right) \ge F_k\left(\hat{\mathbf{z}}_k^{i+1}\right). \qquad (4.35)$$

Based on Lemma 4.1 and the above observations regarding the choice of the step size, we are now ready to establish for linear systems the nominal stability properties of the proposed pMHE iteration scheme in the following theorem.

Theorem 4.1. *Consider system* (4.16) *with* $w_k = 0, v_k = 0, k \in \mathbb{N}$, *and the pMHE iteration scheme in Algorithm 4.1 with a step size in* (4.7b) *satisfying*

$$\eta_k^i \le \frac{\sigma_k}{L_k} \qquad (4.36)$$

for all $i = 0, \cdots, \text{it}(k) - 1$ and $k \in \mathbb{N}$. Let Assumptions 4.1-4.5 hold. Suppose that there exist $M \in \mathbb{N}_+$ and $c \in \mathbb{R}_{++}$ such that the Bregman distance D_{ψ_k} and the a priori estimate operator Φ_k satisfy

$$D_{\psi_k}\left(\Phi_{k-1}(\mathbf{z}), \Phi_{k-1}(\hat{\mathbf{z}})\right) - D_{\psi_{k-1}}(\mathbf{z}, \hat{\mathbf{z}}) \leq -c\left\|\mathbf{z} - \hat{\mathbf{z}}\right\|^2 \tag{4.37}$$

for all $k \geq M$ and $\mathbf{z}, \hat{\mathbf{z}} \in \mathbb{R}^{(N+1)n}$. Then, the estimation error (4.13) is GUES.

Proof. Similar to the stability proofs in the previous chapter, we first show GUES of the pMHE error (4.12) by proving the existence of a continuous time-varying Lyapunov function V_k satisfying conditions (A.6a) and (A.6b) in Theorem A.2 in Appendix A. In this case, we choose V_k as the Bregman distance employed in the iteration procedure (4.7), i.e.,

$$V_k\left(\mathbf{z}_k, \hat{\mathbf{z}}_k^{\text{it}(k)}\right) = D_{\psi_k}\left(\mathbf{z}_k, \hat{\mathbf{z}}_k^{\text{it}(k)}\right). \tag{4.38}$$

By Assumption 4.4, (A.6a) follows with $c_1 = \frac{\sigma}{2}$ and $c_2 = \frac{\gamma}{2}$. Furthermore, by (4.19) in Lemma 4.1, we have

$$\Delta V_k := V_k\left(\mathbf{z}_k, \hat{\mathbf{z}}_k^{\text{it}(k)}\right) - V_{k-1}\left(\mathbf{z}_{k-1}, \hat{\mathbf{z}}_{k-1}^{\text{it}(k-1)}\right) \tag{4.39}$$

$$= D_{\psi_k}\left(\mathbf{z}_k, \hat{\mathbf{z}}_k^{\text{it}(k)}\right) - D_{\psi_{k-1}}\left(\mathbf{z}_{k-1}, \hat{\mathbf{z}}_{k-1}^{\text{it}(k-1)}\right)$$

$$\leq D_{\psi_k}\left(\mathbf{z}_k, \Phi_{k-1}\left(\hat{\mathbf{z}}_{k-1}^{\text{it}(k-1)}\right)\right) - D_{\psi_{k-1}}\left(\mathbf{z}_{k-1}, \hat{\mathbf{z}}_{k-1}^{\text{it}(k-1)}\right) + \frac{1}{2}\sum_{i=0}^{\text{it}(k)-1}\left(\eta_k^i L_k - \sigma_k\right)\left\|\hat{\mathbf{z}}_k^{i+1} - \hat{\mathbf{z}}_k^i\right\|^2.$$

The condition on the step size given in (4.36) implies that $\eta_k^i L_k - \sigma_k \leq 0$, and hence,

$$\frac{1}{2}\sum_{i=0}^{\text{it}(k)-1}\left(\eta_k^i L_k - \sigma_k\right)\left\|\hat{\mathbf{z}}_k^{i+1} - \hat{\mathbf{z}}_k^i\right\|^2 \leq 0. \tag{4.40}$$

Since $\mathbf{z}_k = \Phi_{k-1}\left(\mathbf{z}_{k-1}\right)$ in view of Assumption 4.5, it holds that

$$D_{\psi_k}\left(\mathbf{z}_k, \Phi_{k-1}\left(\hat{\mathbf{z}}_{k-1}^{\text{it}(k-1)}\right)\right) = D_{\psi_k}\left(\Phi_{k-1}\left(\mathbf{z}_{k-1}\right), \Phi_{k-1}\left(\hat{\mathbf{z}}_{k-1}^{\text{it}(k-1)}\right)\right). \tag{4.41}$$

Using the sufficient condition on the Bregman distance and the a priori estimate operator stated in (4.37) as well as (4.40) and (4.41) in (4.39) leads to

$$\Delta V_k \leq D_{\psi_k}\left(\Phi_{k-1}\left(\mathbf{z}_{k-1}\right), \Phi_{k-1}\left(\hat{\mathbf{z}}_{k-1}^{\text{it}(k-1)}\right)\right) - D_{\psi_{k-1}}\left(\mathbf{z}_{k-1}, \hat{\mathbf{z}}_{k-1}^{\text{it}(k-1)}\right) \tag{4.42}$$

$$= -c\left\|\mathbf{z}_{k-1} - \hat{\mathbf{z}}_{k-1}^{\text{it}(k-1)}\right\|^2$$

for all $k \geq M$, which proves that condition (A.6b) holds with $c_3 = c$. Thus, the pMHE error (4.12) is GUES and, according to Definition A.5 in Appendix A, there exist constants $\tilde{\alpha} \in \mathbb{R}_{++}$ and $\beta \in (0, 1)$ such that

$$\left\|\mathbf{z}_k - \hat{\mathbf{z}}_k^{\text{it}(k)}\right\| \leq \tilde{\alpha}\beta^k\left\|\mathbf{z}_0 - \hat{\mathbf{z}}_0^{\text{it}(0)}\right\| = \tilde{\alpha}\beta^k\left\|\mathbf{z}_0 - \hat{\mathbf{z}}_0^0\right\| \tag{4.43}$$

holds for all $k \in \mathbb{N}_+$ and $\mathbf{z}_0, \hat{\mathbf{z}}_0^0 \in \mathbb{R}^{(N+1)n}$. The equality in (4.43) is due to the fact that, by Remark 4.1, it holds that $\hat{\mathbf{z}}_0^{i+1} = \arg\min_{\mathbf{z} \in \mathcal{S}_0} D_{\psi_0}(\mathbf{z}, \hat{\mathbf{z}}_0^i) = \hat{\mathbf{z}}_0^i$. Hence, each pMHE iterate in (4.7b) at time instant $k = 0$ is equal to the warm start, i.e., $\hat{\mathbf{z}}_0^i = \hat{\mathbf{z}}_0^0$ for all $i = 1, \cdots, \text{it}(0)$.

Moreover, evaluating (4.30) at time $k = 0$ implies that $D_{\psi_0}(\mathbf{z}_0, \hat{\mathbf{z}}_0^0) \leq D_{\psi_0}(\mathbf{z}_0, \bar{\mathbf{z}}_0)$, and hence $\|\mathbf{z}_0 - \hat{\mathbf{z}}_0^0\| \leq \sqrt{\frac{\gamma}{\sigma}}\|\mathbf{z}_0 - \bar{\mathbf{z}}_0\|$ due to the uniform bounds in Assumption 4.4. Applying the last inequality in (4.43) yields

$$\left\|\mathbf{z}_k - \hat{\mathbf{z}}_k^{\mathrm{it}(k)}\right\| \leq \alpha\beta^k\left\|\mathbf{z}_0 - \bar{\mathbf{z}}_0\right\| = \alpha\beta^k\left\|x_0 - \bar{x}_0\right\| \tag{4.44}$$

for all $k \in \mathbb{N}_+$ and $x_0, \bar{x}_0 \in \mathbb{R}^n$, where $\bar{x}_0$ denotes the initial guess and $\alpha := \tilde{\alpha}\sqrt{\frac{\gamma}{\sigma}}$. Regarding the stability of the estimation error (4.13), we proceed with similar steps as in the proof of Theorem 3.2. More specifically, by the triangle inequality, we have

$$\left\|x_k - \hat{x}_k\right\| \leq \left\|x\left(k; x_{k-N}, k-N, \mathbf{u}_k\right) - x\left(k; \hat{x}_{k-N}^{\mathrm{it}(k)}, k-N, \mathbf{u}_k\right)\right\| \tag{4.45}$$
$$+ \left\|x\left(k; \hat{x}_{k-N}^{\mathrm{it}(k)}, k-N, \mathbf{u}_k\right) - x\left(k; \hat{x}_{k-N}^{\mathrm{it}(k)}, k-N, \mathbf{u}_k, \hat{\mathbf{w}}_k^{\mathrm{it}(k)}\right)\right\|.$$

By the uniform bound $\|A_k\| \leq c_{\mathrm{f}}$, which holds for all $k \in \mathbb{N}$ due to Assumption 4.1, and recalling (4.12), we can adapt the steps used for deriving (3.35) in order to obtain

$$\left\|x\left(k; x_{k-N}, k-N, \mathbf{u}_k\right) - x\left(k; \hat{x}_{k-N}^{\mathrm{it}(k)}, k-N, \mathbf{u}_k\right)\right\| \leq c_{\mathrm{f}}^N\left\|e_{k-N}\right\| \tag{4.46}$$

and

$$\left\|x\left(k; \hat{x}_{k-N}^{\mathrm{it}(k)}, k-N, \mathbf{u}_k\right) - x\left(k; \hat{x}_{k-N}^{\mathrm{it}(k)}, k-N, \mathbf{u}_k, \hat{\mathbf{w}}_k^{\mathrm{it}(k)}\right)\right\| \leq \sum_{j=k-N}^{k-1} c_{\mathrm{f}}^{k-j}\left\|\hat{w}_j^{\mathrm{it}(k)}\right\|. \tag{4.47}$$

Substituting the above upper bounds in (4.45) leads to

$$\left\|x_k - \hat{x}_k\right\| \leq c_{\mathrm{f}}^N\left\|e_{k-N}\right\| + \sum_{j=k-N}^{k-1} c_{\mathrm{f}}^{k-j}\left\|\hat{w}_j^{\mathrm{it}(k)}\right\| \tag{4.48}$$
$$\leq c_{\mathrm{f}}^N\left\|\mathbf{z}_k - \hat{\mathbf{z}}_k^{\mathrm{it}(k)}\right\| + \sum_{j=k-N}^{k-1} c_{\mathrm{f}}^{k-j}\left\|\mathbf{z}_k - \hat{\mathbf{z}}_k^{\mathrm{it}(k)}\right\|.$$

We finally use (4.44) in (4.48) in order to establish that

$$\left\|x_k - \hat{x}_k\right\| \leq \bar{c}\,\alpha\beta^k\|x_0 - \bar{x}_0\| \tag{4.49}$$

for all $k \in \mathbb{N}_+$ and $x_0, \bar{x}_0 \in \mathbb{R}^n$, where $\bar{c} := c_{\mathrm{f}}^N + \sum_{j=k-N}^{k-1} c_{\mathrm{f}}^{k-j}$. This establishes that the resulting estimation error (4.13) is GUES, which completes the proof. $\square$

Theorem 4.1 implies that, for linear systems, GUES of the estimation error generated by the pMHE iteration scheme in Algorithm 4.1 is guaranteed independently of the executed number of optimization algorithm iterations $\mathrm{it}(k)$. Hence, Algorithm 4.1 can be considered as an anytime MHE algorithm which yields convergent, albeit suboptimal, state estimates after each iteration (4.7b). This desirable property holds also for any horizon length $N \in \mathbb{N}_+$ and any smooth functions r_i, q_i in the MHE stage cost satisfying Assumption 4.3. This includes for instance quadratic functions and the Huber penalty function introduced in (3.114). We also point out that selecting a constant step size given by $\eta_k^i = \frac{\sigma}{L_{\mathrm{F}}}$ for all $i = 0, \cdots, \mathrm{it}(k) - 1$ and $k \in \mathbb{N}$ is sufficient for ensuring GUES of the underlying estimation error.

Remark 4.3. Although GUES of the estimation error holds independently of the number of optimization algorithm iterations it(k), the underlying stability analysis does not establish a direct link between it(k) and how quickly the estimation error approaches zero. Intuitively, one would expect that increasing it(k) leads to a faster convergence and hence an improved performance of the estimator. This observation is also quite apparent in simulation examples as will be demonstrated later in Section 4.5. Nevertheless, a similar observation will be validated through the subsequent regret analysis in Section 4.3.2, in which the derived regret bound characterizing the performance of the pMHE iteration scheme can be rendered smaller by increasing the number of optimization iterations.

Even though the internal iterates are explicitly taken into account in the proof of Theorem 4.1, we require the same sufficient condition (4.37) on the Bregman distance D_{ψ_k} and the a priori estimate operator Φ_k as in (3.28) in Theorem 3.2, in which the exact solution of the pMHE problem (3.19) is considered. This observation is particularity advantageous since it enables us to adopt the key idea of the pMHE framework discussed in the previous chapter, which is to construct Φ_k from an estimator with exponentially stable dynamics and to select D_{ψ_k} as the Lyapunov function with which the stability of this estimator can be verified. Hence, analogous to Section 3.2, we discuss in the following specific design approaches for Φ_k and D_{ψ_k} that are based on the LTV Kalman filter and the Luenberger observer for the special case of LTI systems.

Φ_k and D_{ψ_k} based on the LTV Kalman filter

In the following, we focus on the design of the Bregman distance D_{ψ_k} used in lines 3 an 4 of Algorithm 4.1 and of the a priori estimate operator Φ_k used in line 7 of the algorithm. In particular, we adapt the design approach based on the Kalman filter introduced in Section 3.2 to the pMHE iteration scheme as it is provably shown to satisfy condition (3.28), or equivalently (4.37).

Let $\mathbf{z} = \begin{bmatrix} x^\top & \mathbf{w}^\top \end{bmatrix}^\top$, $\hat{\mathbf{z}} = \begin{bmatrix} \hat{x}^\top & \hat{\mathbf{w}}^\top \end{bmatrix}^\top$, $x, \hat{x} \in \mathbb{R}^n$, $\mathbf{w}, \hat{\mathbf{w}} \in \mathbb{R}^{Nn}$. Given a time instant $k \in \mathbb{N}$ and a horizon length $N \in \mathbb{N}_+$, the a priori estimate operator based on the Kalman filter is

$$\Phi_k(\mathbf{z}) = \begin{bmatrix} A_{k-N}\left(x + K_{k-N}\left(y_{k-N} - C_{k-N}\, x\right)\right) + B_{k-N}\, u_{k-N} \\ \mathbf{0} \end{bmatrix}, \tag{4.50}$$

where $\mathbf{0} \in \mathbb{R}^{Nn}$ refers to the zero vector. In particular, consider the notation introduced for the a priori estimate $\bar{\mathbf{z}}_k$ in (4.6) and the last pMHE iterate $\hat{\mathbf{z}}_k^{\mathrm{it}(k)}$ obtained from (4.7) at time instant k. According to (4.10), the a priori estimate for the next time instant is

$$\bar{\mathbf{z}}_{k+1} = \Phi_k\left(\hat{\mathbf{z}}_k^{\mathrm{it}(k)}\right) = \begin{bmatrix} \bar{x}_{k-N+1}^\top & 0 & \dots & 0 \end{bmatrix}^\top, \tag{4.51a}$$

where we have zero a priori process disturbances, i.e., $\bar{w}_i = 0$, $i = k - N + 1, \dots, k$, and

$$\bar{x}_{k-N+1} = A_{k-N}\left(\hat{x}_{k-N}^{\mathrm{it}(k)} + K_{k-N}\left(y_{k-N} - C_{k-N}\,\hat{x}_{k-N}^{\mathrm{it}(k)}\right)\right) + B_{k-N}\, u_{k-N}. \tag{4.51b}$$

The Bregman distance D_{ψ_k} is a weighted quadratic function given by

$$D_{\psi_k}\left(\mathbf{z}, \hat{\mathbf{z}}\right) = \frac{1}{2}\left\|x - \hat{x}\right\|_{\Pi_{k-N}^-}^2 + \frac{1}{2}\left\|\mathbf{w} - \hat{\mathbf{w}}\right\|_{\bar{W}}^2, \tag{4.52}$$

where $\Pi_{k-N}^- \in \mathbb{S}_{++}^n$ denotes the inverse of the Kalman filter covariance matrix and the weight matrix $\bar{W} \in \mathbb{S}_{++}^{Nn}$ has the form $\bar{W} = \mathrm{diag}(W, \ldots, W)$ with $W \in \mathbb{S}_{++}^n$ arbitrary. The Kalman gain and covariance matrix are computed as

$$K_k = P_k^- C_k^\top \left(C_k P_k^- C_k^\top + R \right)^{-1}, \tag{4.53a}$$

$$P_{k+1}^- = A_k P_k^+ A_k^\top + Q, \qquad \Pi_k^- := \left(P_k^- \right)^{-1}, \tag{4.53b}$$

$$P_k^+ = \left(I - K_k C_k \right) P_k^-, \qquad \Pi_k^+ := \left(P_k^+ \right)^{-1}, \tag{4.53c}$$

where $P_0^- \in \mathbb{S}_{++}^n$, $Q \in \mathbb{S}_{++}^n$ and $R \in \mathbb{S}_{++}^p$.

Assumption 4.6 (Linear system properties). *System* (4.16) *is uniformly completely observable and uniformly completely controllable.*

The definitions of uniform complete observability and uniform complete controllability can be found in Definitions 3.1 and 3.2, respectively. In view of this additional assumption, we can take advantage of Proposition 3.1 to establish that the Bregman distance (4.52) and the a priori estimate operator (4.50) verify Assumptions 4.4 and 4.5 and satisfy the sufficient condition (4.37) in Theorem 4.1. Based thereon, we obtain the following result.

Corollary 4.1. *Consider system* (4.16) *with* $w_k = 0$, $v_k = 0$, $k \in \mathbb{N}$, *and the pMHE iteration scheme in Algorithm 4.1 with a step size in* (4.7b) *satisfying* $\eta_k^i \leq \frac{\sigma_k}{L_k}$ *for all* $i = 0, \cdots, \mathrm{it}(k) - 1$ *and* $k \in \mathbb{N}$, *a priori estimate operator* (4.50) *and Bregman distance* (4.52) *based on the Kalman filter. Let Assumptions 4.1-4.3, 4.6 hold. Then, the estimation error* (4.13) *is GUES.*

Φ_k and D_{ψ_k} based on the Luenberger observer

In the following, we consider the problem setup presented in Section 3.2.4 and focus on the special case of discrete-time LTI systems

$$x_{k+1} = A\,x_k + B\,u_k + w_k, \tag{4.54a}$$

$$y_k = C\,x_k + v_k, \tag{4.54b}$$

where the state and disturbances are known to satisfy

$$x_k \in \mathcal{X} \subseteq \mathbb{R}^n, \quad w_k \in \mathcal{W} \subseteq \mathbb{R}^n, \quad v_k \in \mathcal{V} \subseteq \mathbb{R}^p, \qquad k \in \mathbb{N}. \tag{4.55}$$

Consider line 4 of Algorithm 4.1 and in particular the iteration procedure (4.7). In the cost function F_k given in (4.5a), we fix the stage costs $r_i(\cdot) := r(\cdot)$ and $q_i(\cdot) := q(\cdot)$ for all $i \in \mathbb{N}$. Furthermore, we fix the function $\psi_k(\cdot) := \psi(\cdot)$ for all time instants $k \in \mathbb{N}$ and let D_ψ refer to the associated Bregman distance, which is constructed via (4.8) and whose specific choice will be presented shortly. In line 6 of Algorithm 4.1, based on the last pMHE iterate $\hat{\mathbf{z}}_k^{\mathrm{it}(k)}$, the state estimate $\hat{x}_k$ is obtained via a forward prediction of the dynamics (4.54a) as

$$\hat{x}_k = x\left(k; \hat{x}_{k-N}^{\mathrm{it}(k)}, k-N, \mathbf{u}_k, \hat{\mathbf{w}}_k^{\mathrm{it}(k)}\right) = A^N \hat{x}_{k-N}^{\mathrm{it}(k)} + \sum_{j=k-N}^{k-1} A^{k-1-j}\left(B\,u_j + \hat{w}_j^{\mathrm{it}(k)}\right). \tag{4.56}$$

In line 7 of Algorithm 4.1, analogous to Section 3.2.4, we construct the a priori estimate operator Φ_k based on the Luenberger observer as follows.

Let $\mathbf{z} = \begin{bmatrix} x^\top & \mathbf{w}^\top \end{bmatrix}^\top$, $\hat{\mathbf{z}} = \begin{bmatrix} \hat{x}^\top & \hat{\mathbf{w}}^\top \end{bmatrix}^\top$, $x, \hat{x} \in \mathbb{R}^n$, $\mathbf{w}, \hat{\mathbf{w}} \in \mathbb{R}^{Nn}$. Given a time instant $k \in \mathbb{N}$ and a horizon length $N \in \mathbb{N}_+$, let

$$\Phi_k(\mathbf{z}) = \begin{bmatrix} A\,x + B\,u_{k-N} + L(y_{k-N} - Cx) \\ \mathbf{0} \end{bmatrix}, \tag{4.57}$$

where $\mathbf{0} \in \mathbb{R}^{Nn}$ and the observer gain L is designed such that all the eigenvalues of $A - LC$ are strictly within the unit circle. In particular, given the last pMHE iterate $\hat{\mathbf{z}}_k^{\mathrm{it}(k)}$ obtained from (4.7) at time instant k, we compute the a priori estimate for the next time instant as

$$\bar{\mathbf{z}}_{k+1} = \Phi_k\left(\hat{\mathbf{z}}_k^{\mathrm{it}(k)}\right) = \begin{bmatrix} \bar{x}_{k-N+1}^\top & 0 & \cdots & 0 \end{bmatrix}^\top \tag{4.58a}$$

with zero a priori process disturbances, i.e., $\bar{w}_i = 0$ for all $i = k - N + 1, \ldots, k$ and

$$\bar{x}_{k-N+1} = A\hat{x}_{k-N}^{\mathrm{it}(k)} + B\,u_{k-N} + L\left(y_{k-N} - C\hat{x}_{k-N}^{\mathrm{it}(k)}\right). \tag{4.58b}$$

The Bregman distance in (4.7) can be designed as

$$D_\psi(\mathbf{z}, \hat{\mathbf{z}}) = \frac{1}{2}\|x - \hat{x}\|_P^2 + \frac{1}{2}\|\mathbf{w} - \hat{\mathbf{w}}\|_{\bar{W}}^2, \tag{4.59}$$

where $P \in \mathbb{S}_{++}^n$ and $\bar{W} = \mathrm{diag}(W, \ldots, W) \in \mathbb{S}_{++}^{Nn}$ with $W \in \mathbb{S}_{++}^n$. Following the same line of arguments as in Section 3.2.4, we can show that the Bregman distance (4.59) and the a priori estimate operator (4.58) verify Assumptions 4.4 and 4.5. In addition, a sufficient condition on the weight matrix P given by the linear matrix inequality (LMI) (3.123) can be established such that D_ψ and Φ_k fulfill (4.37) in Theorem 4.1. Global exponential stability (GES) of the estimation error generated from anytime pMHE based on the Luenberger observer can be therefore formalized in the following corollary.

Corollary 4.2. *Consider system* (4.54) *with $w_k = 0$, $v_k = 0$, $k \in \mathbb{N}$, and the pMHE iteration scheme in Algorithm 4.1 with a step size in* (4.7b) *satisfying $\eta_k^i \leq \frac{\sigma_k}{L_k}$ for all $i = 0, \cdots, \mathrm{it}(k) - 1$ and $k \in \mathbb{N}$, a priori estimate operator* (4.57), *and Bregman distance* (4.59) *based on the Luenberger observer. Let Assumptions 4.1-4.3 hold. Suppose that the weight matrix $P \in \mathbb{S}_{++}^n$ in the Bregman distance fulfills the LMI*

$$(A - LC)^\top P(A - LC) - P \preceq -Q. \tag{4.60}$$

Then, the estimation error (4.13) *is GES.*

Corollary 4.2 states that, even though the pMHE iteration scheme performs only a limited number of iterations at each time instant, GES of the underlying estimation error can be guaranteed under very mild assumptions. In particular, as implied by the LMI (4.60), we require detectability instead of observability of the pair (A, C).

Remark 4.4. In principle, we can also employ the design approach based on the steady-state Kalman filter discussed in Remark 3.7. In addition, according to Remark 3.8, we can generate Φ_k from the Luenberger observer (4.57) and choose the Bregman distance (3.130), in which relaxed barrier functions allow to eliminate polytopic state constraints of the form (3.129). In order to be able to ensure GES of the underlying estimation error, the weight matrix $P \in \mathbb{S}_{++}^n$ in the Bregman distance has to fulfill the LMI (3.132).

We pointed out in Section 4.1 that there exists a number (rather small) of MHE iteration schemes in the literature which use single and multi iteration optimization algorithms and derive sufficient conditions for the stability of the resulting estimation error. For LTI systems, we mentioned the work of Alessandri and Gaggero (2017), where the established conditions are comparable to ours, especially those derived for the step size used in the underlying gradient and Newton methods. However, its design depends on the minimum eigenvalue of the observability matrix $\begin{bmatrix} C^\top & (CA)^\top & \cdots & (CA^N)^\top \end{bmatrix}^\top$ and the MHE cost function under consideration is assumed to be quadratic. In contrast to this work, our results hold for detectable linear systems and any convex cost function satisfying Assumption 4.3. Furthermore, in the next section, we additionally provide for the anytime pMHE algorithm novel performance guarantees in terms of a rigorous regret analysis.

4.3.2 Regret analysis

In this section, we study the performance of the proposed anytime pMHE iteration scheme (Algorithm 4.1) for the state estimation problem of linear systems (4.16). Recall the performance criterion of the original MHE problem (4.3) and its reformulation in (4.5), which is to minimize at each time instant k the cost function F_k. In order to characterize the overall performance of Algorithm 4.1, we investigate the accumulation of losses F_k over the considered simulation time $T \in \mathbb{N}_+$ given by

$$\sum_{k=1}^{T} F_k\left(\hat{\mathbf{z}}_k^{\mathrm{it}(k)}\right). \tag{4.61}$$

We consider the last pMHE iterate $\hat{\mathbf{z}}_k^{\mathrm{it}(k)}$ in (4.61) since it is used to generate the state estimate $\hat{x}_k$ and yields the best achieved function value if we choose the step size η_k^i in (4.7b) such that $\eta_k^i \leq \frac{\sigma_k}{L_k}$ for all $i = 0, \cdots, \mathrm{it}(k) - 1$ and $k \in \mathbb{N}_+$. Indeed, in view of Remark 4.2, this choice of the step size guarantees that the sequence $\left\{ F_k\left(\hat{\mathbf{z}}_k^0\right), \cdots, F_k\left(\hat{\mathbf{z}}_k^{\mathrm{it}(k)}\right) \right\}$ generated by the pMHE iteration procedure (4.7) at each time instant k is monotonically nonincreasing, i.e., we have that $F_k\left(\hat{\mathbf{z}}_k^0\right) \geq \cdots \geq F_k\left(\hat{\mathbf{z}}_k^{\mathrm{it}(k)}\right)$.

Our goal is to ensure that (4.61) is not much larger than the total loss $\sum_{k=1}^{T} F_k\left(\mathbf{z}_k^{\mathrm{c}}\right)$ incurred by any comparator sequence $\{\mathbf{z}_1^{\mathrm{c}}, \mathbf{z}_2^{\mathrm{c}}, \ldots, \mathbf{z}_T^{\mathrm{c}}\}$ satisfying

$$\mathbf{z}_k^{\mathrm{c}} := \begin{bmatrix} x_{k-N}^{\mathrm{c}\top} & w_{k-N}^{\mathrm{c}\top} & \cdots & w_{k-1}^{\mathrm{c}\top} \end{bmatrix}^\top \in \mathcal{S}_k. \tag{4.62}$$

In other words, we aim to obtain a low *regret*, which we define as

$$R(T) := \sum_{k=1}^{T} F_k\left(\hat{\mathbf{z}}_k^{\mathrm{it}(k)}\right) - \sum_{k=1}^{T} F_k\left(\mathbf{z}_k^{\mathrm{c}}\right). \tag{4.63}$$

For instance, by computing an upper bound on the regret, we can evaluate how well the pMHE iteration scheme performs compared to an estimation scheme that computes the optimal solutions $\{\mathbf{z}_1^{\mathrm{c}}, \mathbf{z}_2^{\mathrm{c}}, \ldots, \mathbf{z}_T^{\mathrm{c}}\}$. Thus, we measure the real-time regret of our algorithm that carries out only finitely many optimization iterations (due to limited hardware resources and/or minimum required sampling rate) relative to a comparator algorithm that gets instantaneously an optimal solution from some oracle. If Algorithm 4.1 achieves the regret bound $\mathcal{O}(\sqrt{T})$, we obtain for the average regret that $\lim_{T \to \infty} R(T)/T \to 0$. Hence, the proposed algorithm performs well (on average as well as the camparator) if the generated sequence $\left\{ \hat{\mathbf{z}}_1^{\mathrm{it}(1)}, \hat{\mathbf{z}}_2^{\mathrm{it}(2)}, \cdots, \hat{\mathbf{z}}_T^{\mathrm{it}(T)} \right\}$ leads to a sublinear regret (Mokhtari et al. (2016)).

Regret with respect to arbitrary comparator sequences

In the following, we establish upper bounds on the regret $R(T)$ generated by Algorithm 4.1 relative to an arbitrary comparator sequence. Similar to (Hall and Willett (2013)), we derive regret bounds in terms of

$$C_T(\mathbf{z}_1^c, \cdots, \mathbf{z}_T^c) := \sum_{k=1}^{T} \left\| \mathbf{z}_{k+1}^c - \Phi_k\left(\mathbf{z}_k^c\right) \right\|, \tag{4.64}$$

i.e., the deviation of the comparator sequence from the dynamics of the model-based and recursive estimator used to construct the a priori estimate and characterized by the operator Φ_k as specified by (4.10). We define the following notations:

$$\begin{aligned} M_1 &:= \max_{\mathbf{z} \in \mathcal{S}_k, k \in \mathbb{N}_+} \|\nabla \psi_k(\mathbf{z})\|, \quad M_2 := \max_{\mathbf{z} \in \mathcal{S}_k, k \in \mathbb{N}_+} \|\nabla \psi_k(\Phi_k(\mathbf{z}))\|, \quad M := M_1 + M_2, \\ G_{\mathrm{F}} &:= \max_{\mathbf{z} \in \mathcal{S}_k, k \in \mathbb{N}_+} \|\nabla F_k(\mathbf{z})\|, \quad D_{\max} := \max_{\mathbf{z}_1, \mathbf{z}_2 \in \mathcal{S}_k, k \in \mathbb{N}_+} D_{\psi_k}(\mathbf{z}_1, \mathbf{z}_2), \end{aligned} \tag{4.65}$$

where we assume that the maximum in each definition exists and is finite. In the following lemma, we derive a useful inequality satisfied by any pMHE iterate $\hat{\mathbf{z}}_k^i$ resulting from the optimizer update step (4.7b).

Lemma 4.2. *Consider system (4.54) and any comparator sequence* $\{\mathbf{z}_1^c, \mathbf{z}_2^c, \ldots, \mathbf{z}_T^c\}$ *with* $\mathbf{z}_k^c \in \mathcal{S}_k$. *Let* $\left\{ \hat{\mathbf{z}}_k^0, \hat{\mathbf{z}}_k^1, \cdots, \hat{\mathbf{z}}_k^{\mathrm{it}(k)} \right\}$ *be the sequence of pMHE iterates generated by (4.7) at a given time instant* k. *Suppose Assumptions 4.1-4.4 hold. Then,*

$$\eta_k^i \left(F_k(\hat{\mathbf{z}}_k^i) - F_k(\mathbf{z}_k^c) \right) \leq D_{\psi_k}\left(\mathbf{z}_k^c, \hat{\mathbf{z}}_k^i\right) - D_{\psi_k}\left(\mathbf{z}_k^c, \hat{\mathbf{z}}_k^{i+1}\right) + \frac{(\eta_k^i)^2}{2\sigma} \left\| \nabla F_k\left(\hat{\mathbf{z}}_k^i\right) \right\|^2. \tag{4.66}$$

Proof. The following analysis is based on the convergence proof of the MDA presented in Section 2.2.2 and derived in (Beck and Teboulle (2003)).

Since $\mathbf{z}_k^c \in \mathcal{S}_k$, we can set $\mathbf{z} = \mathbf{z}_k^c$ in the first-order optimality condition (4.22) at the minimizer $\hat{\mathbf{z}}_k^{i+1}$ to obtain

$$\left(\eta_k^i \nabla F_k\left(\hat{\mathbf{z}}_k^i\right) + \nabla \psi_k\left(\hat{\mathbf{z}}_k^{i+1}\right) - \nabla \psi_k\left(\hat{\mathbf{z}}_k^i\right) \right)^\top \left(\mathbf{z}_k^c - \hat{\mathbf{z}}_k^{i+1} \right) \geq 0. \tag{4.67}$$

Given that the cost function F_k is convex due to the linear setup and Assumption 4.3, we have that

$$\eta_k^i \left(F_k\left(\hat{\mathbf{z}}_k^i\right) - F_k\left(\mathbf{z}_k^c\right) \right) \leq \eta_k^i \nabla F_k\left(\hat{\mathbf{z}}_k^i\right)^\top \left(\hat{\mathbf{z}}_k^i - \mathbf{z}_k^c \right) = s_1 + s_2 + s_3 \tag{4.68a}$$

with the right-hand side terms

$$s_1 := \left(\nabla \psi_k\left(\hat{\mathbf{z}}_k^i\right) - \nabla \psi_k\left(\hat{\mathbf{z}}_k^{i+1}\right) - \eta_k^i \nabla F_k\left(\hat{\mathbf{z}}_k^i\right) \right)^\top \left(\mathbf{z}_k^c - \hat{\mathbf{z}}_k^{i+1} \right) \tag{4.68b}$$

$$s_2 := \left(\nabla \psi_k\left(\hat{\mathbf{z}}_k^{i+1}\right) - \nabla \psi_k\left(\hat{\mathbf{z}}_k^i\right) \right)^\top \left(\mathbf{z}_k^c - \hat{\mathbf{z}}_k^{i+1} \right) \tag{4.68c}$$

$$s_3 := \eta_k^i \nabla F_k\left(\hat{\mathbf{z}}_k^i\right)^\top \left(\hat{\mathbf{z}}_k^i - \hat{\mathbf{z}}_k^{i+1} \right). \tag{4.68d}$$

By (4.67), it holds that $s_1 \leq 0$. Using the three-points identity (2.33) in Lemma 2.1 (with $a = \hat{\mathbf{z}}_k^{i+1}$, $b = \hat{\mathbf{z}}_k^i$, $c = \mathbf{z}_k^c$) as well as the uniform lower bound on D_{ψ_k}, which arises from the strong convexity of ψ_k in Assumption 4.4, yields

$$
\begin{aligned}
s_2 &= -\left(\nabla\psi_k\left(\hat{\mathbf{z}}_k^i\right) - \nabla\psi_k\left(\hat{\mathbf{z}}_k^{i+1}\right)\right)^\top \left(\mathbf{z}_k^{\mathrm{c}} - \hat{\mathbf{z}}_k^{i+1}\right) \\
&= D_{\psi_k}\left(\mathbf{z}_k^{\mathrm{c}}, \hat{\mathbf{z}}_k^i\right) - D_{\psi_k}\left(\mathbf{z}_k^{\mathrm{c}}, \hat{\mathbf{z}}_k^{i+1}\right) - D_{\psi_k}\left(\hat{\mathbf{z}}_k^{i+1}, \hat{\mathbf{z}}_k^i\right) \\
&\leq D_{\psi_k}\left(\mathbf{z}_k^{\mathrm{c}}, \hat{\mathbf{z}}_k^i\right) - D_{\psi_k}\left(\mathbf{z}_k^{\mathrm{c}}, \hat{\mathbf{z}}_k^{i+1}\right) - \frac{\sigma}{2}\left\|\hat{\mathbf{z}}_k^{i+1} - \hat{\mathbf{z}}_k^i\right\|^2 .
\end{aligned}
\tag{4.69}
$$

Moreover, by Young's inequality,

$$
\begin{aligned}
s_3 &\leq \eta_k^i\left(\frac{\eta_k^i}{2\sigma}\left\|\nabla F_k\left(\hat{\mathbf{z}}_k^i\right)\right\|^2 + \frac{\sigma}{2\eta_k^i}\left\|\hat{\mathbf{z}}_k^{i+1} - \hat{\mathbf{z}}_k^i\right\|^2\right) \\
&= \frac{(\eta_k^i)^2}{2\sigma}\left\|\nabla F_k\left(\hat{\mathbf{z}}_k^i\right)\right\|^2 + \frac{\sigma}{2}\left\|\hat{\mathbf{z}}_k^{i+1} - \hat{\mathbf{z}}_k^i\right\|^2 .
\end{aligned}
\tag{4.70}
$$

Hence, we obtain the result of Lemma 4.2 by substituting the derived bounds (4.69) and (4.70) into (4.68). □

Under suitable conditions on the Bregman distance D_{ψ_k}, the a priori estimate operator Φ_k, and the step size η_k^i, we establish in the next lemma an upper bound on the so-called instantaneous regret (Hall and Willett (2013)), i.e., the difference between the cost function F_k evaluated at the last pMHE iterate $\hat{\mathbf{z}}_k^{\mathrm{it}(k)}$ and F_k evaluated at $\mathbf{z}_k^{\mathrm{c}}$ at a given time instant k.

Lemma 4.3. *Consider system* (4.54) *and any comparator sequence* $\{\mathbf{z}_1^{\mathrm{c}}, \mathbf{z}_2^{\mathrm{c}}, \ldots, \mathbf{z}_T^{\mathrm{c}}\}$ *with* $\mathbf{z}_k^{\mathrm{c}} \in \mathcal{S}_k$. *Let* $\left\{\hat{\mathbf{z}}_k^0, \hat{\mathbf{z}}_k^1, \cdots, \hat{\mathbf{z}}_k^{\mathrm{it}(k)}\right\}$ *be the sequence of pMHE iterates generated by* (4.7) *at a given time instant* k *with a step size satisfying* $\eta_k^i \leq \frac{\sigma_k}{L_k}$. *Suppose Assumptions 4.1-4.4 hold and that the Bregman distance* D_{ψ_k} *and the a priori estimate operator* Φ_k *satisfy*

$$
D_{\psi_k}\left(\Phi_{k-1}(\mathbf{z}), \Phi_{k-1}(\hat{\mathbf{z}})\right) - D_{\psi_{k-1}}(\mathbf{z}, \hat{\mathbf{z}}) \leq 0
\tag{4.71}
$$

for all $k \in \mathbb{N}_+$ *and* $\mathbf{z}, \hat{\mathbf{z}} \in \mathbb{R}^{(N+1)n}$. *Then,*

$$
\begin{aligned}
F_k\left(\hat{\mathbf{z}}_k^{\mathrm{it}(k)}\right) - F_k\left(\mathbf{z}_k^{\mathrm{c}}\right) \leq &\frac{1}{\sum_{i=0}^{\mathrm{it}(k)-1}\eta_k^i}\left(D_{\psi_k}\left(\mathbf{z}_k^{\mathrm{c}}, \hat{\mathbf{z}}_k^0\right) - D_{\psi_{k+1}}\left(\mathbf{z}_{k+1}^{\mathrm{c}}, \hat{\mathbf{z}}_{k+1}^0\right)\right. \\
&\left. + \frac{G_{\mathrm{F}}^2}{2\sigma}\sum_{i=0}^{\mathrm{it}(k)-1}(\eta_k^i)^2 + M\|\mathbf{z}_{k+1}^{\mathrm{c}} - \Phi_k\left(\mathbf{z}_k^{\mathrm{c}}\right)\|\right).
\end{aligned}
\tag{4.72}
$$

Proof. We invoke Lemma 4.2 and evaluate (4.66) at $i = 0$ to obtain

$$
\begin{aligned}
\eta_k^0\left(F_k(\hat{\mathbf{z}}_k^0) - F_k(\mathbf{z}_k^{\mathrm{c}})\right) &\leq D_{\psi_k}\left(\mathbf{z}_k^{\mathrm{c}}, \hat{\mathbf{z}}_k^0\right) - D_{\psi_k}\left(\mathbf{z}_k^{\mathrm{c}}, \hat{\mathbf{z}}_k^1\right) + \frac{(\eta_k^0)^2}{2\sigma}\left\|\nabla F_k\left(\hat{\mathbf{z}}_k^0\right)\right\|^2 \\
&= D_{\psi_k}\left(\mathbf{z}_k^{\mathrm{c}}, \hat{\mathbf{z}}_k^0\right) - D_{\psi_{k+1}}\left(\mathbf{z}_{k+1}^{\mathrm{c}}, \hat{\mathbf{z}}_{k+1}^0\right) + \frac{(\eta_k^0)^2}{2\sigma}\left\|\nabla F_k\left(\hat{\mathbf{z}}_k^0\right)\right\|^2 + T_1 + T_2,
\end{aligned}
\tag{4.73a}
$$

where

$$
T_1 := D_{\psi_{k+1}}\left(\Phi_k\left(\mathbf{z}_k^{\mathrm{c}}\right), \bar{\mathbf{z}}_{k+1}\right) - D_{\psi_k}\left(\mathbf{z}_k^{\mathrm{c}}, \hat{\mathbf{z}}_k^1\right)
\tag{4.73b}
$$

$$
T_2 := D_{\psi_{k+1}}\left(\mathbf{z}_{k+1}^{\mathrm{c}}, \hat{\mathbf{z}}_{k+1}^0\right) - D_{\psi_{k+1}}\left(\Phi_k\left(\mathbf{z}_k^{\mathrm{c}}\right), \bar{\mathbf{z}}_{k+1}\right).
\tag{4.73c}
$$

We can compute an upper bound for each of these two terms as follows. In view of (4.10), we have that $\bar{\mathbf{z}}_{k+1} = \Phi_k\big(\hat{\mathbf{z}}_k^{\mathrm{it}(k)}\big)$, which we plug in T_1 and then use condition (4.71) to get

$$
\begin{aligned}
T_1 &= D_{\psi_{k+1}}\Big(\Phi_k\big(\mathbf{z}_k^{\mathrm{c}}\big), \Phi_k\big(\hat{\mathbf{z}}_k^{\mathrm{it}(k)}\big)\Big) - D_{\psi_k}\big(\mathbf{z}_k^{\mathrm{c}}, \hat{\mathbf{z}}_k^{1}\big) \\
&\leq D_{\psi_k}\big(\mathbf{z}_k^{\mathrm{c}}, \hat{\mathbf{z}}_k^{\mathrm{it}(k)}\big) - D_{\psi_k}\big(\mathbf{z}_k^{\mathrm{c}}, \hat{\mathbf{z}}_k^{1}\big).
\end{aligned}
\tag{4.74}
$$

Moreover, employing (4.66) in Lemma 4.2 for each iteration step starting from $i = \mathrm{it}(k) - 1$ to $i = 2$ yields

$$
\begin{aligned}
T_1 &\leq D_{\psi_k}\big(\mathbf{z}_k^{\mathrm{c}}, \hat{\mathbf{z}}_k^{\mathrm{it}(k)-1}\big) + \frac{(\eta_k^{\mathrm{it}(k)-1})^2}{2\sigma}\Big\|\nabla F_k\big(\hat{\mathbf{z}}_k^{\mathrm{it}(k)-1}\big)\Big\|^2 + \eta_k^{\mathrm{it}(k)-1}\big(F_k\big(\mathbf{z}_k^{\mathrm{c}}\big) - F_k\big(\hat{\mathbf{z}}_k^{\mathrm{it}(k)-1}\big)\big) \\
&\quad - D_{\psi_k}\big(\mathbf{z}_k^{\mathrm{c}}, \hat{\mathbf{z}}_k^{1}\big) \\
&\leq D_{\psi_k}\big(\mathbf{z}_k^{\mathrm{c}}, \hat{\mathbf{z}}_k^{\mathrm{it}(k)-2}\big) + \frac{(\eta_k^{\mathrm{it}(k)-2})^2}{2\sigma}\Big\|\nabla F_k\big(\hat{\mathbf{z}}_k^{\mathrm{it}(k)-2}\big)\Big\|^2 + \frac{(\eta_k^{\mathrm{it}(k)-1})^2}{2\sigma}\Big\|\nabla F_k\big(\hat{\mathbf{z}}_k^{\mathrm{it}(k)-1}\big)\Big\|^2 \\
&\quad + \eta_k^{\mathrm{it}(k)-2}\big(F_k\big(\mathbf{z}_k^{\mathrm{c}}\big) - F_k\big(\hat{\mathbf{z}}_k^{\mathrm{it}(k)-2}\big)\big) + \eta_k^{\mathrm{it}(k)-1}\big(F_k\big(\mathbf{z}_k^{\mathrm{c}}\big) - F_k\big(\hat{\mathbf{z}}_k^{\mathrm{it}(k)-1}\big)\big) - D_{\psi_k}\big(\mathbf{z}_k^{\mathrm{c}}, \hat{\mathbf{z}}_k^{1}\big) \\
&\leq \cdots \\
&\leq \sum_{i=1}^{\mathrm{it}(k)-1} \frac{(\eta_k^i)^2}{2\sigma}\Big\|\nabla F_k\big(\hat{\mathbf{z}}_k^i\big)\Big\|^2 + \sum_{i=1}^{\mathrm{it}(k)-1} \eta_k^i\big(F_k(\mathbf{z}_k^{\mathrm{c}}) - F_k(\hat{\mathbf{z}}_k^i)\big).
\end{aligned}
\tag{4.75}
$$

Let us now compute an upper bound for T_2 in (4.73c). According to (4.7a), the warm start $\hat{\mathbf{z}}_{k+1}^0$ is the Bregman projection of $\bar{\mathbf{z}}_{k+1}$ onto the set $\mathcal{S}_{k+1}$, i.e., $\hat{\mathbf{z}}_{k+1}^0 = \Pi_{\mathcal{S}_{k+1}}^{\psi_{k+1}}(\bar{\mathbf{z}}_{k+1})$ in view of Definition 2.5. Since $\mathbf{z}_{k+1}^{\mathrm{c}} \in \mathcal{S}_{k+1}$, by (2.34) in Lemma 2.2, we have that

$$
0 \leq D_{\psi_{k+1}}\big(\hat{\mathbf{z}}_{k+1}^0, \bar{\mathbf{z}}_{k+1}\big) \leq D_{\psi_{k+1}}\big(\mathbf{z}_{k+1}^{\mathrm{c}}, \bar{\mathbf{z}}_{k+1}\big) - D_{\psi_{k+1}}\big(\mathbf{z}_{k+1}^{\mathrm{c}}, \hat{\mathbf{z}}_{k+1}^0\big)
\tag{4.76}
$$

for all $k \in \mathbb{N}$. Hence, $D_{\psi_{k+1}}(\mathbf{z}_{k+1}^{\mathrm{c}}, \hat{\mathbf{z}}_{k+1}^0) \leq D_{\psi_{k+1}}(\mathbf{z}_{k+1}^{\mathrm{c}}, \bar{\mathbf{z}}_{k+1})$ for all $k \in \mathbb{N}$ and we obtain by using the definition of the Bregman distance in (4.8) that

$$
\begin{aligned}
T_2 &\leq D_{\psi_{k+1}}\big(\mathbf{z}_{k+1}^{\mathrm{c}}, \bar{\mathbf{z}}_{k+1}\big) - D_{\psi_{k+1}}\big(\Phi_k\big(\mathbf{z}_k^{\mathrm{c}}\big), \bar{\mathbf{z}}_{k+1}\big) \\
&= \psi_{k+1}(\mathbf{z}_{k+1}^{\mathrm{c}}) - \nabla\psi_{k+1}(\bar{\mathbf{z}}_{k+1})^{\top}(\mathbf{z}_{k+1}^{\mathrm{c}} - \bar{\mathbf{z}}_{k+1}) - \psi_{k+1}(\Phi_k\big(\mathbf{z}_k^{\mathrm{c}}\big)) \\
&\quad + \nabla\psi_{k+1}(\bar{\mathbf{z}}_{k+1})^{\top}(\Phi_k\big(\mathbf{z}_k^{\mathrm{c}}\big) - \bar{\mathbf{z}}_{k+1}) \\
&= \psi_{k+1}(\mathbf{z}_{k+1}^{\mathrm{c}}) - \psi_{k+1}(\Phi_k\big(\mathbf{z}_k^{\mathrm{c}}\big)) - \nabla\psi_{k+1}(\bar{\mathbf{z}}_{k+1})^{\top}(\mathbf{z}_{k+1}^{\mathrm{c}} - \Phi_k\big(\mathbf{z}_k^{\mathrm{c}}\big)).
\end{aligned}
\tag{4.77}
$$

In addition, convexity of ψ_{k+1} in Assumption 4.4 and the definitions in (4.65) yield

$$
\begin{aligned}
T_2 &\leq \nabla\psi_{k+1}(\mathbf{z}_{k+1}^{\mathrm{c}})^{\top}(\mathbf{z}_{k+1}^{\mathrm{c}} - \Phi_k\big(\mathbf{z}_k^{\mathrm{c}}\big)) - \nabla\psi_{k+1}(\bar{\mathbf{z}}_{k+1})^{\top}(\mathbf{z}_{k+1}^{\mathrm{c}} - \Phi_k\big(\mathbf{z}_k^{\mathrm{c}}\big)) \\
&\leq M_1\big\|\mathbf{z}_{k+1}^{\mathrm{c}} - \Phi_k\big(\mathbf{z}_k^{\mathrm{c}}\big)\big\| + M_2\big\|\mathbf{z}_{k+1}^{\mathrm{c}} - \Phi_k\big(\mathbf{z}_k^{\mathrm{c}}\big)\big\| \\
&= M\big\|\mathbf{z}_{k+1}^{\mathrm{c}} - \Phi_k\big(\mathbf{z}_k^{\mathrm{c}}\big)\big\|.
\end{aligned}
\tag{4.78}
$$

We substitute (4.75) and (4.78) into (4.73) to get

$$
\begin{aligned}
\eta_k^0\big(F_k(\hat{\mathbf{z}}_k^0) - F_k(\mathbf{z}_k^{\mathrm{c}})\big) &\leq D_{\psi_k}\big(\mathbf{z}_k^{\mathrm{c}}, \hat{\mathbf{z}}_k^0\big) - D_{\psi_{k+1}}\big(\mathbf{z}_{k+1}^{\mathrm{c}}, \hat{\mathbf{z}}_{k+1}^0\big) + \frac{(\eta_k^0)^2}{2\sigma}\Big\|\nabla F_k(\hat{\mathbf{z}}_k^0)\Big\|^2 \\
&\quad + \sum_{i=1}^{\mathrm{it}(k)-1} \frac{(\eta_k^i)^2}{2\sigma}\Big\|\nabla F_k\big(\hat{\mathbf{z}}_k^i\big)\Big\|^2 + \sum_{i=1}^{\mathrm{it}(k)-1} \eta_k^i\big(F_k(\mathbf{z}_k^{\mathrm{c}}) - F_k(\hat{\mathbf{z}}_k^i)\big) \\
&\quad + M\big\|\mathbf{z}_{k+1}^{\mathrm{c}} - \Phi_k\big(\mathbf{z}_k^{\mathrm{c}}\big)\big\|.
\end{aligned}
\tag{4.79}
$$

Rearranging the above inequality yields

$$\sum_{i=0}^{\text{it}(k)-1} \eta_k^i (F_k(\hat{\mathbf{z}}_k^i) - F_k(\mathbf{z}_k^c)) \le D_{\psi_k}\left(\mathbf{z}_k^c, \hat{\mathbf{z}}_k^0\right) - D_{\psi_{k+1}}\left(\mathbf{z}_{k+1}^c, \hat{\mathbf{z}}_{k+1}^0\right) \tag{4.80}$$

$$+ \sum_{i=0}^{\text{it}(k)-1} \frac{(\eta_k^i)^2}{2\sigma} \left\|\nabla F_k\left(\hat{\mathbf{z}}_k^i\right)\right\|^2 + M \left\|\mathbf{z}_{k+1}^c - \Phi_k\left(\mathbf{z}_k^c\right)\right\|.$$

Since the condition on the step size $\eta_k^i \le \frac{\sigma_k}{L_k}$ implies by Remark 4.2 that the sequence of function values of the pMHE iterates generated by (4.7) is nonincreasing, it holds that $F_k\left(\hat{\mathbf{z}}_k^{\text{it}(k)}\right) \le F_k\left(\hat{\mathbf{z}}_k^i\right)$ for all $i = 0, \dots, \text{it}(k) - 1$ and $k \in \mathbb{N}_+$. Hence,

$$F_k\left(\hat{\mathbf{z}}_k^{\text{it}(k)}\right) \sum_{i=0}^{\text{it}(k)-1} \eta_k^i \le \sum_{i=0}^{\text{it}(k)-1} \eta_k^i F_k\left(\hat{\mathbf{z}}_k^i\right) \tag{4.81}$$

and we obtain in view of the definition of G_{F} in (4.65) that

$$\left(F_k\left(\hat{\mathbf{z}}_k^{\text{it}(k)}\right) - F_k\left(\mathbf{z}_k^c\right)\right) \sum_{i=0}^{\text{it}(k)-1} \eta_k^i \le D_{\psi_k}\left(\mathbf{z}_k^c, \hat{\mathbf{z}}_k^0\right) - D_{\psi_{k+1}}\left(\mathbf{z}_{k+1}^c, \hat{\mathbf{z}}_{k+1}^0\right) \tag{4.82}$$

$$+ \sum_{i=0}^{\text{it}(k)-1} \frac{(\eta_k^i)^2}{2\sigma} G_{\text{F}}^2 + M \left\|\mathbf{z}_{k+1}^c - \Phi_k\left(\mathbf{z}_k^c\right)\right\|.$$

Dividing the last inequality by $\sum_{i=0}^{\text{it}(k)-1} \eta_k^i > 0$ yields the desired result (4.72). $\qquad\square$

Remark 4.5. Recall that the pMHE iteration procedure (4.7) uses the MDA, whose basic convergence results are presented in Section 2.2.2, as an optimization algorithm. Interestingly, the derived instantaneous regret in Lemma 4.3 reveals how the additional prediction step given by the a priori estimate operator Φ_k can alter the convergence result of the MDA presented in Theorem 2.4. More specifically, we can see that the upper bound in (2.55) is very similar to the upper bound in (4.72) except for the term $\|\mathbf{z}_{k+1}^c - \Phi_k(\mathbf{z}_k^c)\|$. It is also worth noting that, if the comparator sequence is given by the true state, which yields by Assumption 4.5 that $\|\mathbf{z}_{k+1}^c - \Phi_k(\mathbf{z}_k^c)\| = 0$, and if we exploit the fact that $D_{\psi_{k+1}}(\mathbf{z}_{k+1}^c, \hat{\mathbf{z}}_{k+1}^0)$ in (4.72) is nonnegative, we obtain exactly the upper bound in (2.55).

We are in position to state our first main result regarding the regret of Algorithm 4.1.

Theorem 4.2. *Consider any comparator sequence $\{\mathbf{z}_1^c, \mathbf{z}_2^c, \dots, \mathbf{z}_T^c\}$ with $\mathbf{z}_k^c \in \mathcal{S}_k$ and the pMHE iteration scheme in Algorithm 4.1 for system (4.54) with a step size in (4.7b) satisfying $\eta_k^i \le \frac{\sigma_k}{L_k}$ for all $i = 0, \dots, \text{it}(k) - 1$ and $k \in \mathbb{N}_+$ as well as*

$$\sum_{i=0}^{\text{it}(k+1)-1} \eta_{k+1}^i \le \sum_{i=0}^{\text{it}(k)-1} \eta_k^i. \tag{4.83}$$

Let Assumptions 4.1-4.4 hold. Suppose that the Bregman distance D_{ψ_k} and the a priori estimate operator Φ_k satisfy

$$D_{\psi_k}\left(\Phi_{k-1}(\mathbf{z}), \Phi_{k-1}(\hat{\mathbf{z}})\right) - D_{\psi_{k-1}}(\mathbf{z}, \hat{\mathbf{z}}) \le 0 \tag{4.84}$$

for all $k \in \mathbb{N}_+$ and $\mathbf{z}, \hat{\mathbf{z}} \in \mathbb{R}^{(N+1)n}$. Then, Algorithm 4.1 gives the following regret bound

$$R(T) \le \frac{D_{\max}}{\sum_{i=0}^{\text{it}(T)-1} \eta_T^i} + \frac{G_{\text{F}}^2}{2\sigma} \sum_{k=1}^T \frac{\sum_{i=0}^{\text{it}(k)-1} (\eta_k^i)^2}{\sum_{i=0}^{\text{it}(k)-1} \eta_k^i} + \frac{M}{\sum_{i=0}^{\text{it}(T)-1} \eta_T^i} \sum_{k=1}^T \left\|\mathbf{z}_{k+1}^c - \Phi_k(\mathbf{z}_k^c)\right\|. \tag{4.85}$$

Proof. The proof is similar to the proof of (Hall and Willett, 2013, Theorem 4), in which a regret upper bound is derived for the dynamic mirror descent in the context of online convex optimization. For ease of notation, we employ $\sum \eta_k^i$ to refer to the sum of all the step sizes used within the time instant k, i.e., to $\sum_{i=0}^{\mathrm{it}(k)-1} \eta_k^i$.

By Lemma 4.3, the instantaneous regret bound (4.72) holds true. By summing (4.72) over $k = 1, \cdots, T$, we obtain

$$R(T) = \sum_{k=1}^{T} F_k\left(\hat{\mathbf{z}}_k^{\mathrm{it}(k)}\right) - \sum_{k=1}^{T} F_k\left(\mathbf{z}_k^{\mathrm{c}}\right) \tag{4.86}$$

$$\leq \sum_{k=1}^{T} \frac{1}{\sum \eta_k^i} \left(D_{\psi_k}\left(\mathbf{z}_k^{\mathrm{c}}, \hat{\mathbf{z}}_k^0\right) - D_{\psi_{k+1}}\left(\mathbf{z}_{k+1}^{\mathrm{c}}, \hat{\mathbf{z}}_{k+1}^0\right) + \frac{G_{\mathrm{F}}^2}{2\sigma}\sum(\eta_k^i)^2 + M\|\mathbf{z}_{k+1}^{\mathrm{c}} - \Phi_k\left(\mathbf{z}_k^{\mathrm{c}}\right)\| \right).$$

Since the step size satisfies condition (4.83), i.e., $\sum \eta_{k+1}^i \leq \sum \eta_k^i$, it holds that $\frac{1}{\sum \eta_{k+1}^i} \geq \frac{1}{\sum \eta_k^i}$. Exploiting the last inequality and the definition of $D_{\max}$ in (4.65) yields

$$\sum_{k=1}^{T} \frac{1}{\sum \eta_k^i} \left(D_{\psi_k}\left(\mathbf{z}_k^{\mathrm{c}}, \hat{\mathbf{z}}_k^0\right) - D_{\psi_{k+1}}\left(\mathbf{z}_{k+1}^{\mathrm{c}}, \hat{\mathbf{z}}_{k+1}^0\right) \right) \tag{4.87}$$

$$= \frac{D_{\psi_1}\left(\mathbf{z}_1^{\mathrm{c}}, \hat{\mathbf{z}}_1^0\right)}{\sum \eta_1^i} - \frac{D_{\psi_{T+1}}\left(\mathbf{z}_{T+1}^{\mathrm{c}}, \hat{\mathbf{z}}_{T+1}^0\right)}{\sum \eta_T^i} + D_{\psi_2}\left(\mathbf{z}_2^{\mathrm{c}}, \hat{\mathbf{z}}_2^0\right)\left(\frac{1}{\sum \eta_2^i} - \frac{1}{\sum \eta_1^i}\right) + \cdots$$

$$+ D_{\psi_T}\left(\mathbf{z}_T^{\mathrm{c}}, \hat{\mathbf{z}}_T^0\right)\left(\frac{1}{\sum \eta_T^i} - \frac{1}{\sum \eta_{T-1}^i}\right)$$

$$\leq \frac{D_{\max}}{\sum \eta_1^i} + D_{\max}\left(\sum_{k=1}^{T-1} \frac{1}{\sum \eta_{k+1}^i} - \frac{1}{\sum \eta_k^i}\right) = \frac{D_{\max}}{\sum \eta_T^i}.$$

Moreover, since $\frac{1}{\sum \eta_1^i} \leq \cdots \leq \frac{1}{\sum \eta_T^i}$, it holds that

$$\sum_{k=1}^{T} \frac{M}{\sum \eta_k^i} \|\mathbf{z}_{k+1}^{\mathrm{c}} - \Phi_k\left(\mathbf{z}_k^{\mathrm{c}}\right)\| \leq \frac{M}{\sum \eta_T^i} \sum_{k=1}^{T} \|\mathbf{z}_{k+1}^{\mathrm{c}} - \Phi_k\left(\mathbf{z}_k^{\mathrm{c}}\right)\|. \tag{4.88}$$

Hence, substituting the above two upper bounds into (4.86) yields

$$R(T) \leq \frac{D_{\max}}{\sum \eta_T^i} + \frac{G_{\mathrm{F}}^2}{2\sigma} \sum_{k=1}^{T} \frac{\sum(\eta_k^i)^2}{\sum \eta_k^i} + \frac{M}{\sum \eta_T^i} \sum_{k=1}^{T} \|\mathbf{z}_{k+1}^{\mathrm{c}} - \Phi_k\left(\mathbf{z}_k^{\mathrm{c}}\right)\|, \tag{4.89}$$

finishing the proof. □

Notice that D_{ψ_k} and Φ_k satisfy condition (4.84) in Theorem 4.2 if we choose for instance the design approach based on the LTV Kalman filter. More specifically, (4.84) can be satisfied if we choose the a priori estimate operator (4.50) and the Bregman distance (4.52) as discussed in Section 4.3.1, which are provably shown to verify condition (4.37) in Theorem 4.1. Obviously, if (4.37) holds, so does (4.84). For LTI systems, we can consider the design approach based on the Luenberger observer, i.e., with the a priori estimate operator (4.58) and the quadratic Bregman distance (4.59), where the weight matrix $P \in \mathbb{S}_{++}^n$ is designed to fulfill the LMI (4.60).

We discuss in the following an important implication of Theorem 4.2. If we execute a single iteration per time instant, i.e., set $\mathrm{it}(k) = 1$ for all $k \in \mathbb{N}_+$, we get $\sum_{i=0}^{\mathrm{it}(k)-1} \eta_k^i = \eta_k^0 =: \eta_k$ in

(4.85). In this case, the condition (4.83) on the step size becomes $\eta_{k+1} \leq \eta_k$ and the regret bound (4.85) is as follows

$$R(T) \leq \frac{D_{\max}}{\eta_T} + \frac{G_{\mathrm{F}}^2}{2\sigma} \sum_{k=1}^{T} \eta_k + \frac{M}{\eta_T} \sum_{k=1}^{T} \left\| \mathbf{z}_{k+1}^{\mathrm{c}} - \Phi_k\left(\mathbf{z}_k^{\mathrm{c}}\right) \right\|. \tag{4.90}$$

This regret bound is very similar to the bound derived for the dynamic mirror descent (Hall and Willett (2013)) in the context of online convex optimization. Moreover, if we fix the simulation time $T \in \mathbb{N}_+$ and choose $\eta_k = \frac{1}{\sqrt{T}}$ (which fulfills $\eta_{k+1} \leq \eta_k$), Algorithm 4.1 with a single optimization iteration per time instant yields

$$R(T) \leq \sqrt{T} \left(D_{\max} + \frac{G_{\mathrm{F}}^2}{2\sigma} + M \sum_{k=1}^{T} \left\| \mathbf{z}_{k+1}^{\mathrm{c}} - \Phi_k\left(\mathbf{z}_k^{\mathrm{c}}\right) \right\| \right) \tag{4.91}$$

and achieves therefore a regret bound $\mathcal{O}\big(\sqrt{T}(1 + C_T)\big)$, where C_T is defined in (4.64). Furthermore, if the comparator sequence is given by the true system state, i.e., $\mathbf{z}_k^{\mathrm{c}} = \mathbf{z}_k$ for all $k \in \mathbb{N}_+$, and if we choose for instance the Kalman filter to construct the a priori estimate operator Φ_k, it holds that $C_T(\mathbf{z}_1^{\mathrm{c}}, \cdots, \mathbf{z}_T^{\mathrm{c}}) = \sum_{k=1}^{T} \left\| \mathbf{z}_{k+1}^{\mathrm{c}} - \Phi_k\left(\mathbf{z}_k^{\mathrm{c}}\right) \right\| = 0$. Thus, Algorithm 4.1 achieves in this case the desired regret bound $\mathcal{O}(\sqrt{T})$ and the average regret $R(T)/T$ tends to zero when T goes to infinity.

In our second main result, we specify conditions under which Algorithm 4.1 ensures GUES of the underlying estimation error and attains the regret bound $\mathcal{O}\big(\sqrt{T}(1 + C_T)\big)$. For this, we need the following two lemmas, which provide tighter bounds than the bounds established in Lemmas 4.2 and 4.3 thanks to an appropriate choice of the step size which does not require T to be fixed a priori.

Lemma 4.4. *Consider system* (4.54) *and any comparator sequence* $\{\mathbf{z}_1^{\mathrm{c}}, \mathbf{z}_2^{\mathrm{c}}, \ldots, \mathbf{z}_T^{\mathrm{c}}\}$ *with* $\mathbf{z}_k^{\mathrm{c}} \in \mathcal{S}_k$. *Let* $\left\{ \hat{\mathbf{z}}_k^0, \hat{\mathbf{z}}_k^1, \cdots, \hat{\mathbf{z}}_k^{\mathrm{it}(k)} \right\}$ *be the sequence of pMHE iterates generated by* (4.7) *at a given time instant* $k \in \mathbb{N}_+$ *with a step size* $\eta_k^i = \frac{\sigma}{L_{\mathrm{F}}} \frac{1}{\sqrt{k}}$. *Suppose Assumptions 4.1-4.4 hold. Then,*

$$\eta_k^i \left(F_k\left(\hat{\mathbf{z}}_k^{i+1}\right) - F_k\left(\mathbf{z}_k^{\mathrm{c}}\right) \right) \leq D_{\psi_k}\left(\mathbf{z}_k^{\mathrm{c}}, \hat{\mathbf{z}}_k^i\right) - D_{\psi_k}\left(\mathbf{z}_k^{\mathrm{c}}, \hat{\mathbf{z}}_k^{i+1}\right). \tag{4.92}$$

Proof. Since the gradient of F_k is Lipschitz continuous by Assumption 4.3, and given the associated uniform Lipschitz constant L_{F} as well as the uniform strong convexity constant σ in Assumption 4.4, by (4.27), we have

$$\eta_k^i F_k\left(\mathbf{z}\right) \geq \eta_k^i F_k\left(\hat{\mathbf{z}}_k^{i+1}\right) + \frac{1}{2}\left(\sigma - \eta_k^i L_{\mathrm{F}}\right) \left\| \hat{\mathbf{z}}_k^{i+1} - \hat{\mathbf{z}}_k^i \right\|^2 + D_{\psi_k}\left(\mathbf{z}, \hat{\mathbf{z}}_k^{i+1}\right) - D_{\psi_k}\left(\mathbf{z}, \hat{\mathbf{z}}_k^i\right) \tag{4.93}$$

for all $\mathbf{z} \in \mathcal{S}_k$. Moreover, the step size $\eta_k^i = \frac{\sigma}{L_{\mathrm{F}}} \frac{1}{\sqrt{k}}$ yields that $\sigma - \eta_k^i L_{\mathrm{F}} = \sigma - \frac{\sigma}{\sqrt{k}} \geq 0$ for all $k \in \mathbb{N}_+$. Hence, we obtain in (4.93) that

$$\eta_k^i \left(F_k\left(\mathbf{z}\right) - F_k\left(\hat{\mathbf{z}}_k^{i+1}\right) \right) \geq D_{\psi_k}\left(\mathbf{z}, \hat{\mathbf{z}}_k^{i+1}\right) - D_{\psi_k}\left(\mathbf{z}, \hat{\mathbf{z}}_k^i\right). \tag{4.94}$$

Thus, plugging $\mathbf{z} = \mathbf{z}_k^{\mathrm{c}} \in \mathcal{S}_k$ in the above inequality yields the desired result (4.92). $\qquad\square$

Lemma 4.5. *Consider system* (4.54) *and any comparator sequence* $\{\mathbf{z}_1^{\mathrm{c}}, \mathbf{z}_2^{\mathrm{c}}, \ldots, \mathbf{z}_T^{\mathrm{c}}\}$ *with* $\mathbf{z}_k^{\mathrm{c}} \in \mathcal{S}_k$. *Let* $\left\{ \hat{\mathbf{z}}_k^0, \hat{\mathbf{z}}_k^1, \cdots, \hat{\mathbf{z}}_k^{\mathrm{it}(k)} \right\}$ *be the sequence of pMHE iterates generated by* (4.7) *at a* *given time instant $k \in \mathbb{N}_+$ with a step size $\eta_k^i = \frac{\sigma}{L_{\mathrm{F}}} \frac{1}{\sqrt{k}}$. Suppose Assumptions 4.1-4.4 hold and that the Bregman distance D_{ψ_k} and the a priori estimate operator Φ_k satisfy*

$$D_{\psi_k}\left(\Phi_{k-1}(\mathbf{z}), \Phi_{k-1}(\hat{\mathbf{z}})\right) - D_{\psi_{k-1}}(\mathbf{z}, \hat{\mathbf{z}}) \leq 0 \qquad (4.95)$$

for all $k \in \mathbb{N}_+$ and $\mathbf{z}, \hat{\mathbf{z}} \in \mathbb{R}^{(N+1)n}$. Then,

$$F_k\left(\hat{\mathbf{z}}_k^{\mathrm{it}(k)}\right) - F_k\left(\mathbf{z}_k^{\mathrm{c}}\right) \leq \frac{1}{\sum_{i=0}^{\mathrm{it}(k)-1} \eta_k^i} \left(D_{\psi_k}\left(\mathbf{z}_k^{\mathrm{c}}, \hat{\mathbf{z}}_k^0\right) - D_{\psi_{k+1}}\left(\mathbf{z}_{k+1}^{\mathrm{c}}, \hat{\mathbf{z}}_{k+1}^0\right) + M \|\mathbf{z}_{k+1}^{\mathrm{c}} - \Phi_k\left(\mathbf{z}_k^{\mathrm{c}}\right)\| \right).$$
$$(4.96)$$

Proof. Evaluating (4.92) in Lemma 4.4 at $i = 0$ yields

$$\eta_k^0 \left(F_k(\hat{\mathbf{z}}_k^1) - F_k(\mathbf{z}_k^{\mathrm{c}}) \right) \leq D_{\psi_k}(\mathbf{z}_k^{\mathrm{c}}, \hat{\mathbf{z}}_k^0) - D_{\psi_k}(\mathbf{z}_k^{\mathrm{c}}, \hat{\mathbf{z}}_k^1) \qquad (4.97a)$$
$$= D_{\psi_k}\left(\mathbf{z}_k^{\mathrm{c}}, \hat{\mathbf{z}}_k^0\right) - D_{\psi_{k+1}}\left(\mathbf{z}_{k+1}^{\mathrm{c}}, \hat{\mathbf{z}}_{k+1}^0\right) + T_1 + T_2,$$

where

$$T_1 := D_{\psi_{k+1}}\left(\Phi_k\left(\mathbf{z}_k^{\mathrm{c}}\right), \bar{\mathbf{z}}_{k+1}\right) - D_{\psi_k}\left(\mathbf{z}_k^{\mathrm{c}}, \hat{\mathbf{z}}_k^1\right) \qquad (4.97b)$$

$$T_2 := D_{\psi_{k+1}}\left(\mathbf{z}_{k+1}^{\mathrm{c}}, \hat{\mathbf{z}}_{k+1}^0\right) - D_{\psi_{k+1}}\left(\Phi_k\left(\mathbf{z}_k^{\mathrm{c}}\right), \bar{\mathbf{z}}_{k+1}\right). \qquad (4.97c)$$

Similar to the proof of Lemma 4.3, we can compute an upper bound for T_1 and T_2 as follows. Using $\bar{\mathbf{z}}_{k+1} = \Phi_k\left(\hat{\mathbf{z}}_k^{\mathrm{it}(k)}\right)$ in (4.10) as well as condition (4.95) leads to

$$T_1 \leq D_{\psi_k}\left(\mathbf{z}_k^{\mathrm{c}}, \hat{\mathbf{z}}_k^{\mathrm{it}(k)}\right) - D_{\psi_k}\left(\mathbf{z}_k^{\mathrm{c}}, \hat{\mathbf{z}}_k^1\right). \qquad (4.98)$$

Moreover, similar to (4.75), we employ (4.92) in Lemma 4.4 for each iteration step starting from $i = \mathrm{it}(k) - 1$ to $i = 2$ and obtain

$$T_1 \leq D_{\psi_k}\left(\mathbf{z}_k^{\mathrm{c}}, \hat{\mathbf{z}}_k^{\mathrm{it}(k)-1}\right) + \eta_k^{\mathrm{it}(k)-1}\left(F_k(\mathbf{z}_k^{\mathrm{c}}) - F_k\left(\hat{\mathbf{z}}_k^{\mathrm{it}(k)}\right)\right) - D_{\psi_k}\left(\mathbf{z}_k^{\mathrm{c}}, \hat{\mathbf{z}}_k^1\right) \qquad (4.99)$$

$$\leq D_{\psi_k}\left(\mathbf{z}_k^{\mathrm{c}}, \hat{\mathbf{z}}_k^{\mathrm{it}(k)-2}\right) + \sum_{i=\mathrm{it}(k)-1}^{\mathrm{it}(k)} \eta_k^{i-1}\left(F_k(\mathbf{z}_k^{\mathrm{c}}) - F_k(\hat{\mathbf{z}}_k^i)\right) - D_{\psi_k}\left(\mathbf{z}_k^{\mathrm{c}}, \hat{\mathbf{z}}_k^1\right)$$

$$\leq \dots$$

$$\leq \sum_{i=2}^{\mathrm{it}(k)} \eta_k^{i-1}\left(F_k(\mathbf{z}_k^{\mathrm{c}}) - F_k(\hat{\mathbf{z}}_k^i)\right).$$

Notice that an upper bound on T_2, considered also in (4.73c), is derived in (4.78). Hence, substituting (4.78) and the obtained upper bound on T_1 into (4.97) leads to

$$\eta_k^0\left(F_k(\hat{\mathbf{z}}_k^1) - F_k(\mathbf{z}_k^{\mathrm{c}})\right) \leq D_{\psi_k}\left(\mathbf{z}_k^{\mathrm{c}}, \hat{\mathbf{z}}_k^0\right) - D_{\psi_{k+1}}\left(\mathbf{z}_{k+1}^{\mathrm{c}}, \hat{\mathbf{z}}_{k+1}^0\right) + \sum_{i=2}^{\mathrm{it}(k)} \eta_k^{i-1}\left(F_k(\mathbf{z}_k^{\mathrm{c}}) - F_k(\hat{\mathbf{z}}_k^i)\right)$$
$$+ M\|\mathbf{z}_{k+1}^{\mathrm{c}} - \Phi_k\left(\mathbf{z}_k^{\mathrm{c}}\right)\|, \qquad (4.100)$$

and equivalently to

$$\sum_{i=1}^{\mathrm{it}(k)} \eta_k^{i-1}(F_k(\hat{\mathbf{z}}_k^i) - F_k(\mathbf{z}_k^{\mathrm{c}})) \leq D_{\psi_k}\left(\mathbf{z}_k^{\mathrm{c}}, \hat{\mathbf{z}}_k^0\right) - D_{\psi_{k+1}}\left(\mathbf{z}_{k+1}^{\mathrm{c}}, \hat{\mathbf{z}}_{k+1}^0\right) + M\|\mathbf{z}_{k+1}^{\mathrm{c}} - \Phi_k\left(\mathbf{z}_k^{\mathrm{c}}\right)\|.$$
$$(4.101)$$

Since the step size given by $\eta_k^i = \frac{\sigma}{L_F} \frac{1}{\sqrt{k}}$ satisfies $\eta_k^i \leq \frac{\sigma_k}{L_k}$, Remark 4.2 indicates that the sequence of function values of the pMHE iterates generated by (4.7) is nonincreasing. Hence, $F_k\left(\hat{\mathbf{z}}_k^{\mathrm{it}(k)}\right) \leq F_k\left(\hat{\mathbf{z}}_k^i\right)$ for all $i = 0, \ldots, \mathrm{it}(k) - 1$ and $k \in \mathbb{N}_+$. We obtain

$$F_k\left(\hat{\mathbf{z}}_k^{\mathrm{it}(k)}\right) \sum_{i=0}^{\mathrm{it}(k)-1} \eta_k^i = F_k\left(\hat{\mathbf{z}}_k^{\mathrm{it}(k)}\right) \sum_{i=1}^{\mathrm{it}(k)} \eta_k^{i-1} \leq \sum_{i=1}^{\mathrm{it}(k)} \eta_k^{i-1} F_k\left(\hat{\mathbf{z}}_k^i\right) \tag{4.102}$$

and therefore arrive at

$$\left(F_k\left(\hat{\mathbf{z}}_k^{\mathrm{it}(k)}\right) - F_k\left(\mathbf{z}_k^{\mathrm{c}}\right)\right) \sum_{i=0}^{\mathrm{it}(k)-1} \eta_k^i \leq D_{\psi_k}\left(\mathbf{z}_k^{\mathrm{c}}, \hat{\mathbf{z}}_k^0\right) - D_{\psi_{k+1}}\left(\mathbf{z}_{k+1}^{\mathrm{c}}, \hat{\mathbf{z}}_{k+1}^0\right) + M\|\mathbf{z}_{k+1}^{\mathrm{c}} - \Phi_k\left(\mathbf{z}_k^{\mathrm{c}}\right)\|. \tag{4.103}$$

Dividing the above inequality by $\sum_{i=0}^{\mathrm{it}(k)-1} \eta_k^i > 0$ completes the proof. $\qquad\square$

Our second main result summarizes for linear systems the nominal stability and regret properties of Algorithm 4.1.

Theorem 4.3. *Consider system* (4.16) *with* $w_k = 0, v_k = 0, k \in \mathbb{N}$, *and the pMHE iteration scheme in Algorithm 4.1 with* $\mathrm{it}(k+1) \leq \mathrm{it}(k)$ *and a step size in* (4.7b) *satisfying*

$$\eta_k^i = \frac{\sigma}{L_{\mathrm{F}}} \frac{1}{\sqrt{k}} \tag{4.104}$$

for all $i = 0, \ldots, \mathrm{it}(k) - 1$ *and* $k \in \mathbb{N}_+$. *Let Assumptions 4.1-4.5 hold. Suppose that the Bregman distance* D_{ψ_k} *and the a priori estimate operator* Φ_k *satisfy*

$$D_{\psi_k}\left(\Phi_{k-1}(\mathbf{z}), \Phi_{k-1}(\hat{\mathbf{z}})\right) - D_{\psi_{k-1}}(\mathbf{z}, \hat{\mathbf{z}}) \leq -c \|\mathbf{z} - \hat{\mathbf{z}}\|^2 \tag{4.105}$$

for all $k \in \mathbb{N}_+$ *and* $\mathbf{z}, \hat{\mathbf{z}} \in \mathbb{R}^{(N+1)n}$. *Then, the estimation error is GUES.*
Moreover, consider any comparator sequence $\{\mathbf{z}_1^{\mathrm{c}}, \mathbf{z}_2^{\mathrm{c}}, \ldots, \mathbf{z}_T^{\mathrm{c}}\}$ *with* $\mathbf{z}_k^{\mathrm{c}} \in \mathcal{S}_k$. *Then, Algorithm 4.1 gives the following regret bound*

$$R(T) \leq \frac{\sqrt{T}}{\mathrm{it}(T)} \frac{L_{\mathrm{F}}}{\sigma} \left(D_{\max} + M \sum_{k=1}^{T} \|\mathbf{z}_{k+1}^{\mathrm{c}} - \Phi_k\left(\mathbf{z}_k^{\mathrm{c}}\right)\|\right). \tag{4.106}$$

Proof. The first statement of the theorem, i.e., GUES of the underlying estimation error, follows directly from Theorem 4.1. This is due to the fact that the step size η_k^i in (4.104) satisfies $\eta_k^i \leq \frac{\sigma_k}{L_k}$ in (4.36) due to Assumptions 4.3 and 4.4.
Concerning the second statement of the theorem, by Lemma 4.5, i.e., the bound on the instantaneous regret in (4.96), we have

$$R(T) = \sum_{k=1}^{T} F_k\left(\hat{\mathbf{z}}_k^{\mathrm{it}(k)}\right) - F_k\left(\mathbf{z}_k^{\mathrm{c}}\right) \tag{4.107}$$

$$\leq \sum_{k=1}^{T} \frac{1}{\sum_{i=0}^{\mathrm{it}(k)-1} \eta_k^i} \left(D_{\psi_k}\left(\mathbf{z}_k^{\mathrm{c}}, \hat{\mathbf{z}}_k^0\right) - D_{\psi_{k+1}}\left(\mathbf{z}_{k+1}^{\mathrm{c}}, \hat{\mathbf{z}}_{k+1}^0\right) + M\|\mathbf{z}_{k+1}^{\mathrm{c}} - \Phi_k\left(\mathbf{z}_k^{\mathrm{c}}\right)\|\right).$$

Since $\mathrm{it}(k+1) \leq \mathrm{it}(k)$ and $\eta_k^i = \frac{\sigma}{L_F} \frac{1}{\sqrt{k}}$, it holds that

$$\sum_{i=0}^{\mathrm{it}(k)-1} \eta_k^i = \frac{\sigma}{L_F} \frac{\mathrm{it}(k)}{\sqrt{k}} \geq \frac{\sigma}{L_F} \frac{\mathrm{it}(k+1)}{\sqrt{k+1}} = \sum_{i=0}^{\mathrm{it}(k+1)-1} \eta_{k+1}^i. \tag{4.108}$$

Hence, condition (4.83) in Theorem 4.2 holds and as a consequence, we can derive an upper bound analogous to (4.87) to obtain

$$\sum_{k=1}^{T} \frac{1}{\sum_{i=0}^{\mathrm{it}(k)-1} \eta_k^i} \left(D_{\psi_k} \left(\mathbf{z}_k^c, \hat{\mathbf{z}}_k^0 \right) - D_{\psi_{k+1}} \left(\mathbf{z}_{k+1}^c, \hat{\mathbf{z}}_{k+1}^0 \right) \right) \leq \frac{D_{\max}}{\sum_{i=0}^{\mathrm{it}(T)-1} \eta_T^i} \tag{4.109}$$

$$= \frac{D_{\max}}{\sum_{i=0}^{\mathrm{it}(T)-1} \frac{\sigma}{L_F} \frac{1}{\sqrt{T}}} = \frac{\sqrt{T}}{\mathrm{it}(T)} \frac{L_F}{\sigma} D_{\max}.$$

Moreover, due to (4.108), we have that

$$\sum_{k=1}^{T} \frac{M}{\sum_{i=0}^{\mathrm{it}(k)-1} \eta_k^i} \left\| \mathbf{z}_{k+1}^c - \Phi_k \left(\mathbf{z}_k^c \right) \right\| \leq \frac{M}{\sum_{i=0}^{\mathrm{it}(T)-1} \eta_T^i} \sum_{k=1}^{T} \left\| \mathbf{z}_{k+1}^c - \Phi_k \left(\mathbf{z}_k^c \right) \right\| \tag{4.110}$$

$$= \frac{\sqrt{T}}{\mathrm{it}(T)} \frac{L_F}{\sigma} M \sum_{k=1}^{T} \left\| \mathbf{z}_{k+1}^c - \Phi_k \left(\mathbf{z}_k^c \right) \right\|.$$

Substituting (4.109) and (4.110) in (4.107) yields the desired regret bound (4.106). $\qquad\square$

A direct consequence of Theorem 4.3 is that if $C_T(\mathbf{z}_1^c, \cdots, \mathbf{z}_T^c) = \sum_{k=1}^{T} \left\| \mathbf{z}_{k+1}^c - \Phi_k \left(\mathbf{z}_k^c \right) \right\| = 0$, Algorithm 4.1 exhibits a sublinear regret bound $\mathcal{O}(\sqrt{T})$ and $\lim_{T \to \infty} R(T)/T \to 0$. Moreover, we can see that fixing the number of optimization algorithm iterations $\mathrm{it}(k) = \mathrm{it}(k+1) =: \mathrm{it}$ and increasing it lead to a smaller regret bound. In fact, if the comparator sequence $\{\mathbf{z}_1^c, \mathbf{z}_2^c, \ldots, \mathbf{z}_T^c\}$ follows the dynamics described by the a priori estimate operator Φ_k closely, and if we let $\mathrm{it} \to \infty$, then the upper bound (4.106) in Theorem 4.3 vanishes and we obtain an algorithm with zero regret, i.e., $\lim_{\mathrm{it} \to \infty} R(T) \to 0$.

We also remark that the condition $\mathrm{it}(k+1) \leq \mathrm{it}(k)$ requires that we employ a smaller or equal number of optimization iterations each time we receive a new measurement. This condition is in line with the intuitive observation that it is preferable to execute more iterations at the beginning of the pMHE iteration scheme, since our regret measure is aggregated over time and thus memorizes initially poor estimates.

Regret with respect to exponentially stable comparator sequences

As we mentioned before, in general, there is no requirement that the comparator sequence converges to the true state. This being said, it is reasonable to restrict the class of comparator sequences to sequences that converge exponentially fast to the true state. We study this case by imposing the following additional assumption.

Assumption 4.7 (Exponentially stable comparator sequence). *The comparator sequence* $\{\mathbf{z}_1^c, \mathbf{z}_2^c, \ldots, \mathbf{z}_T^c\}$ *with initial guess* $\mathbf{z}_0^c$ *is generated from a state estimator that yields GUES estimation error dynamics. More specifically, there exist positive constants* $\alpha_c \in \mathbb{R}_{++}$ *and* $\beta_c \in (0,1)$ *such that*

$$\left\| \mathbf{z}_k - \mathbf{z}_k^c \right\| \leq \alpha_c \ \beta_c^k \ \left\| \mathbf{z}_0 - \mathbf{z}_0^c \right\| \tag{4.111}$$

holds for all $k \in \mathbb{N}_+$*, where* $\mathbf{z}_k^c$ *is defined in* (4.62).

Notably, when the comparator sequence satisfies the exponential stability assumption, Algorithm 4.1 leads to a constant regret bound, as our next result shows.

Theorem 4.4. *Consider system* (4.16) *with* $w_k = 0, v_k = 0, k \in \mathbb{N}$, *and the pMHE iteration scheme in Algorithm 4.1 with a step size in* (4.7b) *satisfying*

$$\eta_k^i \leq \frac{\sigma_k}{L_k} \tag{4.112}$$

for all $i = 0, \ldots, \mathrm{it}(k) - 1$ *and* $k \in \mathbb{N}_+$. *Let Assumptions 4.1-4.5 hold. Suppose that the Bregman distance* D_{ψ_k} *and the a priori estimate operator* Φ_k *satisfy*

$$D_{\psi_k}\left(\Phi_{k-1}(\mathbf{z}), \Phi_{k-1}(\hat{\mathbf{z}})\right) - D_{\psi_{k-1}}(\mathbf{z}, \hat{\mathbf{z}}) \leq -c \|\mathbf{z} - \hat{\mathbf{z}}\|^2 \tag{4.113}$$

for all $k \in \mathbb{N}_+$ *and* $\mathbf{z}, \hat{\mathbf{z}} \in \mathbb{R}^{(N+1)n}$. *Then, the estimation error is GUES.*
Moreover, suppose that a comparator sequence $\{\mathbf{z}_1^c, \mathbf{z}_2^c, \ldots, \mathbf{z}_T^c\}$ *is generated from an estimator with GUES estimation error dynamics and initial guess* $\mathbf{z}_0^c$, *as in Assumption 4.7. Then, Algorithm 4.1 gives the following constant regret bound*

$$R(T) \leq G_\mathrm{F} \frac{\alpha\,\beta}{1-\beta} \|\mathbf{z}_0 - \bar{\mathbf{z}}_0\| + G_\mathrm{F} \frac{\alpha_\mathrm{c}\,\beta_\mathrm{c}}{1-\beta_\mathrm{c}} \|\mathbf{z}_0 - \mathbf{z}_0^c\|, \tag{4.114}$$

where $\alpha \in \mathbb{R}_{++}$ *and* $\beta \in (0,1)$.

Proof. GUES of the underlying estimation error follows directly from Theorem 4.1.
In the proof of Theorem 4.1, we showed in (4.44) that GUES of the pMHE error (4.12) implies that there exist constants $\alpha \in \mathbb{R}_{++}$ and $\beta \in (0,1)$ such that

$$\left\|\mathbf{z}_k - \hat{\mathbf{z}}_k^{\mathrm{it}(k)}\right\| \leq \alpha\beta^k \left\|\mathbf{z}_0 - \bar{\mathbf{z}}_0\right\| \tag{4.115}$$

holds for all $k \in \mathbb{N}_+$. Recall now the definitions of the regret given in (4.63) and of G_F in (4.65). Exploiting the convexity of the cost function F_k, which is implied by the linear setup and Assumption 4.3, and applying the Cauchy-Schwarz inequality yield

$$F_k\left(\hat{\mathbf{z}}_k^{\mathrm{it}(k)}\right) - F_k\left(\mathbf{z}_k^c\right) \leq \nabla F_k\left(\hat{\mathbf{z}}_k^{\mathrm{it}(k)}\right)^\top \left(\hat{\mathbf{z}}_k^{\mathrm{it}(k)} - \mathbf{z}_k^c\right) \tag{4.116}$$

$$\leq G_\mathrm{F} \left\|\hat{\mathbf{z}}_k^{\mathrm{it}(k)} - \mathbf{z}_k^c\right\|.$$

By the triangle inequality, GUES of the comparator sequence in (4.111) in Assumption 4.7, as well as (4.115), we arrive at

$$F_k\left(\hat{\mathbf{z}}_k^{\mathrm{it}(k)}\right) - F_k\left(\mathbf{z}_k^c\right) \leq G_\mathrm{F} \left\|\mathbf{z}_k - \hat{\mathbf{z}}_k^{\mathrm{it}(k)}\right\| + G_\mathrm{F} \left\|\mathbf{z}_k - \mathbf{z}_k^c\right\| \tag{4.117}$$

$$\leq G_\mathrm{F}\,\alpha\,\beta^k \left\|\mathbf{z}_0 - \bar{\mathbf{z}}_0\right\| + G_\mathrm{F}\,\alpha_\mathrm{c}\,\beta_\mathrm{c}^k \left\|\mathbf{z}_0 - \mathbf{z}_0^c\right\|.$$

Hence, we can bound the regret as follows

$$R(T) \leq G_\mathrm{F}\,\alpha \left\|\mathbf{z}_0 - \bar{\mathbf{z}}_0\right\| \sum_{k=1}^{T} \beta^k + G_\mathrm{F}\,\alpha_\mathrm{c} \left\|\mathbf{z}_0 - \mathbf{z}_0^c\right\| \sum_{k=1}^{T} \beta_\mathrm{c}^k. \tag{4.118}$$

Since $\beta \in (0,1)$ (which holds also for β_c^k), we can plug $\sum_{k=1}^{T} \beta^k = \frac{\beta - \beta^{T+1}}{1-\beta} \leq \frac{\beta}{1-\beta}$ in (4.118) and obtain the desired regret bound (4.114). $\qquad\square$

Table 4.1: Summary of results of Section 4.3. We employ the following notation for abbreviation: $\Delta D_k(\mathbf{z}, \hat{\mathbf{z}}) := D_{\psi_k}(\Phi_{k-1}(\mathbf{z}), \Phi_{k-1}(\hat{\mathbf{z}})) - D_{\psi_{k-1}}(\mathbf{z}, \hat{\mathbf{z}})$ and $\sum_i \eta_k^i := \sum_{i=0}^{\text{it}(k)-1} \eta_k^i$.

Theorem	Assumptions	Step size	Result
Thm. 4.1	A. 4.1-4.5 $\Delta D_k(\mathbf{z}, \hat{\mathbf{z}}) \leq -c\|\mathbf{z} - \hat{\mathbf{z}}\|^2$	$\eta_k^i \leq \frac{\sigma_k}{L_k}$	GUES
Thm. 4.2	A. 4.1-4.4 $\Delta D_k(\mathbf{z}, \hat{\mathbf{z}}) \leq 0$	$\eta_k^i \leq \frac{\sigma_k}{L_k}$ $\sum_i \eta_{k+1}^i \leq \sum_i \eta_k^i$	Regret: $R(T) \leq \frac{D_{\max}}{\sum_i \eta_T^i} + \frac{G_F^2}{2\sigma} \sum_{k=1}^T \frac{\sum_i (\eta_k^i)^2}{\sum_i \eta_k^i} + \frac{M}{\sum_i \eta_T^i} \sum_{k=1}^T \|\mathbf{z}_{k+1}^i - \Phi_k(\mathbf{z}_k^i)\|$
Thm. 4.3	A. 4.1-4.5 $\Delta D_k(\mathbf{z}, \hat{\mathbf{z}}) \leq -c\|\mathbf{z} - \hat{\mathbf{z}}\|^2$ $\text{it}(k+1) \leq \text{it}(k)$	$\eta_k^i = \frac{\sigma}{L_F} \frac{1}{\sqrt{k}}$	GUES + Regret: $R(T) \leq \frac{\sqrt{T}}{\text{it}(T)} \frac{L_F}{\sigma} \left(D_{\max} + M \sum_{k=1}^T \|\mathbf{z}_{k+1}^i - \Phi_k(\mathbf{z}_k^i)\| \right)$
Thm. 4.4	A. 4.1-4.5, 4.7 $\Delta D_k(\mathbf{z}, \hat{\mathbf{z}}) \leq -c\|\mathbf{z} - \hat{\mathbf{z}}\|^2$	$\eta_k^i \leq \frac{\sigma_k}{L_k}$	GUES + Regret (const.): $R(T) \leq \frac{G_F \alpha \beta}{1-\beta} \|\mathbf{z}_0 - \bar{\mathbf{z}}_0\| + \frac{G_F \alpha_c \beta_c}{1-\beta_c} \|\mathbf{z}_0 - \mathbf{z}_0^c\|$

We summarize the main stability and regret results obtained for anytime pMHE of linear systems (4.16) with $w_k = 0, v_k = 0, k \in \mathbb{N}$ in Table 4.1.

Remark 4.6. Note that the obtained stability and regret results were derived under the assumption that the underlying cost function F_k is differentiable, which does not allow for the use of nonsmooth penalty functions such as the sparsity promoting ℓ_1-norm. Hence, it would be worthwhile to extend the proposed approach by replacing the gradient $\nabla F_k(\hat{\mathbf{z}}_k^i)$ in the underlying first-order optimization algorithm (4.7b) with an arbitrary subgradient from the subdifferential $\partial F_k(\hat{\mathbf{z}}_k^i)$ and investigating the resulting stability and regret properties. This can be guided by the convergence analysis of the MDA, which is briefly discussed in Section 2.2.2.

4.4 Anytime proximity MHE for nonlinear systems

In this section, we establish the nominal stability properties of the estimation error generated by the pMHE iteration scheme presented in Algorithm 4.1 for the more general case of nonlinear systems. This section is based on and taken in parts literally from (Gharbi et al. (2021))[2].

We consider nonlinear systems of the form (4.1) satisfying Assumption 4.1. In the MHE problem (4.3), only the initial state in the horizon window $\hat{x}_{k-N}$ and the process disturbances $\hat{w}_i$ are required to lie in the convex sets $\mathcal{X}$ and $\mathcal{W}_i$, respectively, satisfying Assumption 4.2. The resulting feasible set $\mathcal{S}_k$ in the reformulated MHE problem (4.5) is therefore *convex* and has the form

$$\mathcal{S}_k := \left\{ \hat{\mathbf{z}}_k = \begin{bmatrix} \hat{x}_{k-N}^\top & \hat{\mathbf{w}}_k^\top \end{bmatrix}^\top : \ \hat{x}_{k-N} \in \mathcal{X}, \ \hat{w}_i \in \mathcal{W}_i, \quad i = k-N, \ldots, k-1 \right\}. \quad (4.119)$$

In the pMHE iterative procedure (4.7), this implies that the underlying problem to be solved at each iteration is a simple convex optimization problem with a unique solution $\hat{\mathbf{z}}_k^{i+1}$.

[2]M. Gharbi, B. Gharesifard, and C. Ebenbauer. Anytime proximity moving horizon estimation: Stability and regret for nonlinear systems. *In Proc. 60th Conference on Decision and Control (CDC). Accepted.* IEEE, 2021 © 2021 IEEE.

In the previous section, we employed Algorithm 4.1 for the state estimation problem of linear systems and showed in Theorem 4.1 that GUES of the underlying estimation error can be ensured through a proper choice of the Bregman distance D_{ψ_k}, the a priori estimate operator Φ_k, and the step size η_k^i. Thereby, Algorithm 4.1 for linear systems can be considered as an anytime MHE algorithm, in which stability is guaranteed after any number of optimization algorithm iterations. In the following, we extend the nominal stability analysis to the nonlinear case by deriving sufficient conditions on the same design components D_{ψ_k}, Φ_k, and η_k^i for the local uniform exponential stability (UES) of the estimation error. Subsequently, a specific design approach for D_{ψ_k} and Φ_k that is based on the EKF is discussed.

Let us first derive useful properties of the employed optimization algorithm. More specifically, similar to the pMHE operator introduced in (3.59) in the previous chapter, we define in view of (4.7b) the optimizer update operator $\rho_{\eta_k^i} : \mathbb{R}^{Nm_{\mathbf{w}}+n} \to \mathbb{R}^{Nm_{\mathbf{w}}+n}$ as

$$\rho_{\eta_k^i}(\hat{\mathbf{z}}) = \arg\min_{\mathbf{z} \in \mathcal{S}_k} \left\{ \eta_k^i \nabla F_k(\hat{\mathbf{z}})^\top \mathbf{z} + D_{\psi_k}(\mathbf{z}, \hat{\mathbf{z}}) \right\}. \tag{4.120}$$

Notice that $\rho_{\eta_k^i}$ is well-defined since, given the convex constraints (4.119) and the strongly convex Bregman distance, it constitutes the unique solution of a convex optimization problem. In (4.120), setting $\hat{\mathbf{z}} = \hat{\mathbf{z}}_k^i$, i.e., the i-th pMHE iterate, yields the next iterate $\rho_{\eta_k^i}(\hat{\mathbf{z}}_k^i) = \hat{\mathbf{z}}_k^{i+1}$. In the following lemma, we establish that the true system state $\mathbf{z}_k$ with zero disturbances can be interpreted as a fixed point of the operator $\rho_{\eta_k^i}$ for a given time instant k.

Lemma 4.6. *Consider the optimizer update operator $\rho_{\eta_k^i}$ given in (4.120) and let Assumptions 4.1-4.4 hold. In the disturbance-free case where $\mathbf{z}_k = \begin{bmatrix} x_{k-N}^\top & 0 & \dots & 0 \end{bmatrix}^\top$, if $\mathbf{z}_k \in \mathrm{int}(\mathcal{S}_k)$, then $\rho_{\eta_k^i}(\mathbf{z}_k) = \mathbf{z}_k$ for all $k \in \mathbb{N}$.*

Proof. The optimizer update operator at the true system state $\mathbf{z}_k$ can be reformulated as

$$\rho_{\eta_k^i}(\mathbf{z}_k) = \arg\min_{\mathbf{z} \in \mathbb{R}^{Nm_{\mathbf{w}}+n}} \left\{ \eta_k^i I_{\mathcal{S}_k}(\mathbf{z}) + \eta_k^i \nabla F_k(\mathbf{z}_k)^\top \mathbf{z} + D_{\psi_k}(\mathbf{z}, \mathbf{z}_k) \right\}, \tag{4.121}$$

where $I_{\mathcal{S}_k}$ denotes the indicator function of the convex set $\mathcal{S}_k$ as defined in (2.24). In order to prove that $\rho_{\eta_k^i}(\mathbf{z}_k) = \mathbf{z}_k$, we first show that $\mathbf{z}_k$ minimizes the MHE cost

$$F_k(\mathbf{z}) + I_{\mathcal{S}_k}(\mathbf{z}). \tag{4.122}$$

For $\mathbf{z}_k$ to be a local minimizer of (4.122), it must satisfy $-\nabla F_k(\mathbf{z}_k) \in \partial I_{\mathcal{S}_k}(\mathbf{z}_k)$, see (Mine and Fukushima (1981)). Here, $\partial I_{\mathcal{S}_k}(\mathbf{z}_k)$ refers to the subdifferential of the convex function $I_{\mathcal{S}_k}$ at the point $\mathbf{z}_k$ as introduced in Definition 2.3. We show this by employing similar arguments as in the proof of Lemma 3.3. Since $\mathbf{z}_k \in \mathrm{int}(\mathcal{S}_k)$, it holds that $\partial I_{\mathcal{S}_k}(\mathbf{z}_k) = N_{\mathcal{S}_k}(\mathbf{z}_k) = \{0\}$, where $N_{\mathcal{S}_k}$ denotes the normal cone operator defined in (2.25). Moreover, because we have zero true disturbances and due to Assumption 4.3, evaluating the MHE cost function F_k defined in (4.5) at $\mathbf{z}_k$ leads to its minimal value $F_k(\mathbf{z}_k) = \sum_{i=k-N}^{k-1} r_i(0) + q_i(0) = 0$. Hence, $0 \in \partial I_{\mathcal{S}_k}(\mathbf{z}_k) + \nabla F_k(\mathbf{z}_k)$ and we obtain

$$0 \in \partial I_{\mathcal{S}_k}(\mathbf{z}_k) + \nabla F_k(\mathbf{z}_k) \Leftrightarrow 0 \in \eta_k^i (\partial I_{\mathcal{S}_k}(\mathbf{z}_k) + \nabla F_k(\mathbf{z}_k)) + \nabla\psi_k(\mathbf{z}_k) - \nabla\psi_k(\mathbf{z}_k) \tag{4.123}$$

$$\Leftrightarrow \mathbf{z}_k = \arg\min_{\mathbf{z} \in \mathbb{R}^{Nm_{\mathbf{w}}+n}} \eta_k^i I_{\mathcal{S}_k}(\mathbf{z}) + \eta_k^i \nabla F_k(\mathbf{z}_k)^\top \mathbf{z} + D_{\psi_k}(\mathbf{z}, \mathbf{z}_k)$$

$$\Leftrightarrow \mathbf{z}_k = \rho_{\eta_k^i}(\mathbf{z}_k),$$

finishing the proof. $\qquad\square$

With the following lemma, we establish the Lipschitz continuity property of the optimizer update operator $\rho_{\eta_k^i}$.

Lemma 4.7. *Consider the optimizer update operator $\rho_{\eta_k^i}$ given in (4.120) and let Assumptions 4.1-4.4 hold. Then, for any $k \in \mathbb{N}$ and $\mathbf{z}_1, \mathbf{z}_2 \in \mathbb{R}^{Nm_w+n}$,*

$$\left\| \rho_{\eta_k^i}(\mathbf{z}_1) - \rho_{\eta_k^i}(\mathbf{z}_2) \right\| \leq \frac{\eta_k^i}{\sigma_k}\left(L_k + \frac{\gamma_k}{\eta_k^i}\right)\left\| \mathbf{z}_1 - \mathbf{z}_2 \right\|. \tag{4.124}$$

Proof. By optimality of $\rho_{\eta_k^i}(\mathbf{z}_1)$ and $\rho_{\eta_k^i}(\mathbf{z}_2)$ in (4.120) and the definition of the Bregman distance in (4.8), it holds that

$$u \in \partial I_{\mathcal{S}_k}(\rho_{\eta_k^i}(\mathbf{z}_1)), \qquad v \in \partial I_{\mathcal{S}_k}(\rho_{\eta_k^i}(\mathbf{z}_2)) \tag{4.125a}$$

with the subgradients

$$u = -\nabla F_k(\mathbf{z}_1) - \frac{1}{\eta_k^i}\nabla \psi_k(\rho_{\eta_k^i}(\mathbf{z}_1)) + \frac{1}{\eta_k^i}\nabla \psi_k(\mathbf{z}_1), \tag{4.125b}$$

$$v = -\nabla F_k(\mathbf{z}_2) - \frac{1}{\eta_k^i}\nabla \psi_k(\rho_{\eta_k^i}(\mathbf{z}_2)) + \frac{1}{\eta_k^i}\nabla \psi_k(\mathbf{z}_2). \tag{4.125c}$$

Notice that we replaced the constraints specified by the set $\mathcal{S}_k$ with the associated indicator function $I_{\mathcal{S}_k}$ in the cost function. The monotonicity of the subdifferential operator of the convex function $I_{\mathcal{S}_k}$ (Parikh and Boyd (2014)) implies in view of (4.125a) that

$$(u - v)^{\top}(\rho_{\eta_k^i}(\mathbf{z}_1) - \rho_{\eta_k^i}(\mathbf{z}_2)) \geq 0. \tag{4.126}$$

By plugging (4.125b) and (4.125c) in (4.126), we obtain

$$\left(\frac{1}{\eta_k^i}\nabla \psi_k(\mathbf{z}_1) - \frac{1}{\eta_k^i}\nabla \psi_k(\rho_{\eta_k^i}(\mathbf{z}_1))\right) - \nabla F_k(\mathbf{z}_1) \tag{4.127}$$

$$- \frac{1}{\eta_k^i}\nabla \psi_k(\mathbf{z}_2) + \frac{1}{\eta_k^i}\nabla \psi_k(\rho_{\eta_k^i}(\mathbf{z}_2)) + \nabla F_k(\mathbf{z}_2)\bigg)^{\top}\left(\rho_{\eta_k^i}(\mathbf{z}_1) - \rho_{\eta_k^i}(\mathbf{z}_2)\right) \geq 0.$$

Rearranging the above inequality yields

$$\frac{1}{\eta_k^i}\left(\nabla \psi_k(\rho_{\eta_k^i}(\mathbf{z}_1)) - \nabla \psi_k(\rho_{\eta_k^i}(\mathbf{z}_2))\right)^{\top}\left(\rho_{\eta_k^i}(\mathbf{z}_1) - \rho_{\eta_k^i}(\mathbf{z}_2)\right) \tag{4.128}$$

$$\leq \left(\nabla F_k(\mathbf{z}_2) - \nabla F_k(\mathbf{z}_1) + \frac{1}{\eta_k^i}\left(\nabla \psi_k(\mathbf{z}_1) - \nabla \psi_k(\mathbf{z}_2)\right)\right)^{\top}\left(\rho_{\eta_k^i}(\mathbf{z}_1) - \rho_{\eta_k^i}(\mathbf{z}_2)\right).$$

In view of Assumption 4.4, the function ψ_k is strongly convex with constant $\sigma_k \in \mathbb{R}_{++}$, which implies that

$$\left(\nabla \psi_k(\rho_{\eta_k^i}(\mathbf{z}_1)) - \nabla \psi_k(\rho_{\eta_k^i}(\mathbf{z}_2))\right)^{\top}\left(\rho_{\eta_k^i}(\mathbf{z}_1) - \rho_{\eta_k^i}(\mathbf{z}_2)\right) \geq \sigma_k\left\|\rho_{\eta_k^i}(\mathbf{z}_1) - \rho_{\eta_k^i}(\mathbf{z}_2)\right\|^2. \tag{4.129}$$

By applying the Cauchy-Schwarz inequality in (4.128) and using (4.129), we therefore get

$$\frac{\sigma_k}{\eta_k^i}\left\|\rho_{\eta_k^i}(\mathbf{z}_1) - \rho_{\eta_k^i}(\mathbf{z}_2)\right\|^2 \leq \left\|\nabla F_k(\mathbf{z}_2) - \nabla F_k(\mathbf{z}_1)\right\|\left\|\rho_{\eta_k^i}(\mathbf{z}_1) - \rho_{\eta_k^i}(\mathbf{z}_2)\right\| \tag{4.130}$$

$$+ \frac{1}{\eta_k^i}\left\|\nabla \psi_k(\mathbf{z}_1) - \nabla \psi_k(\mathbf{z}_2)\right\|\left\|\rho_{\eta_k^i}(\mathbf{z}_1) - \rho_{\eta_k^i}(\mathbf{z}_2)\right\|.$$

According to Assumptions 4.3 and 4.4, the gradients of the cost function F_k and of ψ_k are Lipschitz continuous with Lipschitz constants $L_k \in \mathbb{R}_{++}$ and $\gamma_k \in \mathbb{R}_{++}$, respectively. Hence,

$$\|\nabla F_k(\mathbf{z}_2) - \nabla F_k(\mathbf{z}_1)\| \leq L_k \|\mathbf{z}_1 - \mathbf{z}_2\|, \tag{4.131}$$

$$\|\nabla \psi_k(\mathbf{z}_1) - \nabla \psi_k(\mathbf{z}_2)\| \leq \gamma_k \|\mathbf{z}_1 - \mathbf{z}_2\|, \tag{4.132}$$

which we use in (4.130) to obtain

$$\frac{\sigma_k}{\eta_k^i} \left\| \rho_{\eta_k^i}(\mathbf{z}_1) - \rho_{\eta_k^i}(\mathbf{z}_2) \right\|^2 \leq \left(L_k + \frac{\gamma_k}{\eta_k^i} \right) \|\mathbf{z}_1 - \mathbf{z}_2\| \left\| \rho_{\eta_k^i}(\mathbf{z}_1) - \rho_{\eta_k^i}(\mathbf{z}_2) \right\|. \tag{4.133}$$

If $\rho_{\eta_k^i}(\mathbf{z}_1) \neq \rho_{\eta_k^i}(\mathbf{z}_2)$, dividing the last inequality by $\|\rho_{\eta_k^i}(\mathbf{z}_1) - \rho_{\eta_k^i}(\mathbf{z}_2)\|$ yields the desired result (4.124) of the lemma. Otherwise, we get $0 \leq 0$ in both inequalities. $\qquad\square$

We require the following strong smoothness assumption on the stage cost and two additional preparatory lemmas before presenting the main stability result.

Assumption 4.8 (Strongly smooth stage cost). *For any $i \in \mathbb{N}$, the gradient of the function r_i is Lipschitz continuous with the uniform Lipschitz constant $L_r \in \mathbb{R}_{++}$, i.e.,*

$$\|\nabla r_i(x) - \nabla r_i(y)\| \leq L_r \|x - y\| \qquad \forall i, \ \forall x, y \in \mathbb{R}^p. \tag{4.134}$$

In the following lemma, we bound the Bregman distance between the true system state $\mathbf{z}_k$ and the last pMHE iterate $\hat{\mathbf{z}}_k^{\mathrm{it}(k)}$ in terms of the distance between $\mathbf{z}_k$ and the stabilizing a priori estimate $\bar{\mathbf{z}}_k = \Phi_{k-1}\big(\hat{\mathbf{z}}_{k-1}^{\mathrm{it}(k-1)}\big)$.

Lemma 4.8. *Consider system (4.1) with $w_k = 0, v_k = 0, k \in \mathbb{N}$, and let $\left\{ \hat{\mathbf{z}}_k^0, \hat{\mathbf{z}}_k^1, \cdots, \hat{\mathbf{z}}_k^{\mathrm{it}(k)} \right\}$ be the sequence of pMHE iterates generated by (4.7) at a given time instant k. Suppose Assumptions 4.1-4.4, 4.8 hold. Then, there exist constants $\delta, \kappa \in \mathbb{R}_{++}$ such that*

$$D_{\psi_k}\big(\mathbf{z}_k, \hat{\mathbf{z}}_k^{i+1}\big) \leq D_{\psi_k}\big(\mathbf{z}_k, \hat{\mathbf{z}}_k^i\big) + \frac{1}{2}\big(\eta_k^i L_k - \sigma_k\big) \left\| \hat{\mathbf{z}}_k^{i+1} - \hat{\mathbf{z}}_k^i \right\|^2 + \eta_k^i \kappa \left\| \mathbf{z}_k - \hat{\mathbf{z}}_k^i \right\|^3 \tag{4.135}$$

holds for any $i \in \{0, ..., \mathrm{it}(k) - 1\}$ and $\mathbf{z}_k, \hat{\mathbf{z}}_k^i$ with $\|\mathbf{z}_k - \hat{\mathbf{z}}_k^i\| \leq \delta$. Moreover, we have that

$$D_{\psi_k}\big(\mathbf{z}_k, \hat{\mathbf{z}}_k^{\mathrm{it}(k)}\big) \leq D_{\psi_k}\big(\mathbf{z}_k, \Phi_{k-1}\big(\hat{\mathbf{z}}_{k-1}^{\mathrm{it}(k-1)}\big)\big) + \frac{1}{2} \sum_{i=0}^{\mathrm{it}(k)-1} \big(\eta_k^i L_k - \sigma_k\big) \left\| \hat{\mathbf{z}}_k^{i+1} - \hat{\mathbf{z}}_k^i \right\|^2 \tag{4.136}$$

$$+ \sum_{i=0}^{\mathrm{it}(k)-1} \eta_k^i \kappa \left\| \mathbf{z}_k - \hat{\mathbf{z}}_k^i \right\|^3,$$

where $\mathbf{z}_k$ denotes the true system state, $\hat{\mathbf{z}}_k^{\mathrm{it}(k)}$ the last pMHE iterate, and Φ_k the a priori estimate operator as introduced in (4.10)

Proof. By optimality of $\hat{\mathbf{z}}_k^{i+1}$ in the convex optimization problem (4.7b), and since the true system state with zero disturbances satisfies the constraints, i.e., $\mathbf{z}_k \in \mathcal{S}_k$, it holds that

$$\big(\eta_k^i \nabla F_k\big(\hat{\mathbf{z}}_k^i\big) + \nabla \psi_k\big(\hat{\mathbf{z}}_k^{i+1}\big) - \nabla \psi_k\big(\hat{\mathbf{z}}_k^i\big)\big)^\top \big(\mathbf{z}_k - \hat{\mathbf{z}}_k^{i+1}\big) \geq 0, \tag{4.137}$$

which we equivalently reformulate as

$$\nabla F_k \left(\hat{\mathbf{z}}_k^i\right)^\top \left(\mathbf{z}_k - \hat{\mathbf{z}}_k^i\right) \geq \nabla F_k \left(\hat{\mathbf{z}}_k^i\right)^\top \left(\hat{\mathbf{z}}_k^{i+1} - \hat{\mathbf{z}}_k^i\right) \tag{4.138}$$
$$+ \frac{1}{\eta_k^i}\left(\nabla \psi_k\left(\hat{\mathbf{z}}_k^i\right) - \nabla \psi_k\left(\hat{\mathbf{z}}_k^{i+1}\right)\right)^\top \left(\mathbf{z}_k - \hat{\mathbf{z}}_k^{i+1}\right).$$

Substituting the Lipschitz continuity property (4.24) of ∇F_k, which is implied by Assumption 4.3, into (4.138) yields

$$\nabla F_k \left(\hat{\mathbf{z}}_k^i\right)^\top \left(\mathbf{z}_k - \hat{\mathbf{z}}_k^i\right) + F_k \left(\hat{\mathbf{z}}_k^i\right) \geq F_k \left(\hat{\mathbf{z}}_k^{i+1}\right) - \frac{L_k}{2}\left\|\hat{\mathbf{z}}_k^{i+1} - \hat{\mathbf{z}}_k^i\right\|^2 \tag{4.139}$$
$$+ \frac{1}{\eta_k^i}\left(\nabla \psi_k\left(\hat{\mathbf{z}}_k^i\right) - \nabla \psi_k\left(\hat{\mathbf{z}}_k^{i+1}\right)\right)^\top \left(\mathbf{z}_k - \hat{\mathbf{z}}_k^{i+1}\right).$$

Similar to the employed notations in the proof of Lemma 3.6 used for establishing local UES of nonlinear pMHE (3.133), we define

$$\tilde{H} := \begin{bmatrix} 0_{[Nm_{\mathrm{w}} \times n]} & I_{[Nm_{\mathrm{w}}]} \end{bmatrix} \in \mathbb{R}^{Nm_{\mathrm{w}} \times (Nm_{\mathrm{w}}+n)} \tag{4.140}$$

and the following vectors in terms of the resulting output residuals and process disturbances obtained at the i-th iteration and for a given time instant k

$$D_{\mathrm{r},k}^i := \begin{bmatrix} \nabla r_{k-N}^\top(\hat{v}_{k-N}^i) & \cdots & \nabla r_{k-1}^\top(\hat{v}_{k-1}^i) \end{bmatrix}^\top \in \mathbb{R}^{Np}, \tag{4.141}$$

$$D_{\mathrm{q},k}^i := \begin{bmatrix} \nabla q_{k-N}^\top(\hat{w}_{k-N}^i) & \cdots & \nabla q_{k-1}^\top(\hat{w}_{k-1}^i) \end{bmatrix}^\top \in \mathbb{R}^{Nm_{\mathrm{w}}}, \tag{4.142}$$

$$\hat{\mathbf{v}}_k^i := \begin{bmatrix} \hat{v}_{k-N}^{i\top} & \cdots & \hat{v}_{k-1}^{i\top} \end{bmatrix}^\top \in \mathbb{R}^{Np}. \tag{4.143}$$

Moreover, we consider the output response function $H_k : \mathbb{R}^{Nm_{\mathrm{w}}+n} \to \mathbb{R}^{Np}$ given in (3.144) and reintroduced here for completeness:

$$H_k(\hat{\mathbf{z}}_k) := \begin{bmatrix} h_{k-N}\left(\hat{x}_{k-N}\right) \\ h_{k-N+1}\left(x\left(k-N+1; \hat{x}_{k-N}, k-N, u_{k-N}, \hat{w}_{k-N}\right)\right) \\ \vdots \\ h_{k-1}\left(x\left(k-1; \hat{x}_{k-N}, k-N, \mathbf{u}, \hat{\mathbf{w}}\right)\right) \end{bmatrix}. \tag{4.144}$$

Following the same line of arguments for deriving (3.146), it holds that

$$\hat{\mathbf{v}}_k^i = H_k(\mathbf{z}_k) - H_k(\hat{\mathbf{z}}_k^i). \tag{4.145}$$

Moreover, by Taylor's theorem (Nocedal and Wright (2006)), we have

$$H_k(\mathbf{z}_k) = H_k(\hat{\mathbf{z}}_k^i) + \nabla H_k(\hat{\mathbf{z}}_k^i)^\top(\mathbf{z}_k - \hat{\mathbf{z}}_k^i) + \mathbf{d}_k(\mathbf{z}_k, \hat{\mathbf{z}}_k^i), \tag{4.146}$$

which we plug in (4.145) to get

$$\hat{\mathbf{v}}_k^i = \nabla H_k(\hat{\mathbf{z}}_k^i)^\top(\mathbf{z}_k - \hat{\mathbf{z}}_k^i) + \mathbf{d}_k(\mathbf{z}_k, \hat{\mathbf{z}}_k^i). \tag{4.147}$$

Here, $\mathbf{d}_k$ denotes the column vector of stacked higher order terms which arise from the Taylor expansion (cf. (3.147b)). According to Assumption 4.3, for any $j \in \mathbb{N}$, the functions r_j and q_j are smooth, convex, and achieve zero at zero. Hence,

$$r_j(\hat{v}_j^i) = r_j(\hat{v}_j^i) - r_j(0) \leq \nabla r_j(\hat{v}_j^i)^\top(\hat{v}_j^i - 0) \tag{4.148a}$$

$$q_j(\hat{w}_j^i) = q_j(\hat{w}_j^i) - q_j(0) \leq \nabla q_j(\hat{w}_j^i)^\top(\hat{w}_j^i - 0). \tag{4.148b}$$

Note that we use the subscript j instead of i for the time-varying stage cost in order to avoid confusion with the iteration index. By (4.148), (4.141), and (4.142), it holds that

$$F_k(\hat{\mathbf{z}}_k^i) = \sum_{j=k-N}^{k-1} r_j(\hat{v}_j^i) + q_j(\hat{w}_j^i) \leq \sum_{j=k-N}^{k-1} \nabla r_j(\hat{v}_j^i)^\top \hat{v}_j^i + \nabla q_j(\hat{w}_j^i)^\top \hat{w}_j^i \tag{4.149}$$
$$= D_{\mathrm{r},k}^{i\top} \hat{\mathbf{v}}_k^i + D_{\mathrm{q},k}^{i\top} \hat{\mathbf{w}}_k^i.$$

Furthermore, in view of (4.5a) and given the introduced notations, we have

$$\nabla F_k(\hat{\mathbf{z}}_k^i) = -\nabla H_k(\hat{\mathbf{z}}_k^i) D_{\mathrm{r},k}^i + \tilde{H}^\top D_{\mathrm{q},k}^i. \tag{4.150}$$

Using (4.150), (4.147), and $\hat{\mathbf{w}}_k^i = \tilde{H}\hat{\mathbf{z}}_k^i = \tilde{H}(\hat{\mathbf{z}}_k^i - \mathbf{z}_k)$ in the (4.149) leads to

$$F_k(\hat{\mathbf{z}}_k^i) \leq D_{\mathrm{r},k}^{i\top} \left(\nabla H_k(\hat{\mathbf{z}}_k^i)^\top (\mathbf{z}_k - \hat{\mathbf{z}}_k^i) + \mathbf{d}_k(\mathbf{z}_k, \hat{\mathbf{z}}_k^i) \right) + D_{\mathrm{q},k}^{i\top} \hat{\mathbf{w}}_k^i \tag{4.151}$$
$$\leq -D_{\mathrm{r},k}^{i\top} \nabla H_k(\hat{\mathbf{z}}_k^i)^\top (\hat{\mathbf{z}}_k^i - \mathbf{z}_k) + D_{\mathrm{q},k}^{i\top} \tilde{H}(\hat{\mathbf{z}}_k^i - \mathbf{z}_k) + D_{\mathrm{r},k}^{i\top} \mathbf{d}_k(\mathbf{z}_k, \hat{\mathbf{z}}_k^i)$$
$$= \nabla F_k(\hat{\mathbf{z}}_k^i)^\top \left(\hat{\mathbf{z}}_k^i - \mathbf{z}_k \right) + D_{\mathrm{r},k}^{i\top} \mathbf{d}_k(\mathbf{z}_k, \hat{\mathbf{z}}_k^i).$$

Substituting (4.139) in (4.151) and exploiting the fact that $F_k\left(\hat{\mathbf{z}}_k^{i+1}\right) \geq 0$ yield

$$D_{\mathrm{r},k}^{i\top} \mathbf{d}_k(\mathbf{z}_k, \hat{\mathbf{z}}_k^i) \geq -\frac{L_k}{2} \left\| \hat{\mathbf{z}}_k^{i+1} - \hat{\mathbf{z}}_k^i \right\|^2 + \frac{1}{\eta_k^i} \left(\nabla \psi_k\left(\hat{\mathbf{z}}_k^i\right) - \nabla \psi_k\left(\hat{\mathbf{z}}_k^{i+1}\right) \right)^\top \left(\mathbf{z}_k - \hat{\mathbf{z}}_k^{i+1} \right). \tag{4.152}$$

In view of the three-points identity of Bregman distances given in Lemma 2.1 (with $a = \hat{\mathbf{z}}_k^{i+1}$, $b = \hat{\mathbf{z}}_k^i$, $c = \mathbf{z}_k$) and due to the lower bound in (4.9) in Assumption 4.4, we have

$$\left(\nabla \psi_k\left(\hat{\mathbf{z}}_k^i\right) - \nabla \psi_k\left(\hat{\mathbf{z}}_k^{i+1}\right) \right)^\top \left(\mathbf{z}_k - \hat{\mathbf{z}}_k^{i+1} \right) = D_{\psi_k}\left(\mathbf{z}_k, \hat{\mathbf{z}}_k^{i+1}\right) + D_{\psi_k}\left(\hat{\mathbf{z}}_k^{i+1}, \hat{\mathbf{z}}_k^i\right) - D_{\psi_k}\left(\mathbf{z}_k, \hat{\mathbf{z}}_k^i\right)$$
$$\geq D_{\psi_k}\left(\mathbf{z}_k, \hat{\mathbf{z}}_k^{i+1}\right) + \frac{\sigma_k}{2}\|\hat{\mathbf{z}}_k^{i+1} - \hat{\mathbf{z}}_k^i\|^2 - D_{\psi_k}\left(\mathbf{z}_k, \hat{\mathbf{z}}_k^i\right). \tag{4.153}$$

Therefore, by employing (4.153) in (4.152), we get

$$D_{\psi_k}\left(\mathbf{z}_k, \hat{\mathbf{z}}_k^{i+1}\right) \leq D_{\psi_k}\left(\mathbf{z}_k, \hat{\mathbf{z}}_k^i\right) + \frac{\eta_k^i}{2}\left(L_k - \frac{\sigma_k}{\eta_k^i}\right) \left\| \hat{\mathbf{z}}_k^{i+1} - \hat{\mathbf{z}}_k^i \right\|^2 + \eta_k^i D_{\mathrm{r},k}^{i\top} \mathbf{d}_k(\mathbf{z}_k, \hat{\mathbf{z}}_k^i). \tag{4.154}$$

The uniform Lipschitz continuity of ∇r_i in Assumption 4.8 implies that

$$D_{\mathrm{r},k}^{i\top} \mathbf{d}_k(\mathbf{z}_k, \hat{\mathbf{z}}_k^i) \leq L_{\mathrm{r}} \|\hat{\mathbf{v}}_k^i\| \|\mathbf{d}_k(\mathbf{z}_k, \hat{\mathbf{z}}_k^i)\|. \tag{4.155}$$

Since f_k, h_k are $\mathcal{C}^2$ functions by Assumption 4.1, the Taylor approximation remainders of their compositions are of second order. Hence, there exist $\kappa_\mathrm{d}, \delta \in \mathbb{R}_{++}$ such that the norm of $\mathbf{d}_k$ in (4.146) can be upper bounded as

$$\|\mathbf{d}_k(\mathbf{z}_k, \hat{\mathbf{z}}_k^i)\| \leq \kappa_\mathrm{d} \|\mathbf{z}_k - \hat{\mathbf{z}}_k^i\|^2 \tag{4.156}$$

for all $\mathbf{z}_k, \hat{\mathbf{z}}_k^i$ with $\|\mathbf{z}_k - \hat{\mathbf{z}}_k^i\| \leq \delta$. Moreover, the assumption that the functions f_k, h_k are Lipschitz continuous uniformly over $k \in \mathbb{N}$ indicates that their compositions in H_k defined in (4.144) is Lipschitz continuous as well with some uniform Lipschitz constant $c_{\mathrm{H}} \in \mathbb{R}_{++}$. Exploiting this property of H_k and using (4.145) and (4.156) in (4.155) lead to

$$D_{\mathrm{r},k}^{i^{\mathsf{T}}} \mathbf{d}_k(\mathbf{z}_k, \hat{\mathbf{z}}_k^i) \leq L_{\mathrm{r}} \|H_k(\mathbf{z}_k) - H_k(\hat{\mathbf{z}}_k^i)\| \|\mathbf{d}_k(\mathbf{z}_k, \hat{\mathbf{z}}_k^i)\| \tag{4.157}$$
$$\leq L_{\mathrm{r}}\, c_{\mathrm{H}}\, \kappa_{\mathrm{d}}\, \|\mathbf{z}_k - \hat{\mathbf{z}}_k^i\|^3$$

for all $\mathbf{z}_k, \hat{\mathbf{z}}_k^i$ with $\|\mathbf{z}_k - \hat{\mathbf{z}}_k^i\| \leq \delta$. Substituting the last inequality in (4.154) yields the first statement (4.135) of the lemma with $\kappa := L_{\mathrm{r}}\, c_{\mathrm{H}}\, \kappa_{\mathrm{d}}$.

Employing similar steps to the derivation of (4.29), i.e., applying (4.135) to each two subsequent iterations i and $i+1$ (where $i = 0, \ldots, \mathrm{it}(k) - 1$) yields

$$D_{\psi_k}\left(\mathbf{z}_k, \hat{\mathbf{z}}_k^{\mathrm{it}(k)}\right) \leq D_{\psi_k}\left(\mathbf{z}_k, \hat{\mathbf{z}}_k^0\right) + \frac{1}{2} \sum_{i=0}^{\mathrm{it}(k)-1} \left(\eta_k^i\, L_k - \sigma_k\right) \left\|\hat{\mathbf{z}}_k^{i+1} - \hat{\mathbf{z}}_k^i\right\|^2 + \sum_{i=0}^{\mathrm{it}(k)-1} \eta_k^i\, \kappa \left\|\mathbf{z}_k - \hat{\mathbf{z}}_k^i\right\|^3 \tag{4.158}$$

for all $\mathbf{z}_k, \hat{\mathbf{z}}_k^i$ with $\|\mathbf{z}_k - \hat{\mathbf{z}}_k^i\| \leq \delta$. Using the fact that $D_{\psi_k}(\mathbf{z}_k, \hat{\mathbf{z}}_k^0) \leq D_{\psi_k}(\mathbf{z}_k, \bar{\mathbf{z}}_k)$ due to (4.30) gives

$$D_{\psi_k}\left(\mathbf{z}_k, \hat{\mathbf{z}}_k^{\mathrm{it}(k)}\right) \leq D_{\psi_k}\left(\mathbf{z}_k, \bar{\mathbf{z}}_k\right) + \frac{1}{2} \sum_{i=0}^{\mathrm{it}(k)-1} \left(\eta_k^i\, L_k - \sigma_k\right) \left\|\hat{\mathbf{z}}_k^{i+1} - \hat{\mathbf{z}}_k^i\right\|^2 + \sum_{i=0}^{\mathrm{it}(k)-1} \eta_k^i\, \kappa \left\|\mathbf{z}_k - \hat{\mathbf{z}}_k^i\right\|^3 \tag{4.159}$$

for all $\mathbf{z}_k, \hat{\mathbf{z}}_k^i$ with $\|\mathbf{z}_k - \hat{\mathbf{z}}_k^i\| \leq \delta$. Setting $\bar{\mathbf{z}}_k = \Phi_{k-1}\left(\hat{\mathbf{z}}_{k-1}^{\mathrm{it}(k-1)}\right)$ as indicated by (4.10) in the above inequality yields the second statement (4.136) of the lemma. $\qquad\square$

The proof of Lemma 4.8 combines tools we used for proving i) Lemma 4.1 in the stability analysis of the pMHE iteration scheme (Algorithm 4.1) for linear systems, and ii) Lemma 3.6 in the stability analysis of the pMHE scheme (Algorithm 3.1) for nonlinear systems, which is carried out based on the optimal solution $\hat{\mathbf{z}}_k^\star$ of the pMHE problem (3.133). In fact, we can observe that the first two terms of the right-hand side of (4.136) appear in (4.19) and that the last term of the right-hand side of (4.136) is of third-order similar to the last term of the right-hand side of (3.135). In both cases, these higher-order terms result from the underlying local analysis based on first-order Taylor approximations of the functions f_k, h_k in the nonlinear system (4.1). Notice also that in this nonlinear case, the proof of Lemma 4.8 allows only state constraints on the initial state $\hat{x}_{k-N}$ as is the case for Lemma 3.6. Nevertheless, a stability-preserving projection of the suboptimal state estimate $\hat{x}_k$, which is obtained via (4.11) based on the last pMHE iterate $\hat{\mathbf{z}}_k^{\mathrm{it}(k)}$, onto the convex set $\mathcal{X}$ can be performed along the lines of Remark 3.10.

The idea behind the following lemma is similar to that of Lemma 3.7 and amounts to bounding the error between the disturbance-free true system state $\mathbf{z}_k$ and any pMHE iterate $\hat{\mathbf{z}}_k^i$ in terms of the error between $\mathbf{z}_k$ and the a priori estimate $\bar{\mathbf{z}}_k$.

Lemma 4.9. *Consider system (4.1) with $w_k = 0, v_k = 0, k \in \mathbb{N}$, and the pMHE iteration scheme in Algorithm 4.1 with $\mathrm{it}(k+1) \leq \mathrm{it}(k)$ and a step size in (4.7b) satisfying $\eta_k^i \leq \frac{\sigma_k}{L_k}$ for all $i = 0, \ldots, \mathrm{it}(k) - 1$ and $k \in \mathbb{N}$. Let Assumptions 4.1-4.4 hold and suppose that the MHE feasible set $\mathcal{S}_k$ given in (4.119) contains the true system state in its interior, i.e., $\mathbf{z}_k \in \mathrm{int}(\mathcal{S}_k)$. Then, there exists a constant $\tilde{c} \in \mathbb{R}_{++}$ such that*

$$\|\mathbf{z}_k - \hat{\mathbf{z}}_k^i\| \leq \tilde{c} \|\mathbf{z}_k - \bar{\mathbf{z}}_k\|, \tag{4.160}$$

where $\mathbf{z}_k$ denotes the true system state, $\hat{\mathbf{z}}_k^i$ a pMHE iterate, and $\bar{\mathbf{z}}_k$ the a priori estimate.

Proof. We invoke (4.124) in Lemma 4.7 and set $\mathbf{z}_1 = \mathbf{z}_k$ and $\mathbf{z}_2 = \hat{\mathbf{z}}_k^i$, i.e.,

$$\left\| \rho_{\eta_k^i}(\mathbf{z}_k) - \rho_{\eta_k^i}(\hat{\mathbf{z}}_k^i) \right\| \leq \frac{\eta_k^i}{\sigma_k} \left(L_k + \frac{\gamma_k}{\eta_k^i} \right) \left\| \mathbf{z}_k - \hat{\mathbf{z}}_k^i \right\|. \tag{4.161}$$

Since $\mathbf{z}_k \in \mathrm{int}(\mathcal{S}_k)$, and due to the absence of disturbances, we have that $\rho_{\eta_k^i}(\mathbf{z}_k) = \mathbf{z}_k$ by Lemma 4.6, which we plug in (4.161) along with $\rho_{\eta_k^i}(\hat{\mathbf{z}}_k^i) = \hat{\mathbf{z}}_k^{i+1}$ to obtain

$$\left\| \mathbf{z}_k - \hat{\mathbf{z}}_k^{i+1} \right\| \leq \left(\frac{\eta_k^i}{\sigma_k} L_k + \frac{\gamma_k}{\sigma_k} \right) \left\| \mathbf{z}_k - \hat{\mathbf{z}}_k^i \right\|. \tag{4.162}$$

Given that the step size satisfies $\eta_k^i \leq \frac{\sigma_k}{L_k}$, it holds that $\frac{\eta_k^i}{\sigma_k} L_k + \frac{\gamma_k}{\sigma_k} \leq 1 + \frac{\gamma_k}{\sigma_k}$ and hence

$$\left\| \mathbf{z}_k - \hat{\mathbf{z}}_k^i \right\| \leq \left(1 + \frac{\gamma_k}{\sigma_k} \right)^i \left\| \mathbf{z}_k - \hat{\mathbf{z}}_k^0 \right\|. \tag{4.163}$$

Since $\mathrm{it}(k+1) \leq \mathrm{it}(k)$, we have for every i-th iteration that $i \leq \mathrm{it}(k) \leq \mathrm{it}(0)$, which yields

$$\left\| \mathbf{z}_k - \hat{\mathbf{z}}_k^i \right\| \leq \left(1 + \frac{\gamma_k}{\sigma_k} \right)^{\mathrm{it}(0)} \left\| \mathbf{z}_k - \hat{\mathbf{z}}_k^0 \right\|. \tag{4.164}$$

By (4.30), it holds that $D_{\psi_k}(\mathbf{z}_k, \hat{\mathbf{z}}_k^0) \leq D_{\psi_k}(\mathbf{z}_k, \bar{\mathbf{z}}_k)$ and hence $\left\| \mathbf{z}_k - \hat{\mathbf{z}}_k^0 \right\| \leq \sqrt{\frac{\gamma_k}{\sigma_k}} \left\| \mathbf{z}_k - \bar{\mathbf{z}}_k \right\|$ due to the lower and upper bounds on the Bregman distance in (4.8) in Assumption 4.4. Using the last inequality in (4.164) leads to

$$\left\| \mathbf{z}_k - \hat{\mathbf{z}}_k^i \right\| \leq \left(1 + \frac{\gamma_k}{\sigma_k} \right)^{\mathrm{it}(0)} \sqrt{\frac{\gamma_k}{\sigma_k}} \left\| \mathbf{z}_k - \bar{\mathbf{z}}_k \right\|. \tag{4.165}$$

In view of the uniform strong convexity and strong smoothness constants σ and γ in Assumption 4.4, we obtain the desired result (4.160) with $\tilde{c} := \left(1 + \gamma/\sigma \right)^{\mathrm{it}(0)} \sqrt{\gamma/\sigma}$. $\qquad \square$

Remark 4.7. In this nonlinear setup, we cannot use the arguments in Remark 4.2 which exploit the convexity of the cost function F_k to establish that the sequence of function values of the pMHE iterates is nonincreasing if $\eta_k^i \leq \frac{\sigma_k}{L_k}$. Nevertheless, we can still arrive to the same statement due to the following. Given that $\hat{\mathbf{z}}_k^{i+1}$ minimizes the convex optimization problem (4.7b), the first-order optimality condition (4.22) holds for any $\mathbf{z} \in \mathcal{S}_k$. Since $\hat{\mathbf{z}}_k^i \in \mathcal{S}_k$, evaluating (4.22) at $\mathbf{z} = \hat{\mathbf{z}}_k^i$ yields

$$\left(\eta_k^i \nabla F_k \left(\hat{\mathbf{z}}_k^i \right) + \nabla \psi_k \left(\hat{\mathbf{z}}_k^{i+1} \right) - \nabla \psi_k \left(\hat{\mathbf{z}}_k^i \right) \right)^\top \left(\hat{\mathbf{z}}_k^i - \hat{\mathbf{z}}_k^{i+1} \right) \geq 0. \tag{4.166}$$

According to Assumption 4.3, the gradient of F_k is Lipschitz continuous with Lipschitz constant $L_k \in \mathbb{R}_{++}$. Hence, (4.24) holds. In particular, we have

$$\eta_k^i \left(F_k \left(\hat{\mathbf{z}}_k^i \right) - F_k \left(\hat{\mathbf{z}}_k^{i+1} \right) \right) + \frac{\eta_k^i L_k}{2} \left\| \hat{\mathbf{z}}_k^{i+1} - \hat{\mathbf{z}}_k^i \right\|^2 \geq \eta_k^i \nabla F_k \left(\hat{\mathbf{z}}_k^i \right)^\top \left(\hat{\mathbf{z}}_k^i - \hat{\mathbf{z}}_k^{i+1} \right). \tag{4.167}$$

We substitute the above inequality in (4.166) to obtain

$$\eta_k^i \left(\nabla F_k \left(\hat{\mathbf{z}}_k^i \right) - \nabla F_k \left(\hat{\mathbf{z}}_k^{i+1} \right) \right) \geq -\frac{\eta_k^i L_k}{2} \left\| \hat{\mathbf{z}}_k^{i+1} - \hat{\mathbf{z}}_k^i \right\|^2 + \left(\nabla \psi_k \left(\hat{\mathbf{z}}_k^{i+1} \right) - \nabla \psi_k \left(\hat{\mathbf{z}}_k^i \right) \right)^\top \left(\hat{\mathbf{z}}_k^{i+1} - \hat{\mathbf{z}}_k^i \right)$$

$$\geq -\eta_k^i L_k \left\| \hat{\mathbf{z}}_k^{i+1} - \hat{\mathbf{z}}_k^i \right\|^2 + \sigma_k \left\| \hat{\mathbf{z}}_k^{i+1} - \hat{\mathbf{z}}_k^i \right\|^2, \tag{4.168}$$

where the second inequality holds by the strong convexity of ψ_k in Assumption 4.4. Thus, also in the nonlinear case, by choosing the step size such that it satisfies $\eta_k^i \leq \frac{\sigma_k}{L_k}$ for all $i = 0, \ldots, \text{it}(k) - 1$ and $k \in \mathbb{N}$, we can make sure that the sequence $\left\{ F_k\left(\hat{\mathbf{z}}_k^0\right), \cdots, F_k\left(\hat{\mathbf{z}}_k^{\text{it}(k)}\right) \right\}$ generated from (4.7) at a given time instant k is nonincreasing.

The following theorem establishes sufficient conditions on $\text{it}(k)$, η_k^i, D_{ψ_k}, and Φ_k for the nominal local UES of the estimation error resulting from applying Algorithm 4.1 to estimate the state of the nonlinear system (4.1).

Theorem 4.5. *Consider system* (4.1) *with* $w_k = 0, v_k = 0, k \in \mathbb{N}$, *and the pMHE iteration scheme in Algorithm 4.1 with* $\text{it}(k+1) \leq \text{it}(k)$ *and a step size in* (4.7b) *satisfying*

$$\eta_k^i \leq \frac{\sigma}{L_F} \tag{4.169}$$

for all $i = 0, \cdots, \text{it}(k) - 1$ *and* $k \in \mathbb{N}$. *Let Assumptions 4.1-4.5 and 4.8 hold and suppose that the MHE feasible set* $\mathcal{S}_k$ *given in* (4.119) *contains the true system state in its interior, i.e.,* $\mathbf{z}_k \in \text{int}(\mathcal{S}_k)$. *Suppose that there exist* $\epsilon, c \in \mathbb{R}_{++}$ *such that the Bregman distance* D_{ψ_k} *and the a priori estimate operator* Φ_k *satisfy*

$$D_{\psi_k}(\Phi_{k-1}(\mathbf{z}), \Phi_{k-1}(\hat{\mathbf{z}})) - D_{\psi_{k-1}}(\mathbf{z}, \hat{\mathbf{z}}) \leq -c\|\mathbf{z} - \hat{\mathbf{z}}\|^2 \tag{4.170}$$

for all $k \in \mathbb{N}_+$ *and* $\mathbf{z}, \hat{\mathbf{z}}$ *with* $\|\mathbf{z} - \hat{\mathbf{z}}\| \leq \epsilon$. *Then, the estimation error* (4.13) *is locally UES, i.e., there exist* $\tilde{\epsilon}, \tilde{\alpha} \in \mathbb{R}_{++}$ *and* $\beta \in (0, 1)$ *such that*

$$\|x_k - \hat{x}_k\| \leq \tilde{\alpha}\beta^k \|x_0 - \bar{x}_0\| \tag{4.171}$$

holds for any $k \in \mathbb{N}_+$ *and* $x_0, \bar{x}_0 \in \mathbb{R}^n$ *with* $\|x_0 - \bar{x}_0\| \leq \tilde{\epsilon}$.

Proof. We follow similar steps as in the proofs of Theorems 3.4 and 4.1. We first prove that the pMHE error (4.12) is locally UES by using

$$V_k\left(\mathbf{z}_k, \hat{\mathbf{z}}_k^{\text{it}(k)}\right) = D_{\psi_k}\left(\mathbf{z}_k, \hat{\mathbf{z}}_k^{\text{it}(k)}\right) \tag{4.172}$$

as a continuous time-varying candidate Lyapunov function. More specifically, we show that V_k satisfies locally the conditions (A.6a) and (A.6b) in Theorem A.2 in Appendix A. Note that by Assumption 4.4, (A.6a) follows with $c_1 = \frac{\sigma}{2}$ and $c_2 = \frac{\gamma}{2}$.

Due to (4.136) in Lemma 4.8, we have

$$\Delta V_k := V_k\left(\mathbf{z}_k, \hat{\mathbf{z}}_k^{\text{it}(k)}\right) - V_{k-1}\left(\mathbf{z}_{k-1}, \hat{\mathbf{z}}_{k-1}^{\text{it}(k-1)}\right) \tag{4.173}$$

$$\leq D_{\psi_k}\left(\mathbf{z}_k, \Phi_{k-1}\left(\hat{\mathbf{z}}_{k-1}^{\text{it}(k-1)}\right)\right) + \frac{1}{2}\sum_{i=0}^{\text{it}(k)-1}\left(\eta_k^i L_k - \sigma_k\right)\left\|\hat{\mathbf{z}}_k^{i+1} - \hat{\mathbf{z}}_k^i\right\|^2 + \sum_{i=0}^{\text{it}(k)-1}\eta_k^i \kappa \left\|\mathbf{z}_k - \hat{\mathbf{z}}_k^i\right\|^3$$

$$- D_{\psi_{k-1}}\left(\mathbf{z}_{k-1}, \hat{\mathbf{z}}_{k-1}^{\text{it}(k-1)}\right)$$

for all $\mathbf{z}_k, \hat{\mathbf{z}}_k^i$ with $\|\mathbf{z}_k - \hat{\mathbf{z}}_k^i\| \leq \delta$. Condition (4.169) on the step size and Assumptions 4.3 and 4.4 imply that (4.40) holds true since $\eta_k^i L_k - \sigma_k \leq \frac{\sigma}{L_F} L_k - \sigma_k \leq \sigma - \sigma_k \leq 0$. Moreover, by Assumption 4.5, it holds for the disturbance-free true system state that $\mathbf{z}_k = \Phi_{k-1}(\mathbf{z}_{k-1})$ which gives (4.41). Using (4.40) and (4.41) in (4.173) leads to

$$\Delta V_k \leq D_{\psi_k}\left(\Phi_{k-1}(\mathbf{z}_{k-1}), \Phi_{k-1}\left(\hat{\mathbf{z}}_{k-1}^{\text{it}(k-1)}\right)\right) - D_{\psi_{k-1}}\left(\mathbf{z}_{k-1}, \hat{\mathbf{z}}_{k-1}^{\text{it}(k-1)}\right) + \sum_{i=0}^{\text{it}(k)-1}\eta_k^i \kappa \left\|\mathbf{z}_k - \hat{\mathbf{z}}_k^i\right\|^3 \tag{4.174}$$

for all $\mathbf{z}_k, \hat{\mathbf{z}}_k^i$ with $\|\mathbf{z}_k - \hat{\mathbf{z}}_k^i\| \leq \delta$. Let us now derive an upper bound on the last term in the right-hand side of (4.174). Since the step size in (4.169) satisfies $\eta_k^i \leq \frac{\sigma}{L_F} \leq \frac{\sigma_k}{L_k}$, we can invoke (4.160) in Lemma 4.9, in which we plug $\mathbf{z}_k = \Phi_{k-1}(\mathbf{z}_{k-1})$ as well as $\bar{\mathbf{z}}_k = \Phi_k\left(\hat{\mathbf{z}}_{k-1}^{\mathrm{it}(k-1)}\right)$ to obtain

$$\|\mathbf{z}_k - \hat{\mathbf{z}}_k^i\| \leq \tilde{c}\|\mathbf{z}_k - \bar{\mathbf{z}}_k\| \tag{4.175}$$
$$= \tilde{c}\left\|\Phi_{k-1}(\mathbf{z}_{k-1}) - \Phi_{k-1}\left(\hat{\mathbf{z}}_{k-1}^{\mathrm{it}(k-1)}\right)\right\|.$$

In view of the sufficient condition (4.170) and the uniform lower and upper bounds in Assumption 4.4, we have that

$$0 \leq \sigma\left\|\Phi_{k-1}(\mathbf{z}_{k-1}) - \Phi_{k-1}\left(\hat{\mathbf{z}}_{k-1}^{\mathrm{it}(k-1)}\right)\right\|^2 \leq (\gamma - 2c)\left\|\mathbf{z}_{k-1} - \hat{\mathbf{z}}_{k-1}^{\mathrm{it}(k-1)}\right\|^2 \tag{4.176}$$

for all $\mathbf{z}_{k-1}, \hat{\mathbf{z}}_{k-1}^{\mathrm{it}(k-1)}$ with $\|\mathbf{z}_{k-1} - \hat{\mathbf{z}}_{k-1}^{\mathrm{it}(k-1)}\| \leq \epsilon$. Using (4.176) in (4.175) yields

$$\|\mathbf{z}_k - \hat{\mathbf{z}}_k^i\| \leq \tilde{c}\sqrt{(\gamma - 2c)/\sigma}\left\|\mathbf{z}_{k-1} - \hat{\mathbf{z}}_{k-1}^{\mathrm{it}(k-1)}\right\| \tag{4.177}$$

for all $\mathbf{z}_{k-1}, \hat{\mathbf{z}}_{k-1}^{\mathrm{it}(k-1)}$ with $\|\mathbf{z}_{k-1} - \hat{\mathbf{z}}_{k-1}^{\mathrm{it}(k-1)}\| \leq \epsilon'$, where

$$\epsilon' := \min\left(\epsilon, \frac{\delta}{\tilde{c}\sqrt{(\gamma - 2c)/\sigma}}\right). \tag{4.178}$$

Note that this choice of ϵ' ensures that $\|\mathbf{z}_k - \hat{\mathbf{z}}_k^i\| \leq \delta$ in (4.177) and $\|\mathbf{z}_{k-1} - \hat{\mathbf{z}}_{k-1}^{\mathrm{it}(k-1)}\| \leq \epsilon$. Moreover, given that $\eta_k^i \leq \frac{\sigma}{L_F}$ and $\mathrm{it}(k+1) \leq \mathrm{it}(k)$,

$$\sum_{i=0}^{\mathrm{it}(k)-1} \eta_k^i \leq \frac{\sigma}{L_F}\mathrm{it}(k) \leq \frac{\sigma}{L_F}\mathrm{it}(0). \tag{4.179}$$

Thus, by (4.177) and the last inequality, we arrive at

$$\sum_{i=0}^{\mathrm{it}(k)-1} \eta_k^i\, \kappa\left\|\mathbf{z}_k - \hat{\mathbf{z}}_k^i\right\|^3 \leq \kappa\tilde{c}^3\sqrt{(\gamma - 2c)/\sigma}^3\left\|\mathbf{z}_{k-1} - \hat{\mathbf{z}}_{k-1}^{\mathrm{it}(k-1)}\right\|^3 \sum_{i=0}^{\mathrm{it}(k)-1} \eta_k^i \tag{4.180}$$
$$\leq \tilde{\kappa}\left\|\mathbf{z}_{k-1} - \hat{\mathbf{z}}_{k-1}^{\mathrm{it}(k-1)}\right\|^3$$

for all $\mathbf{z}_{k-1}, \hat{\mathbf{z}}_{k-1}^{\mathrm{it}(k-1)}$ with $\|\mathbf{z}_{k-1} - \hat{\mathbf{z}}_{k-1}^{\mathrm{it}(k-1)}\| \leq \epsilon'$, where $\tilde{\kappa} := \frac{\sigma}{L_F}\mathrm{it}(0)\,\kappa\,\tilde{c}^3\sqrt{(\gamma - 2c)/\sigma}^3$. Using (4.180) and condition (4.170) in the Lyapunov difference (4.174) yields

$$\Delta V_k \leq -c\left\|\mathbf{z}_{k-1} - \hat{\mathbf{z}}_{k-1}^{\mathrm{it}(k-1)}\right\|^2 + \tilde{\kappa}\left\|\mathbf{z}_{k-1} - \hat{\mathbf{z}}_{k-1}^{\mathrm{it}(k-1)}\right\|^3 \tag{4.181}$$

for any $k \in \mathbb{N}_+$ and $\mathbf{z}_{k-1}, \hat{\mathbf{z}}_{k-1}^{\mathrm{it}(k-1)}$ with $\|\mathbf{z}_{k-1} - \hat{\mathbf{z}}_{k-1}^{\mathrm{it}(k-1)}\| \leq \epsilon'$. We introduce $\bar{\epsilon} := \min\left(\epsilon', \frac{c}{2\tilde{\kappa}}\right)$ and obtain that ΔV_k is locally negative definite, i.e., it satisfies (A.6b) with $c_3 = \frac{c}{2}$ for all $\mathbf{z}_{k-1}, \hat{\mathbf{z}}_{k-1}^{\mathrm{it}(k-1)}$ with $\|\mathbf{z}_{k-1} - \hat{\mathbf{z}}_{k-1}^{\mathrm{it}(k-1)}\| \leq \bar{\epsilon}$. More specifically, in view of the uniform lower bound in Assumption 4.4, we have

$$0 \leq V_k\left(\mathbf{z}_k, \hat{\mathbf{z}}_k^{\mathrm{it}(k)}\right) \leq V_{k-1}\left(\mathbf{z}_{k-1}, \hat{\mathbf{z}}_{k-1}^{\mathrm{it}(k-1)}\right) - \frac{c}{2}\left\|\mathbf{z}_{k-1} - \hat{\mathbf{z}}_{k-1}^{\mathrm{it}(k-1)}\right\|^2$$
$$\leq \tilde{\beta}\, V_{k-1}\left(\mathbf{z}_{k-1}, \hat{\mathbf{z}}_{k-1}^{\mathrm{it}(k-1)}\right) \tag{4.182}$$

for all $k \in \mathbb{N}_+$ and $\mathbf{z}_{k-1}, \hat{\mathbf{z}}_{k-1}^{\text{it}(k-1)}$ with $\left\| \mathbf{z}_{k-1} - \hat{\mathbf{z}}_{k-1}^{\text{it}(k-1)} \right\| \leq \bar{\epsilon}$, where $\tilde{\beta} := 1 - c/\sigma$. Since $V_k = D_{\psi_k}$ is strictly positive for all $\mathbf{z}_k \neq \hat{\mathbf{z}}_k^{\text{it}(k)}$, and given that $c/\sigma \in \mathbb{R}_{++}$, it holds that $\tilde{\beta} \in (0,1)$. Hence, by (4.182) and Assumption 4.4, we get

$$\frac{\sigma}{2} \left\| \mathbf{z}_k - \hat{\mathbf{z}}_k^{\text{it}(k)} \right\|^2 \leq V_k \left(\mathbf{z}_k, \hat{\mathbf{z}}_k^{\text{it}(k)} \right) \leq \tilde{\beta}^k V_0 \left(\mathbf{z}_0, \hat{\mathbf{z}}_0^{\text{it}(0)} \right) \leq \tilde{\beta}^k \frac{\gamma}{2} \left\| \mathbf{z}_0 - \hat{\mathbf{z}}_0^{\text{it}(0)} \right\|^2. \tag{4.183}$$

This yields that, for all $\mathbf{z}_0, \hat{\mathbf{z}}_0^{\text{it}(0)}$ with $\left\| \mathbf{z}_0 - \hat{\mathbf{z}}_0^{\text{it}(0)} \right\| \leq \bar{\epsilon}$,

$$\left\| \mathbf{z}_k - \hat{\mathbf{z}}_k^{\text{it}(k)} \right\| \leq \alpha \beta^k \left\| \mathbf{z}_0 - \hat{\mathbf{z}}_0^{\text{it}(0)} \right\| = \alpha \beta^k \left\| \mathbf{z}_0 - \hat{\mathbf{z}}_0^0 \right\|, \tag{4.184}$$

where $\tilde{\epsilon} := \min(\bar{\epsilon}, \frac{\bar{\epsilon}}{\alpha})$, $\alpha := \sqrt{\gamma/\sigma}$, and $\beta := \sqrt{\tilde{\beta}} \in (0,1)$. Notice that with this choice of $\tilde{\epsilon}$, we can make sure that $\left\| \mathbf{z}_k - \hat{\mathbf{z}}_k^{\text{it}(k)} \right\| \leq \bar{\epsilon}$ in (4.184) and hence can recursively apply (4.182) to obtain (4.183). Moreover, the equality in (4.184) holds due to the fact that $\hat{\mathbf{z}}_0^i = \hat{\mathbf{z}}_0^0$ for all $i = 1, \cdots, \text{it}(0)$ (see the discussion below (4.43) for more detail). Evaluating (4.30) at time instant $k = 0$ and exploiting the uniform bounds in Assumption 4.4 give $\left\| \mathbf{z}_0 - \hat{\mathbf{z}}_0^0 \right\| \leq \sqrt{\frac{\gamma}{\sigma}} \left\| \mathbf{z}_0 - \bar{\mathbf{z}}_0 \right\|$, which we use in (4.184) to obtain

$$\left\| \mathbf{z}_k - \hat{\mathbf{z}}_k^{\text{it}(k)} \right\| \leq \alpha \sqrt{\frac{\gamma}{\sigma}} \beta^k \left\| \mathbf{z}_0 - \bar{\mathbf{z}}_0 \right\| \tag{4.185}$$

for all $\mathbf{z}_0, \bar{\mathbf{z}}_0$ with $\left\| \mathbf{z}_0 - \bar{\mathbf{z}}_0 \right\| \leq \tilde{\epsilon}$. Thus, the pMHE error (4.12) is locally UES.

In the following, we derive an upper bound on the estimation error $x_k - \hat{x}_k$ defined in (4.13) in terms of the pMHE error. We have by the triangular inequality

$$\left\| x_k - \hat{x}_k \right\| \leq \left\| x\left(k; x_{k-N}, k-N, \mathbf{u}_k\right) - x\left(k; \hat{x}_{k-N}^{\text{it}(k)}, k-N, \mathbf{u}_k\right) \right\| \tag{4.186}$$
$$+ \left\| x\left(k; \hat{x}_{k-N}^{\text{it}(k)}, k-N, \mathbf{u}_k\right) - x\left(k; \hat{x}_{k-N}^{\text{it}(k)}, k-N, \mathbf{u}_k, \hat{\mathbf{w}}_k^{\text{it}(k)}\right) \right\|.$$

In view of the uniform Lipschitz continuity of the function f_k in Assumption 4.1 and by recalling (4.12), it holds that

$$\left\| x\left(k; x_{k-N}, k-N, \mathbf{u}_k\right) - x\left(k; \hat{x}_{k-N}^{\text{it}(k)}, k-N, \mathbf{u}_k\right) \right\| \leq c_{\text{f}}^N \left\| e_{k-N} \right\|, \tag{4.187a}$$

$$\left\| x\left(k; \hat{x}_{k-N}^{\text{it}(k)}, k-N, \mathbf{u}_k\right) - x\left(k; \hat{x}_{k-N}^{\text{it}(k)}, k-N, \mathbf{u}_k, \hat{\mathbf{w}}_k^{\text{it}(k)}\right) \right\| \leq \sum_{j=k-N}^{k-1} c_{\text{f}}^{k-j} \left\| \hat{w}_j^{\text{it}(k)} \right\|. \tag{4.187b}$$

Substituting (4.187) in (4.186) yields

$$\left\| x_k - \hat{x}_k \right\| \leq \bar{c} \left\| \mathbf{z}_k - \hat{\mathbf{z}}_k^{\text{it}(k)} \right\|, \tag{4.188}$$

where $\bar{c} := c_{\text{f}}^N + \sum_{j=1}^N c_{\text{f}}^j$. By (4.185) and the fact that $\mathbf{z}_0 = x_0$ and $\bar{\mathbf{z}}_0 = \bar{x}_0$, we get

$$\left\| x_k - \hat{x}_k \right\| \leq \bar{c} \alpha \sqrt{\frac{\gamma}{\sigma}} \beta^k \left\| \mathbf{z}_0 - \bar{\mathbf{z}}_0 \right\| = \bar{c} \alpha \sqrt{\frac{\gamma}{\sigma}} \beta^k \left\| x_0 - \bar{x}_0 \right\| \tag{4.189}$$

for all $x_0, \bar{x}_0$ with $\left\| x_0 - \bar{x}_0 \right\| \leq \tilde{\epsilon}$. This proves (4.171) with $\tilde{\alpha} := \bar{c} \alpha \sqrt{\frac{\gamma}{\sigma}}$ and hence local UES of the estimation error. $\qquad \square$

Theorem 4.5 implies that Algorithm 4.1 possesses the anytime property also for the nonlinear case, in the sense that local exponential stability of the estimation error can be ensured independently of the number of optimization algorithm iterations it(k). A crucial issue for this anytime property to hold is the validity of condition (4.170). It states that, given two vectors $\mathbf{z}, \hat{\mathbf{z}}$ with $\|\mathbf{z} - \hat{\mathbf{z}}\| \leq \epsilon$ and the Bregman distance $D_{\psi_{k-1}}(\mathbf{z}, \hat{\mathbf{z}})$, a prediction from the a priori estimate operator Φ_{k-1} will yield a contracting Bregman distance D_{ψ_k}. Hence, condition (4.170) can be fulfilled by designing the a priori estimate operator from an estimator with locally (exponentially) stable dynamics and selecting the Bregman distance as the Lyapunov function with which the local stability of this estimator can be verified. This includes for instance the computationally efficient discrete-time EKF.

Φ_k and D_{ψ_k} based on the extended Kalman filter

We adapt the design approach based on the EKF discussed in Section 3.3.1 to the pMHE iteration scheme. More specifically, we use it for designing the Bregman distance D_{ψ_k} in lines 3 and 4 of Algorithm 4.1 and for generating the a priori estimate operator Φ_k in line 7 of the algorithm.

Let $\mathbf{z} = \begin{bmatrix} x^\top & \mathbf{w}^\top \end{bmatrix}^\top$, $\hat{\mathbf{z}} = \begin{bmatrix} \hat{x}^\top & \hat{\mathbf{w}}^\top \end{bmatrix}^\top$, $x, \hat{x} \in \mathbb{R}^n$, $\mathbf{w}, \hat{\mathbf{w}} \in \mathbb{R}^{Nm_w}$. We construct the EKF-based a priori estimate operator at the time instant k as follows:

$$x^+ = x + K_{k-N} \left(y_{k-N} - h_{k-N}(x) \right) \tag{4.190a}$$

$$\Phi_k(\mathbf{z}) = \begin{bmatrix} f_{k-N}(x^+, u_{k-N}, 0) \\ 0 \end{bmatrix}, \tag{4.190b}$$

where $0 \in \mathbb{R}^{Nm_w}$. Here, the EKF gain $K_{k-N} \in \mathbb{R}^{n \times p}$ is obtained as

$$K_{k-N} = P_{k-N}^- C_{k-N}^\top \left(C_{k-N} P_{k-N}^- C_{k-N}^\top + R \right)^{-1}, \tag{4.191a}$$

$$P_{k-N+1}^- = A_{k-N} P_{k-N}^+ A_{k-N}^\top + Q, \qquad \Pi_{k-N}^- := \left(P_{k-N}^- \right)^{-1}, \tag{4.191b}$$

$$P_{k-N}^+ = (I - K_{k-N} C_{k-N}) P_{k-N}^-, \qquad \Pi_{k-N}^+ := \left(P_{k-N}^+ \right)^{-1}, \tag{4.191c}$$

where $P_0^- \in \mathbb{S}_{++}^n$, $Q \in \mathbb{S}_{++}^n$, $R \in \mathbb{S}_{++}^p$ and

$$A_{k-N} = \left. \frac{\partial f_{k-N}}{\partial x} \right|_{(x^+, u_{k-N}, 0)}, \qquad C_{k-N} = \left. \frac{\partial h_{k-N}}{\partial x} \right|_x. \tag{4.191d}$$

In particular, consider the last pMHE iterate $\hat{\mathbf{z}}_k^{\mathrm{it}(k)}$ computed at time k. In view of line 7 of Algorithm 4.1, the a priori estimate, whose notation is given in (4.6), is calculated for the next time instant as

$$\bar{\mathbf{z}}_{k+1} = \Phi_k \left(\hat{\mathbf{z}}_k^{\mathrm{it}(k)} \right) = \begin{bmatrix} \bar{x}_{k-N+1}^\top & 0 & \cdots & 0 \end{bmatrix}^\top, \tag{4.192a}$$

where we have zero a priori process disturbances, i.e., $\bar{w}_j = 0$ for $j = k - N + 1, \ldots, k$, and

$$\bar{x}_{k-N}^+ = \hat{x}_{k-N}^{\mathrm{it}(k)} + K_{k-N} \left(y_{k-N} - h_{k-N} \left(\hat{x}_{k-N}^{\mathrm{it}(k)} \right) \right) \tag{4.192b}$$

$$\bar{x}_{k-N+1} = f_{k-N} \left(\bar{x}_{k-N}^+, u_{k-N}, 0 \right). \tag{4.192c}$$

The associated Bregman distance used at time instant k is

$$D_{\psi_k}(\mathbf{z}, \hat{\mathbf{z}}) = \frac{1}{2} \|x - \hat{x}\|_{\Pi_{k-N}^{-}}^2 + \frac{1}{2} \|\mathbf{w} - \hat{\mathbf{w}}\|_{\bar{W}}^2, \qquad (4.193)$$

where $\Pi_{k-N}^{-} \in \mathbb{S}_{++}^n$ given in (4.191b) denotes the inverse of the EKF covariance matrix and $\bar{W} \in \mathbb{S}_{++}^{Nm_w}$ is a weight matrix of the form $\bar{W} = \mathrm{diag}(W, \dots, W)$ with arbitrary $W \in \mathbb{S}_{++}^{m_w}$. In order to be able to exploit the results of Section 3.3.1 and show that the proposed D_{ψ_k} and Φ_k based on the EKF satisfy condition (4.170) in Theorem 4.5, we require the pMHE iteration scheme to inherit the assumptions for ensuring exponential stability of the EKF. These can be found in Assumption 3.8 and are summarized here for completeness.

Assumption 4.9 (Nonlinear system properties). *For any $k \in \mathbb{N}$, A_k is nonsingular, the spectral norms of the Jacobian matrices A_k and C_k defined in (4.191d) are uniformly bounded by $\|A_k\| \leq \bar{a}$, $\|C_k\| \leq \bar{c}$ for some $\bar{a}, \bar{c}, \in \mathbb{R}_{++}$, and the matrices P_k^-, P_k^+ are uniformly bounded as*

$$\underline{p}'I \preceq P_k^- \preceq \bar{p}'I, \quad \underline{p}I \preceq P_k^+ \preceq \bar{p}I, \qquad (4.194)$$

where $\underline{p}', \bar{p}', \underline{p}, \bar{p} \in \mathbb{R}_{++}$.

If Assumption 4.9 holds, we can invoke Proposition 3.3 and establish that the Bregman distance (4.193) and the a priori estimate operator (4.190) satisfy Assumptions 4.4 and 4.5 as well as (4.170) in Theorem 4.5. The stability properties of anytime pMHE based on the EKF can now be easily deduced.

Corollary 4.3. *Consider system (4.1) with $w_k = 0$, $v_k = 0$, $k \in \mathbb{N}$, and the pMHE iteration scheme in Algorithm 4.1 with $\mathrm{it}(k+1) \leq \mathrm{it}(k)$, a step size in (4.7b) satisfying $\eta_k^i \leq \frac{\sigma}{L_F}$ for all $i = 0, \cdots, \mathrm{it}(k) - 1$ and $k \in \mathbb{N}$, a priori estimate operator (4.190) and Bregman distance (4.193) based on the EKF. Let Assumptions 4.1-4.3, 4.8-4.9 hold and suppose that the MHE feasible set $\mathcal{S}_k$ given in (4.119) contains the true system state in its interior, i.e., $\mathbf{z}_k \in \mathrm{int}(\mathcal{S}_k)$. Then, the estimation error (4.13) is locally UES.*

Corollary 4.3 shows that exponential stability of the estimation error generated by the pMHE iteration scheme is ensured independently of the number of internal iterations due to the implicit stabilizing regularization approach of the EKF. This is in contrast to the (explicit) stabilizing regularization of the EKF-based a priori estimate established for the pMHE scheme (3.133) with a priori estimate operator (3.186) and Bregman distance (3.189) in Corollary 3.4. Although stability of the pMHE iteration scheme is induced from the EKF, the latter is used in the a priori estimate only to warm-start the optimization algorithm. Hence, the (suboptimal) bias of the EKF is fading away each time we perform the optimization iteration step (4.7b) and an improved performance can be achieved with each iteration.

Remark 4.8. It is worth mentioning that we can also employ anytime pMHE for nonlinear systems of the form (3.210) which can be transformed into systems (3.211) that are affine in the unmeasured state if Assumptions 4.1-4.3 hold. In particular, we can adapt the design approach discussed in Section 3.3.2 and use the Luenberger observer to construct the a priori estimate $\bar{\mathbf{z}}_{k+1} = \begin{bmatrix} \bar{x}_{k-N+1}^\top & 0 & \dots & 0 \end{bmatrix}^\top$ as follows:

$$\bar{x}_{k-N+1} = A(y_{k-N})\hat{x}_{k-N}^{\mathrm{it}(k)} + \phi(y_{k-N}) + L(y_{k-N})\left(y_{k-N} - C\hat{x}_{k-N}^{\mathrm{it}(k)}\right) + B(y_{k-N}) u_{k-N}. \qquad (4.195)$$

The associated Bregman distance is $D_\psi\left(\mathbf{z}, \hat{\mathbf{z}}\right) = \frac{1}{2}\left\|x - \hat{x}\right\|_P^2 + \frac{1}{2}\left\|\mathbf{w} - \hat{\mathbf{w}}\right\|_{\bar{W}}^2$, where $P \in \mathbb{S}_{++}^n$ and $\bar{W} = \mathrm{diag}(W, \ldots, W) \in \mathbb{S}_{++}^{Nn}$. As discussed in Section 3.3.2, the resulting MHE cost function defined in (4.5) is convex in the decision variables, and we can hence use the steps of Lemma 4.1 to obtain that

$$D_\psi\left(\mathbf{z}_k, \hat{\mathbf{z}}_k^{\mathrm{it}(k)}\right) \leq D_\psi\left(\mathbf{z}_k, \bar{\mathbf{z}}_k\right) + \frac{1}{2}\sum_{i=0}^{\mathrm{it}(k)-1}\left(\eta_k^i L_k - \sigma_k\right)\left\|\hat{\mathbf{z}}_k^{i+1} - \hat{\mathbf{z}}_k^i\right\|^2. \tag{4.196}$$

We assume that the step size in (4.7b) satisfies $\eta_k^i \leq \frac{\sigma_k}{L_k}$ for all $i = 0, \cdots, \mathrm{it}(k) - 1$ and $k \in \mathbb{N}$ and that the weight matrix P in the Bregman distance fulfills the LMI

$$\left(A(y_k) - L(y_k)C\right)^\top P\ \left(A(y_k) - L(y_k)C\right) - P \preceq -Q. \tag{4.197}$$

By considering $V\left(\mathbf{z}_k, \hat{\mathbf{z}}_k^{\mathrm{it}(k)}\right) = D_\psi\left(\mathbf{z}_k, \hat{\mathbf{z}}_k^{\mathrm{it}(k)}\right)$ as a candidate Lyapunov function, we can therefore show that the pMHE error (4.12) is GUES, i.e., we have that

$$V\left(\mathbf{z}_k, \hat{\mathbf{z}}_k^{\mathrm{it}(k)}\right) - V\left(\mathbf{z}_{k-1}, \hat{\mathbf{z}}_{k-1}^{\mathrm{it}(k-1)}\right) \leq D_\psi\left(\mathbf{z}_k, \bar{\mathbf{z}}_k\right) - D_\psi\left(\mathbf{z}_{k-1}, \hat{\mathbf{z}}_{k-1}^{\mathrm{it}(k-1)}\right) \tag{4.198}$$

$$\leq -\lambda_{\min}(Q)\left\|e_{k-N-1}\right\|^2 - \lambda_{\min}(\bar{W})\left\|\hat{\mathbf{w}}_{k-1}^{\mathrm{it}(k-1)}\right\|^2,$$

from which GUES of the estimation error $x_k - \hat{x}_k$ follows.

4.5 Numerical examples

In this section, we consider four academic examples with which we demonstrate the stability and performance properties of the anytime pMHE iteration scheme. First, in order to estimate the state of the linearized system of a batch reactor, we employ Algorithm 4.1 with a priori estimate operator (4.57) and Bregman distance (4.59) based on the Luenberger observer. Second, for the state estimation problem of the nonlinear system of a batch reactor, we employ Algorithm 4.1 with a priori estimate operator (4.190) and Bregman distance (4.193) based on the EKF. For both cases, we illustrate the resulting regret with different numbers of optimization algorithm iterations. Finally, we briefly discuss in the last two examples a special application where the proposed anytime pMHE iteration scheme is combined with so-called relaxed barrier function based linear MPC algorithms in a certainty equivalence output feedback fashion. More specifically, we consider the double integrator with which we investigate, amongst others, the computational efficiency of anytime pMHE, and the three tank system with which we demonstrate how to tackle outliers and ensure an overall satisfactory performance of the output feedback controller given an appropriate choice of the MHE cost. This section is based on and taken in parts literally from (Gharbi et al. (2020b))[3], (Gharbi et al. (2021))[4], and (Gharbi and Ebenbauer (2021))[5].

[3]M. Gharbi, B. Gharesifard, and C. Ebenbauer. Anytime proximity moving horizon estimation: Stability and regret. In *arXiv preprint arXiv:2006.14303*, 2020b.

[4]M. Gharbi, B. Gharesifard, and C. Ebenbauer. Anytime proximity moving horizon estimation: Stability and regret for nonlinear systems. *In Proc. 60th Conference on Decision and Control (CDC), Accepted.* IEEE, 2021 © 2021 IEEE.

[5]M. Gharbi and C. Ebenbauer. Anytime MHE-based output feedback MPC. In *Proc. 7th IFAC Conference on Nonlinear Model Predictive Control*, volume 54, pages 264–271, 2021 © 2021 Elsevier Ltd.

4.5.1 Anytime pMHE for linear systems: Constant volume batch reactor

We consider again the example of a well-mixed, constant volume, isothermal batch reactor, which we introduced in Section 3.4.2. The associated discretized LTI system with a sampling time of $T_s = 0.25$ s is of the form (4.54) with matrices A, B, C in (3.236) and (A, C) observable. Given that the states represent concentrations, they are constrained to be nonnegative, i.e., $x_k \geq 0$. We employ the anytime pMHE iteration scheme in Algorithm 4.1 based on the Luenberger observer, i.e., with a priori estimate operator (4.57) and Bregman distance (4.59). In the MHE cost function F_k given in (4.5a), we set the process disturbances $\hat{w}_i$ as well as $q_i = q$ to zero, such that we only consider the first state in the horizon window $\hat{x}_{k-N}$ as decision variable. The function $r_i = r$ in F_k is chosen as $r(v) = \frac{1}{2}\|v\|_R^2$ with $R = 0.01$. The resulting cost function at time k is then

$$F_k(\hat{x}_{k-N}) = \frac{1}{2} \sum_{i=k-N}^{k-1} \left\| y_i - CA^{i-k+N} \hat{x}_{k-N} \right\|_R^2. \tag{4.199}$$

We choose the horizon length $N = 2$ and design the estimator such that the assumptions and conditions of Theorem 4.3 (or Corollary 4.2) are fulfilled. In particular, in order to satisfy the sufficient condition (4.105) on the Bregman distance D_ψ and a priori estimate operator Φ_k, we design the observer gain L in (4.57) such that the eigenvalues of $A - LC$ are strictly in the unit circle. More specifically, we have that $\lambda_i(A - LC) = \begin{bmatrix} 0.4754 & 0.8497 & 0.9727 \end{bmatrix}^\top$. Furthermore, in the quadratic Bregman distance $D_\psi(x, \hat{x}) = \frac{1}{2}\|x - \hat{x}\|_P^2$, we design the weight matrix $P \in \mathbb{S}_{++}^3$ such that the LMI (4.60) is satisfied. In addition, we use at each time instant k the same number of optimization iterations it(k), i.e., we fix it$(k) = $ it$(k+1) =:$ it. The step size is set to $\eta_k^i = \frac{\sigma}{L_F} \frac{1}{\sqrt{k}}$ in accordance with (4.104). In Assumption 4.4, σ denotes the uniform strong convexity constant of the function ψ used to construct the Bregman distance D_ψ and given by $\sigma = \min(\lambda_i(P))$. In Assumption 4.3, L_F denotes the uniform Lipschitz constant of the gradient of F_k. It can be computed as

$$L_F = R \sum_{i=k-N}^{k-1} \left\| CA^{i-k+N} \right\|^2. \tag{4.200}$$

The true initial state and the initial guess are given by $x_0 = [0.5\ 0.05\ 0]^\top$ and $\bar{x}_0 = [0.1\ 0.1\ 6]^\top$, respectively.

Stability results   Let us start by illustrating the stability properties of the resulting estimation error. For the sake of comparison, we also employ the Luenberger observer designed with the same matrix L and the pMHE scheme based on the Luenberger observer (Algorithm 3.1), where the pMHE optimization problem (3.117) is solved at each time instant k. For this estimator, we choose the same design parameters of the anytime pMHE iteration scheme given by N, F_k, D_ψ and L. The resulting estimation errors for each estimation strategy are shown in Figure 4.2.

All estimators exhibit GES of the estimation errors. This includes the case where we execute a single iteration per time instant k, i.e., it $= 1$. Note that for a small number of iterations, the choice of the observer gain L affects the performance of the estimator. In this case, it is useful to tune L such that a satisfactory performance is attained. Nevertheless,

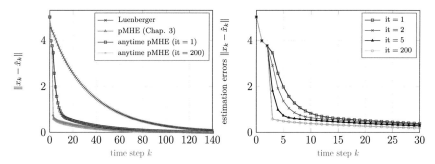

Figure 4.2: *Left:* Estimation errors of the Luenberger observer, the pMHE scheme in Algorithm 3.1, and the pMHE iteration scheme in Algorithm 4.1 with it = 1 and it = 200. *Right:* Estimation errors generated by the pMHE iteration scheme with different numbers of optimization algorithm iterations.

if we perform many iterations and set it = 200 for instance, the choice of L does not have much impact on the performance of the iteration scheme. In fact, we can observe that for the aforementioned scenario, Algorithm 4.1 yields smaller estimation errors than those generated by Algorithm 3.1. We can also observe the effect of increasing the number of iterations executed in the pMHE iteration scheme on the speed of convergence of the estimation error, where the more we iterate, the faster is the convergence of the estimation error to zero.

Regret results In the following, we investigate the regret (4.63) generated by anytime pMHE relative to a comparator sequence given by the true states x_{k-N}. Note that

$$F_k(x_{k-N}) = \frac{1}{2} \sum_{i=k-N}^{k-1} \left\| y_i - CA^{i-k+N} x_{k-N} \right\|_R^2 = 0. \tag{4.201}$$

In particular, we study the effect of increasing the number of iterations on the resulting regret. The results are depicted in Figure 4.3, where we compute the regrets $R(T)$ as well as the average regrets $R(T)/T$ after each simulation time T. Moreover, we plot the regret associated to the pMHE scheme in Algorithm 3.1 relative to this optimal sequence of true states.

Observe that, by increasing the number of optimization algorithm iterations, we can achieve lower regrets $R(T)$ as well as average regrets $R(T)/T$ that tend to zero for $T \to \infty$. This observation is in line with the regret upper bound (4.106) obtained in Theorem 4.3. Furthermore, we can see that the regret obtained by anytime pMHE with it = 20 is lower than that obtained by the pMHE scheme in Algorithm 3.1. This can be explained by the fact that the effect of the Luenberger observer, which might degenerate performance, becomes less significant each time we perform the optimization iteration step (4.7b). In fact, in order to illustrate how the Luenberger observer might impact the performance of the estimator, we also compute the regret $R(T)$ generated by the pMHE iteration scheme in which, instead of centering the Bregman distance around the previous iterate (see (4.7b)),

135

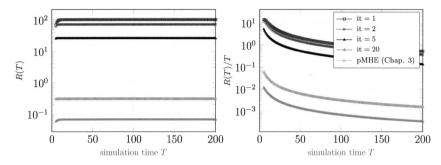

Figure 4.3: *Left:* The resulting regrets of anytime pMHE with different numbers of optimization iterations. *Right:* The resulting average regrets of anytime pMHE with different numbers of optimization iterations.

we let

$$\hat{\mathbf{z}}_k^{i+1} = \arg\min_{\mathbf{z} \in \mathcal{S}_k} \left\{ \eta_k^i \nabla F_k \left(\hat{\mathbf{z}}_k^i \right)^\top \mathbf{z} + D_\psi(\mathbf{z}, \bar{\mathbf{z}}_k) \right\}. \tag{4.202}$$

In this case, the Bregman distance is always centered around the current a priori estimate $\bar{\mathbf{z}}_k$ given by the Luenbeger observer. The results are depicted in Figure 4.4.

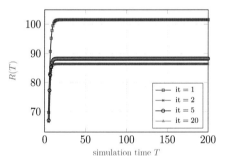

Figure 4.4: The resulting regret for the pMHE iteration scheme with update step (4.202) and with different numbers of optimization iterations.

As demonstrated, increasing the number of iterations per time instant in this case does not necessarily lead to lower regrets.

4.5.2 Anytime pMHE for nonlinear systems: Constant volume batch reactor

In the following, we illustrate the stability and performance properties of the pMHE iteration scheme in Algorithm 4.1 when applied to the nonlinear estimation problem of a constant volume batch reactor, whose continuous-time model is given in (3.237). Recall that the state consists of the partial pressures $x = \begin{bmatrix} p_A & p_B \end{bmatrix}^\top$ and is constrained to be nonnegative. The initial condition of the system is $x_0 = [3\ 1]^\top$.

The goal is to estimate the state x_k by employing the pMHE iteration scheme based on the EKF, where we choose the a priori estimate operator (4.190) and the Bregman distance (4.193). Since we consider only $\hat{x}_{k-N}$ as decision variable by setting $q = 0$ in (4.5a) and ignoring the process disturbances, the Bregman distance in the iteration procedure (4.7) takes the form $D_{\psi_k}(x, \hat{x}) = \frac{1}{2}\|x - \hat{x}\|^2_{\Pi^-_{k-N}}$. Furthermore, we select $r_i(v) = \frac{1}{2}\|v\|^2_{R^{-1}}$ in the cost function (4.5a), choose the horizon length $N = 3$, and fix the number of optimization algorithm iterations by setting $\text{it} := \text{it}(k) = \text{it}(k+1)$ for all time instants k.

Stability results In the following, we focus on the stability properties of the resulting estimation error. We compare the obtained results with the EKF designed accordingly and the nonlinear pMHE scheme (3.133) with Bregman distance and a priori estimate based on the EKF. The corresponding design parameters are $R = 0.01$, $Q = \text{diag}(10^{-4}, 0.01)$, and $P_0 = I_2$. We initialize the considered estimators with a poor initial guess $\bar{x}_0 = \begin{bmatrix} 0.1 & 4.5 \end{bmatrix}^\top$. The resulting state estimates and estimation errors are depicted in Figures 4.5 and 4.6, respectively.

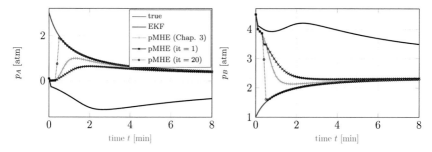

Figure 4.5: Resulting estimates of p_A, p_B generated by the EKF, the pMHE scheme in Algorithm 3.1, and the pMHE iteration scheme in Algorithm 4.1.

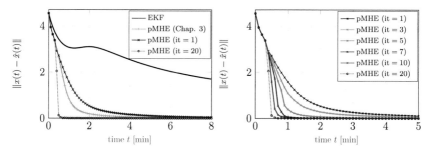

Figure 4.6: Resulting estimation errors generated by the EKF, the pMHE scheme in Algorithm 3.1, and the pMHE iteration scheme in Algorithm 4.1.

Figure 4.5 shows that the EKF estimates of the partial pressure p_A violate the inequality constraints and are negative. In contrast, the employed pMHE approaches yield positive estimates and exponentially stable estimation errors as depicted in Figure 4.6. Moreover, on the left of the latter figure, we can see that using the pMHE iteration scheme in Algorithm 4.1 with it = 20 yields estimation errors smaller than those generated by the pMHE scheme in Algorithm 3.1, which are in turn smaller that those of Algorithm 4.1 with a single iteration it = 1. On the right of Figure 4.6, we can observe that increasing the number of optimization algorithms iterations in the pMHE iteration scheme yields smaller estimation errors. This observation is validated by the computed root mean square errors (RMSE) defined in (3.238) for different values of optimization iterations, where $T_{sim} = 100$ denotes the simulation time for the discretized system. These are reported in Table 4.2, along with the obtained RMSE with other values of the horizon length N.

Table 4.2: RMSE of the pMHE iteration scheme for different numbers of optimization algorithm iterations it and values of the horizon length N.

RMSE	$N = 2$	$N = 3$	$N = 5$	$N = 10$
it = 1	1.0301	0.7747	0.1724	0.0254
it = 5	0.8903	0.5985	0.1560	0.0203
it = 10	0.7583	0.4790	0.1453	0.0193
it = 20	0.5698	0.4011	0.1410	0.0193

As expected, the RMSE decreases with increasing values of optimization algorithm iterations as well as with increasing values of N.

Regret results We investigate for Algorithm 4.1 as well as for the nonlinear pMHE scheme in Algorithm 3.1 the resulting regret (4.63) with respect to the comparator sequence given by the true system state, which yields zero cost F_k. More specifically, after each simulation time T, we compute and plot the resulting regrets $R(T)$ and average regrets $R(T)/T$ in Figure 4.7.

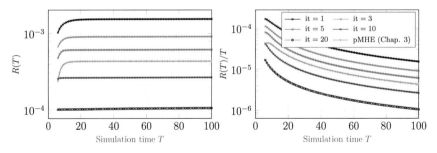

Figure 4.7: Resulting regrets and average regrets of Algorithm 4.1 with different numbers of optimization algorithms iterations.

Although the underlying MHE problem is nonconvex and hence the results of Theorem

4.3 cannot be applied in this case, we can see that the regret decreases with an increasing number of iterations and that the average regret converges to zero for $T \to \infty$.

4.5.3 Output feedback control: Double integrator

In the following, we design an output feedback controller (OFC) for the example of a double integrator, whose model is given by

$$x_{k+1} = \begin{bmatrix} 1 & T_s \\ 0 & 1 \end{bmatrix} x_k + \begin{bmatrix} T_s^2 \\ T_s \end{bmatrix} u_k, \qquad T_s = 0.1, \tag{4.203a}$$

$$y_k = \begin{bmatrix} 1 & 0 \end{bmatrix} x_k. \tag{4.203b}$$

The control goal is to steer the state of the system to the origin. We assume input and state constraints of the form $\mathcal{U} := \{u \in \mathbb{R} | -1 \le u \le 1\}$ and $\mathcal{X} := \{x \in \mathbb{R}^2 | -2 \le x_1 \le 3, -1 \le x_2 \le 1\}$. Note that the pair (A, B) is controllable and that the pair (A, C) is observable. The employed OFC algorithm combines the so-called anytime relaxed barrier function based MPC algorithm (rbMPC) introduced in (Feller and Ebenbauer (2017b)) with the proposed anytime pMHE approach in a certainty equivalence fashion. The stabilizing rbMPC iteration scheme is tailored to a relaxed barrier function based MPC formulation that replaces the state and input constraints by suitable barrier functions in the cost function (cf. Remark 3.8). Furthermore, it performs a limited number of optimization algorithm iterations and is shown in case of full state feedback to possess the anytime property, where asymptotic stability of the closed-loop system is guaranteed after any number of optimization iterations. In (Gharbi and Ebenbauer, 2021, Corollary 1), it is shown that the state of the overall closed-loop system resulting from the OFC algorithm remains bounded and converges to the origin under mild assumptions and for any number of rbMPC and pMHE optimization algorithm iterations. We refer the reader to (Gharbi and Ebenbauer (2021)) for further details on the considered output feedback control strategy. In this example, the underlying pMHE and rbMPC iterations have to be executed within the system sampling time $T_s = 0.1$ s. More specifically, at each time instant, a limited number of pMHE iterations are performed to compute a state estimate, based on which a limited number of rbMPC iterations, referred to as $\mathrm{it}_c \in \mathbb{R}_{++}$, are executed to deliver the control input. In the rbMPC iteration scheme (Feller and Ebenbauer, 2017b, Section 3), the stage cost of the associated optimization problem has the form

$$l(x, u) := \|x\|_{Q_c}^2 + \|u\|_{R_c}^2 + \varepsilon B_\mathrm{x}(x) + \varepsilon B_\mathrm{u}(u), \tag{4.204}$$

where the functions $B_\mathrm{x} : \mathbb{R}^n \to \mathbb{R}_+$ and $B_\mathrm{u} : \mathbb{R}^m \to \mathbb{R}_+$ refer to recentered relaxed logarithmic barrier functions for the polytopic sets $\mathcal{X}$ and $\mathcal{U}$. We choose the design parameters as $Q_c = \mathrm{diag}(1, 0.1)$, $R_c = 0.1$, and the relaxed barrier function weighting parameter as $\varepsilon = 10^{-3}$. Moreover, we choose the horizon length $N_c = 30$ and set the number of rbMPC optimization iterations to $\mathrm{it}_c = 10$, since we observed in simulations that these choices yield a satisfactory control performance. In the pMHE iteration scheme in Algorithm 4.1, we employ the Luenberger observer to construct the a priori estimate operator (4.57) and the associated Bregman distance in (4.59). Moreover, we consider for simplicity only $\hat{x}_{k-N}$ as a decision variable and omit the state constraints. This yields a pMHE update step (4.7b) of the form (4.15). In addition, we choose the sage cost functions in (4.5a) as

$r_i(v) = 1/2\|v\|_R^2$ with $R = 10^3$ and $q_i(w) = 0$. In order to satisfy the conditions of Corollary 4.2, we design the weight matrix P in the Bregman distance $D_\psi(x, \hat{x}) = \frac{1}{2}\|x - \hat{x}\|_P^2$ such that the LMI (4.60) is satisfied and set the step size $\eta_k^i = \lambda_{\min}(P)/L_F$. Since the setup is similar to the first example in Section 4.5.1, the constant L_F can be computed as in (4.200). We fix the number of pMHE iterations, i.e., we let it $:= \text{it}(k) = \text{it}(k+1)$ for all time instants k. In the following simulations, we vary the horizon length N, the initial guess $\bar{x}_0$, and the number of optimization algorithm iterations.

In Figure 4.8, we vary the pMHE horizon length N and plot the resulting RMSE defined in (3.238) as an estimation performance index. We also plot as a closed-loop performance index the averaged cumulative stage cost

$$\bar{L} = \frac{1}{T_{\text{sim}} - N + 1} \sum_{k=N}^{T_{\text{sim}}} l(x, u), \tag{4.205}$$

where the rbMPC stage cost l is defined in (4.204) and the simulation time is set to $T_{\text{sim}} = 300$. In Figure 4.9, we depict the average pMHE computation time per time instant k and the averaged cumulative stage cost $\bar{L}$ over the number of pMHE optimization iterations. For both aforementioned figures, the initial condition is set to $x_{0,1} = [2.5, -0.65]$. We can see in Figure 4.8 that a longer estimation horizon N can decrease the RMSE and improve the closed-loop performance. This effect can be also observed in Figure 4.9 when increasing the number of pMHE optimization algorithm iterations, at the cost of a higher computation time. However, even for a large number of iterations, the resulting average computation time remains very small.

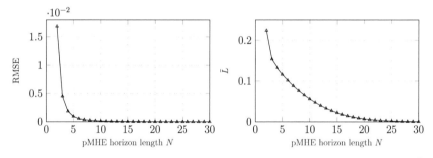

Figure 4.8: The resulting RMSE defined in (3.238) and averaged cumulative stage cost $\bar{L}$ defined in (4.205) over the pMHE horizon length. Here, we choose the initial condition $x_{0,1} = [2.5, -0.65]$ and the initial guess $\bar{x}_0 = [0, 0]$, and set the number of pMHE optimization iterations as it $= 1$.

In Figure 4.10, we consider different initial guesses and plot the corresponding closed-loop state trajectories as well as the evolution of the state x_2 and its estimate over time. Note that the initial condition $x_{0,3} = [1.25, 1.25]$ in this case lies outside the constraint set $\mathcal{X}$. The results of the nominal rbMPC iteration scheme with full state measurement (SFC), i.e., with $C = I_2$, are also depicted.

In all scenarios, the state of the closed-loop system converges successfully to zero. Moreover, we can see that the resulting trajectories are able to remain within the boundaries of the

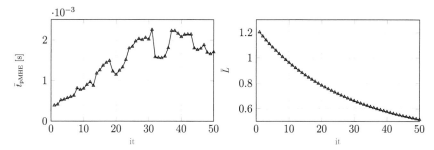

Figure 4.9: The pMHE average computation time per time instant and the resulting averaged cumulative stage cost $\bar{L}$ defined in (4.205) over the number of pMHE optimization iterations. We choose the initial condition $x_{0,1} = [2.5, -0.65]$ and initial guess $\bar{x}_0 = [0, 0]$ and set the pMHE parameters $N = 5$ and $R = 10$.

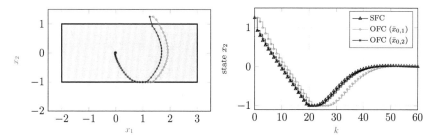

Figure 4.10: Closed-loop state trajectories and the resulting state x_2 for the initial condition $x_{0,3} = [1.25, 1.25]$. We set the pMHE tuning parameters $N = 5$ and it $= 1$, and choose the initial guesses $\bar{x}_{0,1} = [0, 0]$ with initial error $\|x_0 - \bar{x}_{0,1}\| = 1.6971$ as well as $\bar{x}_{0,2} = [2, 0.7]$ with initial error $\|x_0 - \bar{x}_{0,2}\| = 0.9434$.

constraint set given by -1 and 1 even in the absence of constraints in the pMHE iteration scheme. Observe also that a better state estimate is obtained for the more accurate initial guess $\bar{x}_{0,2}$ with a smaller initial error.

4.5.4 Output feedback control: Three tank system

For the simulation example of a three tank system, the control task is to regulate the water heights h in a cascaded three tank system where the control variable consists of the input flow v entering the top tank. The system is linearized around the desired set points $h_s = [11.19, 10.34, 10.80]^\top$ cm and $v_s = 40.47$ ml/s and discretized with a sampling time of $T_s = 5$ s. We define $x = h - h_s \in \mathbb{R}^3$ and $u = v - v_s \in \mathbb{R}$ and consider the resulting discrete-time linear system of the form (4.54) with

$$A = \begin{bmatrix} 0.8913 & 0 & 0 \\ 0.1540 & 0.8233 & 0 \\ 0.0148 & 0.1608 & 0.8300 \end{bmatrix}, \qquad B = \begin{bmatrix} 0.0601 \\ 0.0052 \\ 0.0003 \end{bmatrix}$$
$$C = \begin{bmatrix} 0 & 0 & 1 \end{bmatrix}, \tag{4.206}$$

where (A, B) is controllable and (A, C) is observable. The constraints on the water heights and the input flow are given by $0 \leq h_{1,2,3} \leq 28$ and $0 \leq v \leq 60$, respectively. This results in polytopic input and state constraints of the form $x_k \in \mathcal{X} := \{x \in \mathbb{R}^3 : C_x x \leq d_x\}$ and $u_k \in \mathcal{U} := \{u \in \mathbb{R} : C_u u \leq d_u\}$. Furthermore, we assume that the measurements are subject to two outliers which occur at the times $t = 75s$ and $t = 100s$ with values 250 and 100, respectively. The initial condition is given by $x_0 = h_0 - h_s$ where $h_0 = \begin{bmatrix} 27.5 & 15 & 15 \end{bmatrix}^\top$. We apply the OFC algorithm introduced for the double integrator in Section 4.5.3 to the linearized system with matrices (4.206). The parameters for the rbMPC iteration scheme are $Q_c = \operatorname{diag}(1, 1, 10)$, $R_c = 10^{-2}$ and $\varepsilon = 10^{-3}$ in the stage cost (4.204). In addition, we set the horizon length $N_c = 15$ and the number of rbMPC iterations $it_c = 5$. Concerning the pMHE iteration scheme, we consider a setup similar to the previous example. In particular, we design Algorithm 4.1 based on the Luenberger observer, i.e., with a priori estimate operator (4.57) and Bregman distance (4.59). Furthermore, we choose the horizon length $N = 10$. We set the initial guess $\bar{x}_0 = \begin{bmatrix} 0 & 0 & 0 \end{bmatrix}^\top$ and perform a single iteration per time instant, i.e., we set $it = 1$ for all time instants k.

Given that one further advantage of pMHE is its flexibility of designing the stage cost based on prior knowledge on the measurement disturbances, we choose instead of a quadratic stage cost the more appropriate function $r(v) = R \, \Phi_{\text{hub}}(v)$ with $R = 100$. Here, Φ_{hub} refers to the Huber penalty function which is known to be robust to outliers and verifies Assumption 4.3. Its definition can be found in (3.114), where $M \in \mathbb{R}_{++}$ is a given design parameter. We obtain the results in Figure 4.11, where we also plot the nominal rbMPC with full state measurement.

Note that the trajectory of the water height h_2 in the top right plot results from simulating the continuous-time original nonlinear system with the piecewise constant control input obtained from the OFC iteration scheme. We can see that for the value $M = 10^3$ with which the Huber penalty function mimics the ℓ_2-norm, the resulting estimation error becomes very large when the outliers take place. This leads to a poor performance of the OFC algorithm, which can be observed in the resulting evolutions of the applied control input and water height h_2. We can observe a much better closed-loop performance for the Huber parameter $M = 10$. This can be explained by the fact that for smaller values of M, the effect of the ℓ_1-norm, which is robust against outliers, is more dominant. For all the considered scenarios in this example, the input and state constraints are satisfied despite the underlying relaxation of the barrier functions and the estimation errors.

Hence, the two previous examples show that anytime pMHE employed in the considered OFC approach can be real-time feasible by adjusting the number of optimization algorithm iterations and that it ensures an improved overall performance through an appropriate choice of the MHE cost, which can be guided based on prior knowledge on the measurement disturbances.

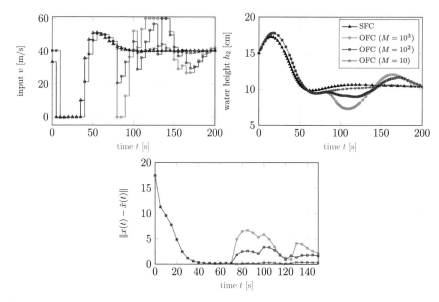

Figure 4.11: Results of the OFC iteration scheme in the case where the measurements are subject to two outliers. *Top left:* The evolution of the control input. *Top right:* The evolution of the water level in the second tank when applying the OFC inputs to the original nonlinear model. *Bottom:* The evolution of the estimation error between the state of the linearized model and the estimates from the pMHE iteration scheme with stage cost $r(v) = 100\,\Phi_{\text{hub}}(v)$, $N = 10$ and it $= 1$.

4.6 Summary

In this chapter, we proposed a novel pMHE iteration scheme in which a state estimate is computed at each time instant after an arbitrary number of optimization algorithm iterations. The optimization algorithm consists of the mirror descent method which generalizes the computationally cheap gradient descent and can be therefore executed quickly. Under suitable assumptions on the Bregman distance, the a priori estimate operator, and the step size, desirable stability properties of the underlying estimation errors were established. Overall, we obtained an anytime MHE algorithm where convergent (and suboptimal) state estimates are generated after any number of optimization algorithm iterations as well as for any horizon length, and whose design is based on flexible and rather general convex stage cost functions that can handle outliers.

In the first main part of the chapter, we considered linear systems and proved global uniform exponential stability of the estimation error. We also discussed specific design procedures of the Bregman distance and the a priori estimate operator that are based on the Kalman filter and the Luenberger observer. In addition, the performance of the pMHE iteration scheme was characterized by the resulting real-time regret for which upper bounds were derived. A combined result shows that the proposed iteration scheme provides stable estimation errors after each optimization iteration and possesses a sublinear regret which can be rendered arbitrarily small by increasing the number of iterations.

In the second main part of the chapter, we considered nonlinear systems and proved that (local) uniform exponential stability of the estimation error can be ensured for any number of optimization algorithm iterations. Subsequently, we presented a design approach of the Bregman distance and the a priori estimate operator that is based on the EKF.

Finally, we demonstrated the theoretical properties and efficacy of anytime pMHE using numerical examples, in which a rather high accuracy of the resulting state estimates was observed even with a small number of iterations per time instant. Moreover, we applied the proposed algorithm in the context of MHE-based output feedback MPC. The considered examples illustrated an overall satisfactory control performance and a real-time capable implementation.

Chapter 5

Conclusions

In this chapter, we summarize the main results of this thesis and provide an outlook on possible directions for future work.

5.1 Summary

In this thesis, we addressed the state estimation problem of linear and nonlinear constrained discrete-time systems within the novel framework of proximity moving horizon estimation, or pMHE. The proposed framework is based on the general conceptual idea of employing a stabilizing analytical a priori estimate and combining it via a proximity measure given by a suitable Bregman distance with an online optimization in order to obtain an improved performance. We derived for the proposed proximity-based MHE approaches desirable theoretical properties including stability, robustness, and performance guarantees, which hold for any horizon length and irrespectively of the convex stage cost being used in the underlying optimization problem. From a technical point of view, the formulation of the presented pMHE approaches as well as the obtained theoretical results were achieved by linking MHE to proximal methods and exploiting the rich theory therein.

In Chapter 3, we formulated the novel pMHE scheme and derived sufficient conditions on the Bregman distance and the a priori estimate operator for the nominal exponential stability of the underlying estimation error, i.e., global uniform exponential stability for LTV systems, global exponential stability for LTI systems, and local uniform exponential stability for nonlinear systems. These stability properties were proven by employing the Bregman distance used in the pMHE problem as a Lyapunov function and investigated under the assumption that a solution of the optimization problem is available at each time instant. Based on the established sufficient conditions, we provided specific design approaches for the Bregman distance and the a priori estimate operator from which stability can be provably inherited. These design approaches are based on the Kalman filter in the LTV case, the Luenberger observer in the LTI case, and the extended Kalman filter in the nonlinear case. In addition, we proved for LTV systems that the pMHE scheme based on the Kalman filter yields an estimation error which is input-to-state stable with respect to additive process and measurement disturbances. Moreover, we showed under suitable assumptions that pMHE can be interpreted as a Bayesian estimator with stability guarantees and gave some insight into the design and tuning of the components of the pMHE problem which can be guided by the statistics of the disturbances.

In Chapter 4, we introduced a novel pMHE iteration scheme which allows to reduce computational effort by performing only a limited number of optimization algorithm iterations each time a new measurement is received. As another added advantage over the proximity-based formulation in Chapter 3, we emphasized that the implicit stabilizing regularization of the a priroi estimate in the iteration scheme ensures that its (suboptimal) bias is reduced with each optimization iteration since it is used only as a warm start strategy. By taking the dynamics of the employed optimization algorithm into account in the stability analysis, we established nominal exponential stability of the resulting estimation error. More specifically, we derived sufficient conditions on the Bregman distance, the a priori estimate operator, and the step size used in the underlying optimization algorithm for the global uniform exponential stability in the LTV case, global exponential stability in the LTI case, and local uniform exponential stability in the nonlinear case. In addition, we adapted the design procedures presented in Chapter 3 to the associated pMHE iteration schemes. As an overall main result, we obtained *anytime MHE algorithms* for which these guarantees are ensured independently of the executed number of optimization algorithm iterations. Furthermore, for LTV systems, we investigated the performance properties of the pMHE iteration scheme in terms of a regret analysis, which allows to characterize the real-time regret of the proposed algorithm that performs only finitely many optimization iterations relative to an estimation scheme that computes the optimal solutions. In particular, we derived upper bounds on the regret with respect to arbitrary as well as exponentially stable comparator sequences based on suitable conditions on the Bregman distance, the a priori estimate operator, and the step size. Our results ensure both global uniform exponential stability and a sublinear regret which can be rendered smaller by increasing the number of optimization algorithm iterations.

In conclusion, the overall contribution of this thesis is to provide a unified framework for the flexible design, analysis, and implementation of theoretically sound and computationally efficient MHE approaches.

5.2 Outlook

Based on the results obtained in this thesis, we point out further research topics which include additional extensions of the developed theory and the validation of the proposed pMHE approaches in practical applications.

The stability results presented in this thesis hold if pMHE inherits the assumptions for ensuring exponential stability of the recursive state estimation strategy that is used to generate the a priori estimate and the Bregman distance. Especially for the state estimation problem of nonlinear systems where the EKF is employed due to its local stability properties, the underlying assumptions can be rather strong and hard to verify offline. Thus, although pMHE performs at least as good as the EKF and is shown to exhibit a far superior performance in simulations, searching for alternative recursive and stable estimators with less strong assumptions would be valuable. More generally speaking, exploiting the general applicability of the proposed proximity-based framework in order to investigate different design approaches other than the ones used in this thesis could be of interest. For instance, we could consider employing the recursive update equations of the unscented Kalman filter to generate the a priori solution and test whether pMHE yields an improved performance due to the online optimization. Note that this filter was considered in the context of MHE to approximate the arrival cost (Qu and Hahn (2009)). In addition, exploring other approximate nonlinear filters that extend the EKF, such as the iterated EKF or second-order filters based on a second-order Taylor expansion of the nonlinear system (Jazwinski

(1970); Robertson et al. (1996)), would be interesting.

Regrading the computational aspects of the pMHE iteration scheme proposed in Chapter 4, a further exploration of the computation time and estimation performance of the proposed algorithm compared to real-time MHE approaches from the literature deserves further research. Furthermore, since there exist accelerated versions of proximal methods such as the fast proximal gradient method (Beck (2017)), another interesting future research direction may focus on developing even faster pMHE algorithms with stability and performance guarantees.

From a practical point of view, given the reduced computation complexity and the desirable theoretical guarantees of the anytime pMHE algorithm, a rather straightforward next step is to implement the proposed approach in real-world applications. Particularly interesting are safety-critical engineering problems, such as automotive applications, aircraft flight control, and the design of medical devices. The goal is to demonstrate the efficient numerical implementation and the reliability of the presented algorithm despite the complexity of the considered application. Moreover, it might be also interesting to apply anytime pMHE to the state estimation problem of large-scale systems, where the advantages of the proposed approach are expected to be even more evident since the underlying optimization algorithm is a simple first-order method well-suited for large-scale optimization (Beck (2017)).

On a more general note, we focused in this thesis on the state estimation problem of dynamical systems. A natural extension is to consider combined state and parameter estimation similar to the MHE approaches in (Kühl et al. (2011); Sui and Johansen (2011)). If the parameter is assumed to be constant or slowly time-varying, a common approach is to augment the state space with the unknown parameter and model the latter as a discrete random walk process (Robertson et al. (1996)). In pMHE, although the augmented model can be directly incorporated into the underlying optimization problem without any modifications, the design of the a priori estimate and the Bregman distance such that stability of the estimator can be ensured is not straightforward and deserves further investigation. Furthermore, extending the proximity-based framework to the estimation problem of external disturbances is both theoretically and practically an interesting research direction.

Appendix A

Stability properties of discrete-time systems

In the following, we present useful stability properties of discrete-time systems that we employ throughout the thesis, and which can be found in the standard textbook of Rawlings et al. (2017). For the stability properties of continuous-time systems, we refer the reader to the textbook by Khalil (2002).

Stability of autonomous time-invariant systems

We first consider discrete-time systems of the form

$$x_{k+1} = f(x_k), \qquad x_0 \in \mathbb{R}^n, \tag{A.1}$$

where $x_k \in \mathbb{R}^n$ refers to the current state. We assume the transition function $f : \mathbb{R}^n \to \mathbb{R}^n$ to be continuous and let $x(k; x_0)$ denote the solution of (A.1) at time k with initial condition x_0. A point $\bar{x} \in \mathbb{R}^n$ is an equilibrium point of (A.1) if it satisfies $\bar{x} = f(\bar{x})$. Without loss of generality, we assume in the following that $\bar{x} = 0$ is an equilibrium point of system (A.1).

Definition A.1 (Asymptotic stability). *The origin $\bar{x} = 0$ is stable for system* (A.1) *if, for all $\varepsilon \in \mathbb{R}_{++}$, there exists $\delta \in \mathbb{R}_{++}$ such that $\|x_0\| \leq \delta$ implies that $\|x(k; x_0)\| \leq \varepsilon$ for all $k \in \mathbb{N}$. It is asymptotically stable if it is stable and $\lim_{k \to \infty} \|x(k; x_0)\| = 0$ for all x_0 in a neighborhood of the origin. It is globally asymptotically stable if it is asymptotically stable for all $x_0 \in \mathbb{R}^n$.*

Definition A.2 (Exponential stability). *The origin $\bar{x} = 0$ is exponentially stable for system* (A.1) *if there exist $\alpha \in \mathbb{R}_{++}$ and $\beta \in (0, 1)$ such that $\|x(k; x_0)\| \leq \alpha \|x_0\| \beta^k$ for all $k \in \mathbb{N}$ and for any x_0 in a neighborhood of the origin. It is globally exponentially stable if it is exponentially stable for all $x_0 \in \mathbb{R}^n$.*

By a slight abuse of language, we say that a system is asymptotically (exponentially) stable if its origin $\bar{x} = 0$ is asymptotically (exponentially) stable. In the following, we use Lyapunov functions to characterize the stability properties of system (A.1).

Definition A.3 (Lyapunov function for time-invariant systems). *A continuous function $V : \mathbb{R}^n \to \mathbb{R}_+$ is a Lyapunov function for system* (A.1) *if there exist K_∞-functions α_1, α_2 and a continuous positive definite function $\alpha_3 : \mathbb{R} \to \mathbb{R}_+$ such that, for any $x \in \mathbb{R}^n$*

$$\alpha_1(\|x\|) \leq V(x) \leq \alpha_2(\|x\|) \tag{A.2a}$$
$$V(f(x)) - V(x) \leq -\alpha_3(\|x\|). \tag{A.2b}$$

Theorem A.1. *System* (A.1) *is globally asymptotically stable if it admits a Lyapunov function. Moreover, it is globally exponentially stable if it admits a continuous Lyapunov function satisfying for some $c_1, c_2, c_3 \in \mathbb{R}_{++}$ the following inequalities*

$$c_1 \|x\|^2 \leq V(x) \leq c_2 \|x\|^2 \tag{A.3a}$$

$$V(f(x)) - V(x) \leq -c_3 \|x\|^2 \tag{A.3b}$$

for any $x \in \mathbb{R}^n$.

Stability of autonomous time-varying systems

We consider now discrete-time time-varying systems of the form

$$x_{k+1} = f_k(x_k), \tag{A.4}$$

with $k \geq k_0 \in \mathbb{N}$ and initial condition $x_0 = x_0(k_0) \in \mathbb{R}^n$. We assume that the origin $\bar{x} = 0$ is an equilibrium point, i.e., $0 = f_k(0)$ for all $k \in \mathbb{N}$ and that the function f_k is continuous uniformly over $k \in \mathbb{N}$. We let $x(k; x_0, k_0)$ denote the solution of (A.4) at time k with initial condition x_0 at time k_0. For time-varying systems, we have the following stability definitions (Zhou (2017)).

Definition A.4 (Uniform asymptotic stability). *The origin $\bar{x} = 0$ is uniformly stable for system* (A.4) *if, for all $\varepsilon \in \mathbb{R}_{++}$, there exists $\delta \in \mathbb{R}_{++}$ independent of k_0 such that $\|x_0\| \leq \delta$ implies that $\|x(k; x_0, k_0)\| \leq \varepsilon$ for all $k \geq k_0 \in \mathbb{N}$. It is uniformly asymptotically stable if it is uniformly stable and $\lim_{k \to \infty} \|x(k; x_0, k_0)\| = 0$ for all $\|x_0\| < c$, where c is a positive constant that is independent of k_0. It is globally uniformly asymptotically stable if it is uniformly asymptotically stable for all $x_0 \in \mathbb{R}^n$.*

Definition A.5 (Uniform exponential stability). *The origin $\bar{x} = 0$ is uniformly exponentially stable for system* (A.4) *if there exist $\alpha \in \mathbb{R}_{++}$ and $\beta \in (0, 1)$ such that $\|x(k; x_0, k_0)\| \leq \alpha \|x_0\| \beta^{k-k_0}$ for all $k \geq k_0 \in \mathbb{N}$ and for any $\|x_0\| < c$, where the constants α and c are independent of k_0. It is globally uniformly exponentially stable if it is uniformly exponentially stable for all $x_0 \in \mathbb{R}^n$.*

The following definition of a Lyapunov function for discrete-time time-varying systems is based on (Jazwinski, 1970, p. 240).

Definition A.6 (Lyapunov function for time-varying systems). *A continuous time-varying function $V_k : \mathbb{R}^n \to \mathbb{R}_+$ is a Lyapunov function for system* (A.4) *if there exist K_∞-functions α_1, α_2 and a continuous positive definite function $\alpha_3 : \mathbb{R} \to \mathbb{R}_+$ such that, for some positive integer $M \in \mathbb{N}_+$, the following holds*

$$\alpha_1(\|x\|) \leq V_k(x) \leq \alpha_2(\|x\|) \tag{A.5a}$$

$$V_k(f_{k-1}(x)) - V_{k-1}(x) \leq -\alpha_3(\|x\|), \tag{A.5b}$$

for any $k \geq M$ and $x \in \mathbb{R}^n$.

Theorem A.2. *System* (A.4) *is globally uniformly asymptotically stable if it admits a time-varying Lyapunov function. Moreover, it is (locally) uniformly exponentially stable if it admits a time-varying Lyapunov function satisfying for some $c_1, c_2, c_3, r \in \mathbb{R}_{++}$ and $M \in \mathbb{N}_+$ the following inequalities*

$$c_1 \|x\|^2 \leq V_k(x) \leq c_2 \|x\|^2 \tag{A.6a}$$

$$V_k(f_{k-1}(x)) - V_k(x) \leq -c_3 \|x\|^2, \tag{A.6b}$$

for any $k \geq M$ and $x \in \mathbb{R}^n$ with $\|x\| \leq r$. It is globally uniformly exponentially stable if the above assumptions hold globally.

Input-to-state stability

We now introduce some basics regarding the concept of input-to-state stability (ISS) for discrete-time systems (Jiang and Wang (2001); Lin et al. (2005)). More specifically, we consider discrete-time systems of the form

$$x_{k+1} = f_k(x_k, w_k), \tag{A.7}$$

with $k \geq k_0 \in \mathbb{N}$ and initial condition $x_0 = x_0(k_0) \in \mathbb{R}^n$, and where $x_k \in \mathbb{R}^n$ and $w_k \in \mathbb{R}^{m_w}$ refer to the state and input vectors. We assume that $0 = f_k(0,0)$ for all $k \in \mathbb{N}$ and that the function f_k is continuous in both of its arguments and uniformly over $k \in \mathbb{N}$. We let $x(k; x_0, k_0, \mathbf{w}_k)$ denote the solution of (A.7) at time k with initial condition x_0 at time k_0 and with input sequence $\mathbf{w}_k \coloneqq \{w_0, w_1, \ldots w_{k-1}\}$. We base our discussion on the results in (Laila and Astolfi (2005)), which is one of the few works that focus on the ISS property and its Lyapunov characterization for the case of discrete-time time-varying systems. Note that to the best of our knowledge, in the literature, the notion of ISS for time-varying systems is used to characterize *uniform* ISS. In words, input-to-state stability states that, for any bounded input sequence $\mathbf{w}_k$, the state remains bounded with a well-defined gain between the size of the state bound and the input magnitude. This is stated more accurately in the following definition.

Definition A.7 ((Uniform) Input-to-state stability). *System* (A.7) *is ISS with respect to w_k if there exist a $\mathcal{KL}$-function β and a $\mathcal{K}$-function γ such that, for each bounded input sequence $\mathbf{w}_k$ and each initial condition $x_0 \in \mathbb{R}^n$, it holds that*

$$\|x(k, x_0, k_0, \mathbf{w}_k)\| \leq \beta(\|x_0\|, k - k_0) + \gamma(\|\mathbf{w}_{[k_0:k-1]}\|) \tag{A.8}$$

for any $k \geq k_0 \in \mathbb{N}$, where $\|\mathbf{w}_{[i:k]}\| \coloneqq \sup_{i \leq j \leq k}\{\|w_j\|\}$.

In particular, the state of an ISS system decays for a decaying input and the origin $\bar{x} = 0$ is globally asymptotically stable whenever $w_k = 0$ for all $k \geq k_0$.

In the following, we present the concept of an ISS-Lyapunov function which can be used to prove that system (A.7) is ISS (Jiang and Wang (2001)).

Definition A.8 (ISS-Lyapunov function). *A continuous time-varying function $V_k : \mathbb{R}^n \to \mathbb{R}_+$ is an ISS-Lyapunov function for system* (A.7) *if there exist $\mathcal{K}_\infty$-functions $\alpha_1, \alpha_2, \alpha_3$ and a $\mathcal{K}$-function σ such that*

$$\alpha_1(\|x\|) \leq V_k(x) \leq \alpha_2(\|x\|) \tag{A.9a}$$

$$V_k(f_{k-1}(x,w)) - V_k(x) \leq -\alpha_3(\|x\|) + \sigma(\|w\|) \tag{A.9b}$$

hold for any $k \geq k_0 \in \mathbb{N}_+$, $x \in \mathbb{R}^n$ and $w \in \mathbb{R}^{m_w}$.

Theorem A.3. *If system (A.7) admits an ISS-Lyapunov function, then it is ISS.*

Notation

Sets

$\mathbb{N}$	set of natural numbers
$\mathbb{N}_+$	set of strictly positive natural numbers
$\mathbb{R}$	set of real numbers
$\mathbb{R}_+$	set of nonnegative real numbers
$\mathbb{R}_{++}$	set of strictly positive real numbers
$\mathbb{R}^n$	set of real-valued n-dimensional vectors
$\mathbb{R}^n_+$	set of real-valued n-dimensional vectors with nonnegative entries
$\mathbb{R}^{n \times m}$	set of real-valued $n \times m$ matrices
$\mathbb{S}^n$	set of $n \times n$ symmetric real matrices
$\mathbb{S}^n_{++}$	set of $n \times n$ positive definite matrices
$\mathcal{C}^k$	set of k-times continuously differentiable functions, $k \in \mathbb{N}$
$[a,b)$	interval $\{x \in \mathbb{R} : \ a \leq x < b\}$ for $a, b \in \mathbb{R}$
$\text{int}(\mathcal{X})$	interior of a set $\mathcal{X} \subset \mathbb{R}^n$

Vectors, matrices, and norms

$I_{[n]}$	identity matrix with dimension $n \times n$
$0_{[n \times m]}$	zero matrix with dimension $n \times m$
$(\cdot)^\top$	transpose of a real vector or matrix
$(\cdot)^i$	i-th element or row of a real vector or matrix
$\lambda_i(A)$	eigenvalues $[\lambda_1, \cdots, \lambda_n]$ of a symmetric matrix $A \in \mathbb{S}^n$
$\lambda_{\max}(A)$	maximum eigenvalue of a symmetric matrix $A \in \mathbb{S}^n$
$\lambda_{\min}(A)$	minimum eigenvalue of a symmetric matrix $A \in \mathbb{S}^n$
$\|x\|$	Euclidean norm of a vector $x \in \mathbb{R}^n$
$\|x\|_Q^2$	weighted Euclidean norm $x^\top Q x$ for $x \in \mathbb{R}^n$ and $Q \in \mathbb{S}^n_{++}$
$\|A\|$	spectral norm of a matrix $A \in \mathbb{R}^{n \times m}$: $\|A\| = \|A\|_2 = \sqrt{\lambda_{\max}(A^\top A)}$
$\text{diag}(A_1, \ldots, A_r)$	block-diagonal matrix with main diagonal blocks $A_1, \ldots, A_r, \ r \in \mathbb{N}_+$
$Q \succ 0 \ (Q \succeq 0)$	matrix $Q \in \mathbb{S}^n$ is positive definite (positive semidefinite), i.e., $x^\top Q x > 0 \ (x^\top Q x \geq 0)$ for all $x \in \mathbb{R}^n$ with $x \neq 0$
$Q \prec 0 \ (Q \preceq 0)$	matrix $Q \in \mathbb{S}^n$ is negative definite (negative semidefinite), i.e., the matrix $-Q$ is positive definite (positive semidefinite)
(x, y)	stacked vector $\begin{bmatrix} x^\top & y^\top \end{bmatrix}^\top \in \mathbb{R}^{n+m}$ for $x \in \mathbb{R}^n$ and $y \in \mathbb{R}^m$

Functions and derivatives

$f : \mathcal{D} \to C$ the function $f(\cdot)$ maps from its domain $\mathcal{D}$ into the set C

$\mathcal{K}$ A function $\alpha : \mathbb{R}_+ \mapsto \mathbb{R}_+$ is a class $\mathcal{K}$-function, i.e., $\alpha \in \mathcal{K}$, if it is continuous, strictly increasing, and $\alpha(0) = 0$.

$\mathcal{K}_\infty$ A function $\alpha : \mathbb{R}_+ \mapsto \mathbb{R}_+$ is a class $\mathcal{K}_\infty$-function, i.e., $\alpha \in \mathcal{K}_\infty$, if $\alpha \in \mathcal{K}$ and α is radially unbounded, i.e., $\alpha(r) \to \infty$ for $r \to \infty$.

$\mathcal{K}\mathcal{L}$ A function $\beta : \mathbb{R}_+ \times \mathbb{R}_+ \mapsto \mathbb{R}_+$ is a class $\mathcal{K}\mathcal{L}$-function, i.e., $\beta \in \mathcal{K}\mathcal{L}$, if $\beta(\cdot, t) \in \mathcal{K}$ for each fixed $t \in \mathbb{R}_+$ and the function $\beta(r, \cdot)$ is decreasing with $\lim_{t\to\infty} \beta(r, t) = 0$ for each fixed $r \in \mathbb{R}_+$.

$\nabla f(\bar{x})$ gradient of a function $f \in \mathcal{C}^1 : \mathcal{D} \to \mathbb{R}$ evaluated at $\bar{x} \in \mathcal{D}$

$\nabla^2 f(\bar{x})$ Hessian of a function $f \in \mathcal{C}^2 : \mathcal{D} \to \mathbb{R}$ evaluated at $\bar{x} \in \mathcal{D}$

Acronyms and abbreviations

EKF   extended Kalman filter

FIE   full information estimation

GAS   global asymptotic stability

GES   global exponential stability

GUES   global uniform exponential stability

i-IOSS   incremental input/output-to-state stability

ISS   input-to-state stability

KKT   Karush-Kuhn-Tucker

LMI   linear matrix inequality

LTI   linear time-invariant

LTV   linear time-varying

MDA   mirror descent algorithm

MHE   moving horizon estimation

MPC   model predictive control

PMD   proximal minimization with D-functions

PPA   proximal point algorithm

RGAS   robust global asymptotic stability

RMSE   root mean square error

UES   uniform exponential stability

Bibliography

M. Abdollahpouri, G. Takács, and B. Rohal'-Ilkiv. Real-time moving horizon estimation for a vibrating active cantilever. *Mechanical Systems and Signal Processing*, 86:1–15, 2017.

A. Alessandri. Design of observers for Lipschitz nonlinear systems using LMI. *Proc. IFAC Nonlinear Control Systems*, 37(13):459–464, 2004.

A. Alessandri and M. Awawdeh. Moving-horizon estimation with guaranteed robustness for discrete-time linear systems and measurements subject to outliers. *Automatica*, 67:85–93, 2016.

A. Alessandri and M. Gaggero. Fast moving horizon state estimation for discrete-time systems using single and multi iteration descent methods. *IEEE Transactions on Automatic Control*, 62(9):4499–4511, 2017.

A. Alessandri and M. Gaggero. Fast moving horizon state estimation for discrete-time systems with linear constraints. *International Journal of Adaptive Control and Signal Processing*, 34(6):706–720, 2020.

A. Alessandri, M. Baglietto, and G. Battistelli. Receding-horizon estimation for discrete-time linear systems. *IEEE Transactions on Automatic Control*, 48(3):473–478, 2003.

A. Alessandri, M. Baglietto, and G. Battistelli. Moving-horizon state estimation for nonlinear discrete-time systems: New stability results and approximation schemes. *Automatica*, 44 (7):1753–1765, 2008.

A. Alessandri, M. Baglietto, G. Battistelli, and V. Zavala. Advances in moving horizon estimation for nonlinear systems. In *Proc. 49th Conference on Decision and Control (CDC)*, pages 5681–5688. IEEE, 2010.

A. Alessandri, M. Baglietto, and G. Battistelli. Min-max moving-horizon estimation for uncertain discrete-time linear systems. *SIAM Journal on Control and Optimization*, 50 (3):1439–1465, 2012.

D. A. Allan, J. B. Rawlings, and A. R. Teel. Nonlinear detectability and incremental input/output-to-state stability. *Texas–Wisconsin–California Control Consortium (TWCCC), Tech. Rep*, 1, 2020.

A. Andersson and T. Thiringer. Motion sensorless IPMSM control using linear moving horizon estimation with Luenberger observer state feedback. *IEEE transactions on transportation electrification*, 4(2):464–473, 2018.

A. Aravkin, J. V. Burke, L. Ljung, A. Lozano, and G. Pillonetto. Generalized Kalman smoothing: Modeling and algorithms. *Automatica*, 86:63–86, 2017.

H. Bae and J.-H. Oh. Humanoid state estimation using a moving horizon estimator. *Advanced Robotics*, 31(13):695–705, 2017.

A. Banerjee, S. Merugu, I. S. Dhillon, and J. Ghosh. Clustering with Bregman divergences. *Journal of machine learning research*, 6:1705–1749, 2005.

G. Battistelli, L. Chisci, and S. Gherardini. Moving horizon estimation for discrete-time linear systems with binary sensors: Algorithms and stability results. *Automatica*, 85: 374–385, 2017.

K. Baumgärtner, A. Zanelli, and M. Diehl. A gradient condition for the arrival cost in moving horizon estimation. In *Proc. European Control Conference (ECC)*, pages 1286–1291. IEEE, 2020.

F. Bayer. *Proximity moving horizon estimation for nonlinear systems*. Master thesis, University of Stuttgart, 2019.

A. Beck. *First-order methods in optimization*. SIAM, 2017.

A. Beck and M. Teboulle. Mirror descent and nonlinear projected subgradient methods for convex optimization. *Operations Research Letters*, 31(3):167–175, 2003.

M. A. Belabbas. On implicit regularization: Morse functions and applications to matrix factorization. *arXiv preprint arXiv:2001.04264*, 2020.

I. Ben-Gal. Outlier detection. In *Data mining and knowledge discovery handbook*, pages 131–146. Springer, 2005.

M. Boutayeb and D. Aubry. A strong tracking extended Kalman observer for nonlinear discrete-time systems. *IEEE Transactions on Automatic Control*, 44(8):1550–1556, 1999.

S. Boyd and L. Vandenberghe. *Convex optimization*. Cambridge university press, 2004.

L. M. Bregman. The relaxation method of finding the common point of convex sets and its application to the solution of problems in convex programming. *USSR computational mathematics and mathematical physics*, 7(3):200–217, 1967.

E. J. Candes, M. B. Wakin, and S. P. Boyd. Enhancing sparsity by reweighted ℓ_1 minimization. *Journal of Fourier Analysis and Applications*, 14(5-6):877–905, 2008.

Y. Censor and S. A. Zenios. Proximal minimization algorithm with D-functions. *Journal of Optimization Theory and Applications*, 73(3):451–464, 1992.

G. Chen and M. Teboulle. Convergence analysis of a proximal-like minimization algorithm using Bregman functions. *SIAM Journal on Optimization*, 3(3):538–543, 1993.

T. Chen, N. F. Kirkby, and R. Jena. Optimal dosing of cancer chemotherapy using model predictive control and moving horizon state/parameter estimation. *Computer methods and programs in biomedicine*, 108(3):973–983, 2012.

E. Chu, A. Keshavarz, D. Gorinevsky, and S. Boyd. Moving horizon estimation for staged QP problems. In *Proc. 51st Conference on Decision and Control (CDC)*, pages 3177–3182. IEEE, 2012.

S.-T. Chung and J. Grizzle. Sampled-data observer error linearization. *Automatica*, 26(6): 997–1007, 1990.

M. L. Darby and M. Nikolaou. A parametric programming approach to moving-horizon state estimation. *Automatica*, 43(5):885–891, 2007.

C. De Souza, M. Gevers, and G. Goodwin. Riccati equations in optimal filtering of nonstabilizable systems having singular state transition matrices. *IEEE Transactions on Automatic control*, 31(9):831–838, 1986.

I. S. Dhillon and J. A. Tropp. Matrix nearness problems with Bregman divergences. *SIAM Journal on Matrix Analysis and Applications*, 29(4):1120–1146, 2008.

D. L. Donoho. Compressed sensing. *IEEE Transactions on Information Theory*, 52(4): 1289–1306, 2006.

C. Ebenbauer and F. Allgöwer. Analysis and design of polynomial control systems using dissipation inequalities and sum of squares. *Computers & chemical engineering*, 30(10-12): 1590–1602, 2006.

M. Farina, G. Ferrari-Trecate, and R. Scattolini. Moving-horizon partition-based state estimation of large-scale systems. *Automatica*, 46(5):910–918, 2010.

A. Favato, F. Toso, P. G. Carlet, M. Carbonieri, and S. Bolognani. Fast moving horizon estimator for induction motor sensorless control. In *Proc. 10th International Symposium on Sensorless Control for Electrical Drives (SLED)*, pages 1–6. IEEE, 2019.

C. Feller and C. Ebenbauer. Relaxed logarithmic barrier function based model predictive control of linear systems. *IEEE Transactions on Automatic Control*, 62(3):1223–1238, 2017a.

C. Feller and C. Ebenbauer. A stabilizing iteration scheme for model predictive control based on relaxed barrier functions. *Automatica*, 80:328–339, 2017b.

C. Feller and C. Ebenbauer. Sparsity-exploiting anytime algorithms for model predictive control: A relaxed barrier approach. *IEEE Transactions on Control Systems Technology*, 2018.

G. Ferrari-Trecate, D. Mignone, and M. Morari. Moving horizon estimation for hybrid systems. *IEEE Transactions on Automatic Control*, 47(10):1663–1676, 2002.

R. Findeisen, K. Graichen, and M. Mönnigmann. Eingebettete Optimierung in der Regelungstechnik–Grundlagen und Herausforderungen. *at-Automatisierungstechnik*, 66 (11):877–902, 2018.

M. A. Gandhi and L. Mili. Robust Kalman filter based on a generalized maximum-likelihood-type estimator. *IEEE Transactions on Signal Processing*, 58(5):2509–2520, 2009.

K. Geebelen, A. Wagner, S. Gros, J. Swevers, and M. Diehl. Moving horizon estimation with a Huber penalty function for robust pose estimation of tethered airplanes. In *Proc. American Control Conference (ACC)*, pages 6169–6174. IEEE, 2013.

M. Gharbi and C. Ebenbauer. A proximity approach to linear moving horizon estimation. In *Proc. 6th IFAC Conference on Nonlinear Model Predictive Control*, volume 51, pages 549–555. Elsevier, 2018.

M. Gharbi and C. Ebenbauer. A proximity moving horizon estimator based on Bregman distances and relaxed barrier functions. In *Proc. 18th European Control Conference (ECC)*, pages 1790–1795. IEEE, 2019a.

M. Gharbi and C. Ebenbauer. Proximity moving horizon estimation for linear time-varying systems and a Bayesian filtering view. In *Proc. 58th Conference on Decision and Control (CDC)*, pages 3208–3213. IEEE, 2019b.

M. Gharbi and C. Ebenbauer. A proximity moving horizon estimator for a class of nonlinear systems. *International Journal of Adaptive Control and Signal Processing*, 34(6):721–742, 2020.

M. Gharbi and C. Ebenbauer. Anytime MHE-based output feedback MPC. In *Proc. 7th IFAC Conference on Nonlinear Model Predictive Control*, volume 54, pages 264–271. Elsevier, 2021.

M. Gharbi, F. Bayer, and C. Ebenbauer. Proximity moving horizon estimation for discrete-time nonlinear systems. *IEEE Control Systems Letters*, 5(6):2090–2095, 2020a.

M. Gharbi, B. Gharesifard, and C. Ebenbauer. Anytime proximity moving horizon estimation: Stability and regret. *arXiv preprint arXiv:2006.14303*, 2020b.

M. Gharbi, B. Gharesifard, and C. Ebenbauer. Anytime proximity moving horizon estimation: Stability and regret for nonlinear systems. In *Proc. 60th Conference on Decision and Control (CDC). Accepted*. IEEE, 2021.

Y. Gu, Y. Chou, J. Liu, and Y. Ji. Moving horizon estimation for multirate systems with time-varying time-delays. *Journal of the Franklin Institute*, 356(4):2325–2345, 2019.

O. Güler. On the convergence of the proximal point algorithm for convex minimization. *SIAM Journal on Control and Optimization*, 29(2):403–419, 1991.

A. Haber and M. Verhaegen. Moving horizon estimation for large-scale interconnected systems. *IEEE Transactions on Automatic Control*, 58(11):2834–2847, 2013.

E. Hall and R. Willett. Dynamical models and tracking regret in online convex programming. In *Proc. International Conference on Machine Learning*, pages 579–587. PMLR, 2013.

E. L. Haseltine and J. B. Rawlings. Critical evaluation of extended Kalman filtering and moving-horizon estimation. *Industrial & engineering chemistry research*, 44(8):2451–2460, 2005.

N. Haverbeke. *Efficient numerical methods for moving horizon estimation*. PhD thesis, Katholieke Universiteit Leuven, Heverlee, Belgium, 2011.

N. Haverbeke, M. Diehl, and B. De Moor. A structure exploiting interior-point method for moving horizon estimation. In *Proc. 48h IEEE Conference on Decision and Control (CDC) held jointly with 28th Chinese Control Conference*, pages 1273–1278. IEEE, 2009.

D. M. Hawkins. *Identification of outliers*, volume 11. Springer, 1980.

X. Hu, D. Cao, and B. Egardt. Condition monitoring in advanced battery management systems: Moving horizon estimation using a reduced electrochemical model. *IEEE/ASME Transactions on Mechatronics*, 23(1):167–178, 2017.

J. Huang, G. Zhao, and X. Zhang. MEMS gyroscope/TAM-integrated attitude estimation based on moving horizon estimation. *Proc. Institution of Mechanical Engineers, Part G: Journal of Aerospace Engineering*, 231(8):1451–1459, 2017.

R. Huang, S. C. Patwardhan, and L. T. Biegler. Robust stability of nonlinear model predictive control based on extended Kalman filter. *Journal of Process Control*, 22(1): 82–89, 2012.

P. J. Huber. Robust estimation of a location parameter. In *Breakthroughs in statistics*, pages 492–518. Springer, 1992.

A. Jazwinski. Limited memory optimal filtering. *IEEE Transactions on Automatic Control*, 13(5):558–563, 1968.

A. Jazwinski. *Stochastic processes and filtering theory*. Elsevier, 1970.

L. Ji, J. B. Rawlings, W. Hu, A. Wynn, and M. Diehl. Robust stability of moving horizon estimation under bounded disturbances. *IEEE Transactions on Automatic Control*, 61 (11):3509–3514, 2015.

Z.-P. Jiang and Y. Wang. Input-to-state stability for discrete-time nonlinear systems. *Automatica*, 37(6):857–869, 2001.

J. B. Jørgensen, J. B. Rawlings, and S. B. Jørgensen. Numerical methods for large-scale moving horizon estimation and control. In *Proc. Int. Symposium on Dynamics and Control Process Systems (DYCOPS)*, volume 7, 2004.

T. Kailath. A view of three decades of linear filtering theory. *IEEE Transactions on information theory*, 20(2):146–181, 1974.

R. E. Kalman. On the general theory of control systems. In *Proc. First International Conference on Automatic Control, Moscow*, pages 481–492, 1960a.

R. E. Kalman. A new approach to linear filtering and prediction problems. *Journal of Basic Engineering*, 82(1):35–45, 1960b.

H. K. Khalil. *Nonlinear systems; 3rd edition*. Prentice-Hall, 2002.

S. Knüfer and M. A. Müller. Robust global exponential stability for moving horizon estimation. In *Proc. Conference on Decision and Control (CDC)*, pages 3477–3482. IEEE, 2018.

H. Kong and S. Sukkarieh. Suboptimal receding horizon estimation via noise blocking. *Automatica*, 98:66–75, 2018.

D. Kouzoupis, R. Quirynen, F. Girrbach, and M. Diehl. An efficient SQP algorithm for moving horizon estimation with Huber penalties and multi-rate measurements. In *Proc. Conference on Control Applications (CCA)*, pages 1482–1487. IEEE, Sept 2016.

A. J. Krener and A. Isidori. Linearization by output injection and nonlinear observers. *Systems & Control Letters*, 3(1):47–52, 1983.

P. Kühl, M. Diehl, T. Kraus, J. P. Schlöder, and H. G. Bock. A real-time algorithm for moving horizon state and parameter estimation. *Computers & chemical engineering*, 35 (1):71–83, 2011.

A. Küpper, M. Diehl, J. P. Schlöder, H. G. Bock, and S. Engell. Efficient moving horizon state and parameter estimation for SMB processes. *Journal of Process Control*, 19(5): 785–802, 2009.

D. S. Laila and A. Astolfi. Input-to-state stability for discrete-time time-varying systems with applications to robust stabilization of systems in power form. *Automatica*, 41(11): 1891–1903, 2005.

F. Lauer, G. Bloch, and R. Vidal. A continuous optimization framework for hybrid system identification. *Automatica*, 47(3):608–613, 2011.

W. C. Lee and K. H. Nam. Observer design for autonomous discrete-time nonlinear systems. *Systems & Control Letters*, pages 49–58, 1991.

W. Lin and C. I. Byrnes. Remarks on linearization of discrete-time autonomous systems and nonlinear observer design. *Systems & Control Letters*, 25(1):31–40, 1995.

Y. Lin, Y. Wang, and D. Cheng. On nonuniform and semi-uniform input-to-state stability for time varying systems. *Proc. 16th IFAC World Congress*, 38(1):312–317, 2005.

K. V. Ling and K. W. Lim. Receding horizon recursive state estimation. *IEEE Transactions on Automatic Control*, 44(9):1750–1753, 1999.

A. Liu, W.-A. Zhang, M. Z. Chen, and L. Yu. Moving horizon estimation for mobile robots with multirate sampling. *IEEE Transactions on Industrial Electronics*, 64(2):1457–1467, 2016.

R. López-Negrete, S. C. Patwardhan, and L. T. Biegler. Constrained particle filter approach to approximate the arrival cost in moving horizon estimation. *Journal of Process Control*, 21(6):909–919, 2011.

H. Michalska and D. Mayne. Moving horizon observers. In *Proc. Nonlinear Control Systems Design*, pages 185–190. Elsevier, 1993.

H. Michalska and D. Q. Mayne. Moving horizon observers and observer-based control. *IEEE Transactions on Automatic Control*, 40(6):995–1006, 1995.

H. Mine and M. Fukushima. A minimization method for the sum of a convex function and a continuously differentiable function. *Journal of Optimization Theory and Applications*, 33(1):9–23, 1981.

A. Mokhtari, S. Shahrampour, A. Jadbabaie, and A. Ribeiro. Online optimization in dynamic environments: Improved regret rates for strongly convex problems. In *Proc. 55th Conference on Decision and Control (CDC)*, pages 7195–7201. IEEE, 2016.

P. Moraal and J. Grizzle. Observer design for nonlinear systems with discrete-time measurements. *IEEE Transactions on automatic control*, 40(3):395–404, 1995.

B. Morabito, M. Kögel, E. Bullinger, G. Pannocchia, and R. Findeisen. Simple and efficient moving horizon estimation based on the fast gradient method. *Proc. 5th IFAC Conference on Nonlinear Model Predictive Control*, 48(23):428–433, 2015.

B. Morabito, R. Klein, and R. Findeisen. Real time feasibility and performance of moving horizon estimation for Li-ion batteries based on first principles electrochemical models. In *Proc. American Control Conference (ACC)*, pages 3457–3462. IEEE, 2017.

J.-J. Moreau. Proximité et dualité dans un espace hilbertien. *Bulletin de la Société mathématique de France*, 93:273–299, 1965.

M. A. Müller. Nonlinear moving horizon estimation in the presence of bounded disturbances. *Automatica*, 79:306–314, 2017.

K. R. Muske, J. B. Rawlings, and J. H. Lee. Receding horizon recursive state estimation. In *Proc. American Control Conference*, pages 900–904. IEEE, 1993.

A. Nemirovskiĭ and D. Yudin. *Problem Complexity and Method Efficiency in Optimization.* Wiley, 1983.

J. Nocedal and S. Wright. *Numerical optimization.* Springer Science & Business Media, 2006.

N. Parikh and S. Boyd. Proximal algorithms. *Foundations and Trends in optimization*, 1 (3):127–239, 2014.

T. Polóni, A. Aas Eielsen, B. Rohal-Ilkiv, and T. Arne Johansen. Adaptive model estimation of vibration motion for a nanopositioner with moving horizon optimized extended Kalman filter. *Journal of Dynamic Systems, Measurement, and Control*, 135(4), 2013.

H. Qin and W. Chen. Application of the constrained moving horizon estimation method for the ultra-short baseline attitude determination. *Acta Geodaetica et Geophysica*, 48 (1):27–38, 2013.

C. C. Qu and J. Hahn. Computation of arrival cost for moving horizon estimation via unscented Kalman filtering. *Journal of Process Control*, 19(2):358–363, 2009.

S. V. Raković and W. S. Levine. *Handbook of model predictive control.* Springer, 2018.

C. V. Rao. *Moving horizon strategies for the constrained monitoring and control of nonlinear discrete-time systems.* PhD thesis, University of Wisconsin–Madison, 2000.

C. V. Rao, J. B. Rawlings, and J. H. Lee. Constrained linear state estimation – a moving horizon approach. *Automatica*, 37(10):1619–1628, 2001.

C. V. Rao, J. B. Rawlings, and D. Q. Mayne. Constrained state estimation for nonlinear discrete-time systems: Stability and moving horizon approximations. *IEEE Transactions on Automatic Control*, 48(2):246–258, 2003.

J. B. Rawlings, D. Q. Mayne, and M. Diehl. *Model Predictive Control: Theory, Computation, and Design*. Nob Hill Publishing, 2017.

K. Reif and R. Unbehauen. The extended Kalman filter as an exponential observer for nonlinear systems. *IEEE Transactions on Signal processing*, 47(8):2324–2328, 1999.

K. Reif, S. Gunther, E. Yaz, and R. Unbehauen. Stochastic stability of the discrete-time extended Kalman filter. *IEEE Transactions on Automatic control*, 44(4):714–728, 1999.

D. G. Robertson, J. H. Lee, and J. B. Rawlings. A moving horizon-based approach for least-squares estimation. *AIChE Journal*, 42(8):2209–2224, 1996.

R. T. Rockafellar. *Convex Analysis*. Princeton University Press, 1970.

R. T. Rockafellar. Monotone operators and the proximal point algorithm. *SIAM journal on control and optimization*, 14(5):877–898, 1976.

P. J. Rousseeuw and A. M. Leroy. *Robust regression and outlier detection*. John wiley & sons, 2005.

E. K. Ryu and S. Boyd. Primer on monotone operator methods. *Applied and Computational Mathematics*, 15(1):3–43, 2016.

S. Särkkä. *Bayesian filtering and smoothing*, volume 3. Cambridge University Press, 2013.

J. D. Schiller, S. Knüfer, and M. A. Müller. Robust stability of suboptimal moving horizon estimation using an observer-based candidate solution. *arXiv preprint arXiv:2011.08723*, 2020.

R. Schneider, R. Hannemann-Tamás, and W. Marquardt. An iterative partition-based moving horizon estimator with coupled inequality constraints. *Automatica*, 61:302–307, 2015.

P. Segovia, L. Rajaoarisoa, F. Nejjari, E. Duviella, and V. Puig. Model predictive control and moving horizon estimation for water level regulation in inland waterways. *Journal of Process Control*, 76:1–14, 2019.

J.-N. Shen, Y.-J. He, Z.-F. Ma, H.-B. Luo, and Z.-F. Zhang. Online state of charge estimation of lithium-ion batteries: A moving horizon estimation approach. *Chemical Engineering Science*, 154:42–53, 2016.

D. Simon. Kalman filtering with state constraints: A survey of linear and nonlinear algorithms. *IET Control Theory & Applications*, 4(8):1303–1318, 2010.

D. Simon and D. L. Simon. Aircraft turbofan engine health estimation using constrained Kalman filtering. *J. Eng. Gas Turbines Power*, 127(2):323–328, 2005.

E. D. Sontag and Y. Wang. Output-to-state stability and detectability of nonlinear systems. *Systems & Control Letters*, 29(5):279–290, 1997.

H. W. Sorenson. Least-squares estimation: from Gauss to Kalman. *IEEE spectrum*, 7(7): 63–68, 1970.

D. A. Sprott. Gauss's contributions to statistics. *Historia Mathematica*, 5(2):183–203, 1978.

D. Sui and T. A. Johansen. Moving horizon observer with regularisation for detectable systems without persistence of excitation. *International journal of control*, 84(6):1041–1054, 2011.

D. Sui and T. A. Johansen. Linear constrained moving horizon estimator with pre-estimating observer. *Systems & Control Letters*, 67:40–45, 2014.

D. Sui, T. A. Johansen, and L. Feng. Linear moving horizon estimation with pre-estimating observer. *IEEE Transactions on Automatic Control*, 55(10):2363–2368, 2010.

V. Sundarapandian. Local observer design for nonlinear systems. *Mathematical and computer modelling*, 35(1-2):25–36, 2002.

M. Teboulle. Entropic proximal mappings with applications to nonlinear programming. *Mathematics of Operations Research*, 17(3):670–690, 1992.

M. Teboulle. A simplified view of first order methods for optimization. *Mathematical Programming*, pages 1–30, 2018.

M. J. Tenny and J. B. Rawlings. Efficient moving horizon estimation and nonlinear model predictive control. In *Proc. American Control Conference*, volume 6, pages 4475–4480. IEEE, 2002.

S. Ungarala. Computing arrival cost parameters in moving horizon estimation using sampling based filters. *Journal of Process Control*, 19(9):1576–1588, 2009.

J. Vandersteen, M. Diehl, C. Aerts, and J. Swevers. Spacecraft attitude estimation and sensor calibration using moving horizon estimation. *Journal of Guidance, Control, and Dynamics*, 36(3):734–742, 2013.

A. Wynn, M. Vukov, and M. Diehl. Convergence guarantees for moving horizon estimation based on the real-time iteration scheme. *IEEE Transactions on Automatic Control*, 59 (8):2215–2221, 2014.

V. M. Zavala. Stability analysis of an approximate scheme for moving horizon estimation. *Computers & Chemical Engineering*, 34(10):1662–1670, 2010.

V. M. Zavala, C. D. Laird, and L. T. Biegler. A fast moving horizon estimation algorithm based on nonlinear programming sensitivity. *Journal of Process Control*, 18(9):876–884, 2008.

A. Zemouche, M. Boutayeb, and G. I. Bara. Observer design for nonlinear systems: An approach based on the differential mean value theorem. *Proc. 44th Conference on Decision and Control*, pages 6353–6358, 2005.

J. Zhang and J. Liu. Distributed moving horizon state estimation for nonlinear systems with bounded uncertainties. *Journal of Process Control*, 23(9):1281–1295, 2013.

B. Zhou. Stability analysis of non-linear time-varying systems by Lyapunov functions with indefinite derivatives. *IET Control Theory & Applications*, 11(9):1434–1442, 2017.

L. Zou, Z. Wang, J. Hu, and Q.-L. Han. Moving horizon estimation meets multi-sensor information fusion: Development, opportunities and challenges. *Information Fusion*, 60: 1–10, 2020.